LINEAR
CONTROL SYSTEMS

LINEAR
CONTROL SYSTEMS:
MODELING,
ANALYSIS,
AND DESIGN

JAMES R. ROWLAND

Professor of Electrical and Computer Engineering
University of Kansas

JOHN WILEY & SONS

New York   Chichester   Brisbane   Toronto   Singapore

Library of Congress Cataloging in Publication Data:

Rowland, James R.
  Linear control systems.

  Includes bibliography and index.
  1. Feedback control systems.   2. Linear systems.
I. Title.
TJ216.R67   1986   629.8'3   85-22612
ISBN 0-471-03276-X

Printed in the United States of America

10   9   8   7   6   5   4   3   2   1

PREFACE

Dynamic in coverage and emphasis, the field of control engineering has progressed through several distinct stages in recent decades. At one time dealing primarily with the analysis and design of servomechanisms, feedback control applications have expanded to such areas as economics, sociology, transportation, energy, and ecology. Concurrently, a shift in emphasis has taken place from classical transfer function procedures to state variable techniques. Renewed interests in the transfer function approach and its relationship to state variable design have been exhibited recently. This book presents an integrated treatment of linear control system modeling, analysis, and design based on relating and interpreting concepts from both transfer function and state variable viewpoints.

This book is intended for a one-semester or one-quarter introductory feedback course at the undergraduate level for seniors (or possibly advanced juniors) in electrical engineering and related engineering fields. It should also be useful for self-study, or as a reference, for graduate engineers on the job or in graduate school. Although both transform and state variable techniques are presented and integrated throughout the book, a conscious effort has been made to develop the transform procedures somewhat more completely. This choice is based on the realization that subsequent graduate courses usually emphasize the state variable approach and that introductory feedback control courses traditionally emphasize classical transform procedures. Relating and interpreting concepts from both viewpoints with a firm development of the classical approach appears to be a suitable blend for a modern textbook at the senior undergraduate level.

A prerequisite knowledge of the analysis of linear circuits is assumed. A background including the use of Laplace transforms to solve linear ordinary differential equations and some familiarity with frequency response methods and matrix fundamentals are also recommended, though these concepts are covered as needed within the book.

This book is organized into parts on modeling (Part I: Chapters 1, 2, and 3), analysis (Part II: Chapters 4, 5, and 6), and design (Part III: Chapters 7 and 8). Points of integration between classical transform and state variable viewpoints appear in Chapters 2 to 8 as alternative approaches within the same section (e.g., Sections 2.5, 3.3, or 3.4), in neighboring sections (e.g., Sections 2.4 and 2.6), and in related sections in different chapters (e.g., Sections 7.3 and 8.3). Computational methods within the book include the use of iterative formulas for determining the roots of a polynomial (e.g., Sections 4.2 and 6.6) and step-by-step procedures for designing series compensation networks (Chapter 7). These computational methods emphasize the numerical algorithms being used and not their specialization to a particular calculator or digital computer.

I wish to express my appreciation to the reviewers for their detailed evaluations in helping to shape the scope, content, and tone of this book. Initial reviewers were Dean K. Frederick, Rensselaer Polytechnic Institute; J. B. Cruz, Jr., University of Illinois; Robert N. Clark, University of Washington; and Richard Roberts, University of Colorado. Final reviewers were Arild Thowsen, Iowa State University; Alexander Nauda, Bucknell University; and Paul E. Russell, Arizona State University. Special thanks go to Charles M. Bacon, Oklahoma State University, for his personal and administrative encouragement; to Daniel D. Lingelbach, Gary E. Young, James H. Taylor, and Lynn R. Ebbesen for helpful suggestions while teaching from manuscripts for this book at Oklahoma State University; and to Robert J. Mulholland, J. Mark Richardson, and John M. Acken for reviewing selected chapters.

Among the many students who offered suggestions, I am particularly indebted to Jessy W. Grizzle, Keith A. Teague, Andrea S. Johnson, W. Craig Henry, Daniel C. Easley, Beverly Rainwater, D. Mark Anderson, Ronald J. Franco, and William H. Hensley. I am pleased to acknowledge Ms. Charlene Fries for typing the final manuscript, and I thank my wife, Jonell, for typing much of the first draft and especially for her support and understanding.

James R. Rowland

ABOUT THE AUTHOR

James R. Rowland, Ph.D., P.E., received his doctorate in electrical engineering from Purdue University. He is currently Professor and Chairman of the Department of Electrical and Computer Engineering at the University of Kansas. He has previous faculty experience at Oklahoma State University and Georgia Institute of Technology and consulting and industrial experience at Lockheed-Georgia, the U.S. Army Missile Command, and Sandia National Laboratories. Teaching and research interests have included control systems theory, stochastic modeling, digital signal processing, and estimation theory. He is a Senior Member of the Institute of Electrical and Electronics Engineers and was the recipient of an IEEE Centennial Medal. He has served the IEEE as Education Society President and as Awards and Recognition Chairman of the Educational Activities Board. A member of the American Society for Engineering Education and the National Society of Professional Engineers, he has served as a program accreditation visitor for the Accreditation Board for Engineering and Technology.

CONTENTS

PART I. INTRODUCTION AND CONTROL SYSTEM MODELING

PART II. CONTROL SYSTEM ANALYSIS

PART III.   CONTROL SYSTEM DESIGN

INTRODUCTION AND CONTROL SYSTEM MODELING

1

Introduction to Automatic Control

Ingenuity in devising new and better controllers for physical dynamic systems has helped to stimulate rapid technological progress in this century. These improvements have been enhanced by the continual development of new devices for use as system components and by the expanding capabilities of computers for use in modeling, analysis, and design tasks. The marketplace of the world has come to depend on the products of the control engineer. Familiar control applications range from autopilots to air conditioners, elevators to ecosystems, spacecraft to solar trackers, and from nuclear submarines to sensor-equipped industrial robots. Some of these systems that employ feedback are shown in Figure 1.1.

A control engineer's first step is the formation of a suitable model of the dynamic system or process to be controlled. This model may be validated by analyzing its performance for realistic input conditions and then by comparing with field test data taken from the dynamic system in its operating environment. Model validation is often achieved with the aid of computer simulations in which the effects of varying parameters can be determined more easily. Further analysis of the simulated model is usually necessary to obtain the model response for different feedback configurations and parameter settings. Once an acceptable controller has been designed and tested on the model, the feedback control strategy is then applied to the actual system to be controlled. In this chapter we present introductory concepts as a motivation for a further appreciation and understanding of these modeling, analysis, and design steps.

1.1 FEEDBACK AND ITS SIGNIFICANCE

Regulatory processes in nature can be regarded as natural feedback that provides an automatic control feature for the natural system. Often assumed to be a relatively simple structure, the underlying regulatory process and its relationship to the total natural system might be quite complex. Familiar examples are skin temperature control through perspiration cooling (homeostasis) and the ecological

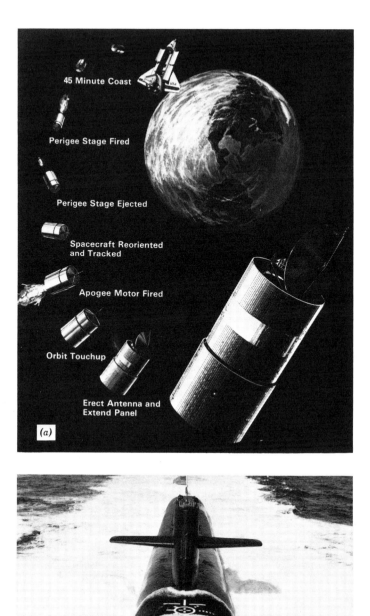

Figure 1.1 Common control system applications include (a) spacecraft control, (b) the Trident submarine USS Alabama.

Source: Courtesy of (a) Hughes Aircraft Company, (b) McDonnell Douglas.

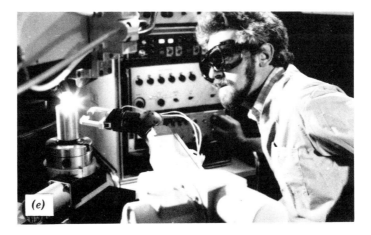

Figure 1.1 (*Continued*) (*c*) the F / A-18 Hornet fighter aircraft, (*d*) PRŌVOX®
boiler control instrumentation, and (*e*) a prototype robot arm used in conjunction
with leaser welding.

(*c*) General Dynamics, (*d*) Fisher Controls, and (*e*) Northrop Corporation.

balance between predator and prey. In the first case, the perspiration, which appears on the skin when the body becomes overheated, begins to evaporate and lower the skin temperature. The perspiration that forms and then cools makes possible the regulatory process, and the body itself is the natural system that needs temperature control. In the second case, if a large number of prey (e.g., rabbits) is available in a wildlife system, the predators (e.g., wolves) increase in number and consequently consume more of the prey. Since fewer of the prey are then available, the number of predators automatically decreases, causing a larger

TABLE 1.1 Typical Feedback Applications

Categories	Specific Applications
Ecological	Wildlife management and control; control of plant chemical wastes via monitoring lakes and rivers; air pollution abatement; water control and distribution; flood control via dams and reservoirs; forest growth management
Medical	Medical instrumentation for monitoring and control; artificial limbs (prosthesis)
Home appliances	Home heating, refrigeration, and airconditioning via thermostatic control; electronic sensing and control in clothes dryers; humidity controllers; temperature control of ovens
Power/energy	Power system control and planning; feedback instrumentation in oil recovery; optimal control of windmill blade and solar panel surfaces; optimal power distribution via power factor control
Transportation	Control of roadway vehicle traffic flows using sensors; automatic speed control devices on automobiles; propulsion control in rail transit systems; building elevators and escalators
Manufacturing	Sensor-equipped robots for cutting, drilling, die casting, forging, welding, packaging, and assembling; chemical process control; tension control windup processes in textile mills; conveyor speed control with optical pyrometer sensing in hot steel rolling mills
Aerospace and military	Missile guidance and control; automatic piloting; spacecraft control; tracking systems; nuclear submarine navigation and control; fire-control systems (artillery)

number of prey because of a lower predator threat, and the cycle is repeated. If the number of prey are controlled, we may regard the number of predators as the natural feedback controller, and vice versa. In each of these cases, both the system to be controlled and the regulatory process exist as inherent parts of the total natural system.

A second category of feedback applications is comprised of natural systems to be controlled with human-devised feedback controllers. Examples are control of a disease epidemic by inoculations, increased agricultural production by the proper use of fertilizers, irrigation, and insecticides, containment of forest fires set by lightning or other natural means, and intentional wildlife management for a desired ecological balance. In each of these cases, definite human actions, often on a large scale, are employed to bring the materials and forces of nature under control.

A third grouping of feedback applications includes human-devised systems and human-devised controllers. Examples are sensor-equipped industrial robots used in manufacturing for welding, assembling, painting, and packaging; home heating, refrigeration, and air-conditioning systems; missile, aircraft, spacecraft, and ship guidance and control; and chemical process control for the blending or separation of concentrates. Even automatic "cruise" control is a human-devised controller designed to maintain a preset vehicle speed on flat or hilly terrain.

The fourth possibility, human-devised systems having natural feedback, completes the spectrum of feedback applications. Specific examples for this category incorporate technological, ecological, and other aspects within very broad multidisciplinary areas. For example, the earth's system of energy usage may be considered a predominantly human-devised system that has a natural feedback control dependent on the inherent capability of the planet to sustain life. Medical and agricultural advancements influence the effectiveness of natural feedback in supporting the earth's population and defining its distribution. World simulations that have internal human-devised and natural feedback have been used to study these complex large-scale interactions.

Several of these examples, as well as some additional ones, are listed in Table 1.1.

1.2 OPEN-LOOP AND CLOSED-LOOP SYSTEMS

Feedback is defined as the use of the system output to modify its input for control purposes. The term *plant* designates the dynamic system or process to be controlled. Without feedback, the plant and its controller form an *open-loop system*, as shown in the schematic block diagram of Figure 1.2*a*. In contrast, a *closed-loop system* consists of the plant, one or more sensors, and the controller in a feedback configuration, as shown in Figure 1.2*b*. Block diagrams are generally either schematic block diagrams, which indicate only the functional identity of each block (e.g., "plant," "sensor," or "controller"), or transfer-function block diagrams, which provide mathematical descriptions within blocks.

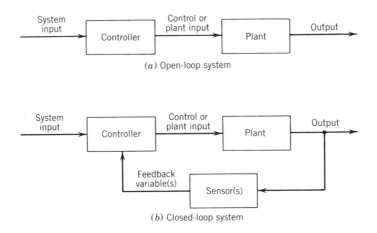

Figure 1.2 Schematic block diagrams of open-loop and closed-loop systems.

Among the advantages of an open-loop system are its simplicity, lower cost, and lower weight. Such systems are used in situations where only known inputs are applied and plant parameters are well defined. Examples are toasters, electric fans, beverage dispensing machines, home lighting systems, and lawn sprinklers. These systems operate under preset conditions without taking into account the actual output. The toaster operates for a fixed time; the electric fan has a set motor speed; the beverage machine has a prearranged schedule of events; and the lights and sprinklers operate at set conditions until turned off physically by some person or by timing mechanisms. The results from such open-loop systems could be suitable enough to preclude a need for more refined closed-loop operations. Figure 1.3 shows other systems for which open-loop operations are adequate.

The advantage of closed-loop operation is improved performance in the presence of unknown disturbance inputs and unknown system parameters. The output response of properly designed closed-loop systems is less sensitive to these unwanted disturbances and system parameter variations. Examples of closed-loop systems are tracking systems, rudder control of a ship, thermostatically controlled ovens, home heating and air-conditioning systems, and building elevators. Other closed-loop systems are shown in Figure 1.4.

Some systems may be classified as open-loop or closed-loop only upon close inspection. An electric clothes dryer operates in an open-loop fashion (preset period of drying time) unless it has an electronic sensor that determines the moisture content of the clothes and shuts itself off when they are dry (closed loop). "Automatic" dishwashers and clothes washing machines operate as open-loop systems (for their primary goals) according to a preset timing routine for proceeding through their cycles. Once cycle options have been selected, these operations are unaffected by the amount of food left on the dishes at any time or the cleanliness of the clothes. However, as shown in Figure 1.5, water-fill subsystems within dishwashers do perform in a closed-loop manner. Float devices

Figure 1.3 Open-loop control applications include (*a*) a computer-controlled tester for checking Very High Speed Integrated Circuit chips for use in future tactical aircraft radars, (*b*) a small insulin pump carried by diabetics, which continuously trickles a predetermined amount of insulin into the body over a 24-hour period, and (*c*) Weyerhaeuser's log merchandiser, which forces small pine logs into the teeth of rotary saws for processing.

Source: Courtesy of (*a*) Hughes Aircraft Company, (*b*) Sandia National Laboratories, and (*c*) Weyerhaeuser Company.

are used as water-level sensors that provide feedback signals to cut off the input flow when the desired amount of dishwater has entered the water circulation area.

Control systems, whether open loop or closed loop, may be single-input, single-output systems, or they may be multivariable control systems, having multiple inputs and multiple outputs. An example of a multivariable control system, an electronic engine control system, is shown in Figure 1.6. The sensed output variables include crankshaft and throttle positions, the exhaust gas recirculation valve position, inlet air and coolant temperatures, barometric pressure, and absolute pressure of the manifold. These variables are fed back to a microprocessor controller that forms engine control inputs, ignition timing and duty cycle, exhaust gas recirculation flow rate, throttle air bypass, fuel flow, and fuel injector sequencing. Each sensed output variable contains key information about the engine's performance. Feeding back a large number of these outputs for control permits the design of a highly efficient engine with reduced emissions and improved fuel economy. Another multivariable control system which senses and controls other variables is shown in Figure 1.7.

The control of nuclear power plants provides insight into the choices for open-loop and closed-loop operations within a large-scale system. Closed-loop operations are used to control the plant when disturbances with a time scale of a

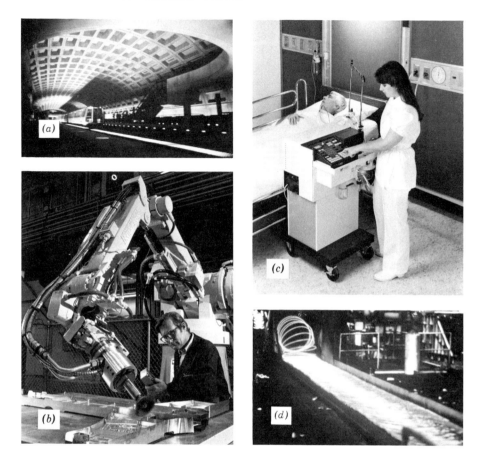

Figure 1.4 Closed-loop control applications include (*a*) both air-conditioning and operating controls for this mass-transit vehicle of the Washington, D.C., Metropolitan Area Transit Authority, (*b*) the aerospace industry's first robot capable of deburring complex machine parts, (*c*) microprocessor control of both pneumatic and patient monitoring by providing increased ease of breathing and better control over ventilatory parameters, and (*d*) the computer control of machines that mold loops of steel cable.

Source: Courtesy of (*a*) The Trane Company, La Crosse, Wisconsin, (*b*) Northrop Corporation, (*c*) Puritan-Bennett Corporation, and (*d*) Hewlett-Packard Company.

few seconds or a few minutes are encountered. Examples of such disturbances are process noise and changes in process parameters. Open-loop control is usually implemented by the power plant operator for disturbances occurring within a larger time scale, such as the human regulation of the control rods for core power distribution. This human/machine interface between the operator and the many mechanical indicators on the power plant control panels presents problems in human engineering. For additional closed-loop control, minor retrofits for the

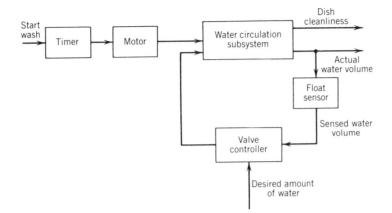

Figure 1.5 Dishwasher performance in open-loop (primary goal) and closed-loop (water-fill) operations.

Figure 1.6 Motorola's electronic engine control for passenger cars and light trucks maximizes fuel economy within emission constraints with acceptable levels of performance, reliability, and cost. The unit senses all critical operating parameters and computes realtime commands to control engine functions such as ignition timing and fuel flow. A custom designed 16-bit microprocessor and companion read-only memory are used in this two-sided printed circuit board design that makes extensive use of surface mounted devices and automated manufacturing techniques.

Source: Courtesy of Motorola, Inc.

Figure 1.7 The USAF F-16 aircraft (*a*) and its fly-by-wire advanced flight control computer system (*b*) which processes sensor information such as altitude and acceleration to provide electrical signals for the surface actuators.

Source: Courtesy of (*a*) General Dynamics and (*b*) Lear Siegler Inc., Astronics Division.

single-input, single-output control of separate plant components are possible in current operating plants. Major design improvements for the multivariable control of highly interacting components and subsystems are expected for the next generation of nuclear plants.

1.3 TRANSITIONS AND TRENDS

The earliest known feedback mechanisms are the waterclock of Ktesibios and the oil lamp of Philon, both invented in the third century B.C. [1].[1] Designed to maintain desired liquid levels, these first float regulators were followed in the first century A.D. by the more refined feedback devices of Heron, who separated the functions of sensing (float) and control (valve). By the end of the seventeenth century, temperature and steam-pressure regulators had been invented. Watt's flyball governor for controlling the speed of a steam engine in the eighteenth century apparently marks the beginning of automatic control as a science [1,2]. Within the next century, the feedback design of "regulators" expanded to the position control of "servomechanisms," including the use of proportional, derivative, and integral control in closed-loop system designs.

Analytical techniques, which now serve as the fundamentals of the classical approach to feedback analysis and design, were developed by Bode, Black, Nyquist, Evans, and others in the 1920s, 1930s, and 1940s [3–7]. These methods

---

[1] References are listed at the ends of chapters.

apply to linear, constant-parameter, continuous-time systems and are based on the formation of transfer functions in terms of the Laplace transform variable s. Feedback analysis and design procedures were applied to servomechanisms in fire-control systems (artillery) and in the rudder control of ships [8,9], as well as to the design of feedback amplifiers.

The state variable representation for system modeling became popular among control engineers in the late 1950s and early 1960s [10]. Compact vector-matrix notation was introduced to describe nonlinear time-varying systems in the same state-space format as linear time-invariant (constant-parameter) systems. This modern technique was applied for stability analysis and for the design of systems that were optimal according to some prescribed time-integral performance measure. The term *automatic control* replaced the restrictive and outdated term *servomechanism theory*. However, the control engineer was faced with a gap between theory and application caused ostensibly by feedback applications in practice being designed by the classical transfer function approach while most researchers had shifted to the modern state variable techniques [11]. In the 1960s and 1970s optimal filtering, prediction, and smoothing procedures were developed for systems having noisy input signals. Combined estimation and control techniques were developed for closed-loop system design, and the emergence of the ultrafast digital computer made possible the solution of specific control problems,

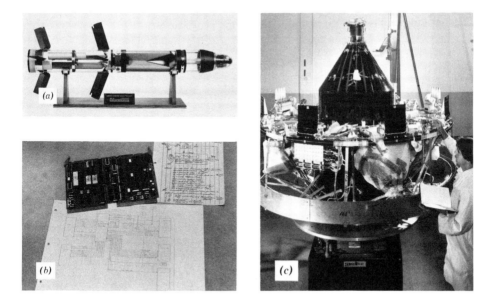

Figure 1.8 Modern estimation and control algorithms can help to (*a*) identify software / hardware tradeoffs for a laser-guided, cannon-launched missile (Copperhead), (*b*) design a realtime microprocessor controller for a voice-coil disk drive, and (*c*) determine the control subsystem design architecture and operations for the spacecraft in Figure 1.1*a*.

Source: (*a*) Martin Marietta Aerospace, (*b*) Magnetic Peripherals Inc., and (*c*) Hughes Aircraft Company.

TABLE 1.2   Transitions and Trends in Control Engineering

Circa 1965	Circa 1985
Theory	Mathematical system theory
↑	Algorithm development
(Gap)	Software engineering
↓	
Application	Hardware engineering

such as quantifying the effects of noise disturbances in ballistic calculations of long-range missile trajectories.

Today, the enormous speed and capacity of computers have opened up new vistas in control engineering for large-scale analysis and design and for the parallel processing of sensed data to provide on-line estimation and control, as illustrated in Figure 1.8. An emphasis on algorithm development (open-loop and closed-loop solutions) and software engineering (computer program development) has now entered the spectrum between abstract mathematical system theory (existence and uniqueness proofs) and hardware engineering (physical mechanization of control laws), as shown in Table 1.2. The previous gap between theory and practice has been sharply narrowed by a renewed interest on the part of the control engineer to relate and interpret classical transform and state variable techniques and to use computational procedures for their efficient implementation in practice [12–14]. This trend in control engineering toward an increased emphasis on computational methods will evidently parallel the continuing development in computers from microprocessors to large-scale computer facilities [15–20].

1.4 SYSTEM CLASSIFICATION AND LINEARIZATION

Systems may be classified according to the alternatives listed in Table 1.3. The scope of this book is limited to a consideration of linear, lumped-parameter, continuous-time, dynamic systems having only deterministic parameters and inputs. Both time-varying and time-invariant systems are modeled in Chapter 2, but the analysis and design emphasis of subsequent chapters is on time-invariant cases. Systems with single inputs and single outputs are considered primarily, but with some extensions to systems having multiple inputs and multiple outputs (multivariable systems).

Nonlinear dynamic systems may often be linearized about either static or dynamic operating points. *Static operating points*, referred to also as *equilibrium points*, occur from the application of a constant input, such as a constant force (gravity) or a constant voltage (battery). *Dynamic operating points* occur when the input varies with time to yield a nominal response (or solution) as a function of time. Examples are a nominal flight path of an aircraft and a nominal trajectory

TABLE 1.3 Alternatives for System Classification

STATIC OR DYNAMIC SYSTEMS

Static systems are composed of simple linear gains or nonlinear devices and described by algebraic equations, and dynamic systems are described by differential or difference equations.

CONTINUOUS-TIME OR DISCRETE-TIME SYSTEMS

Continuous-time dynamic systems are described by differential equations, and discrete-time dynamic systems by difference equations.

LINEAR OR NONLINEAR SYSTEMS

Linear dynamic systems are described by differential (or difference) equations having solutions that are linearly related to their inputs. Equations describing nonlinear dynamic systems contain one or more nonlinear terms.

LUMPED OR DISTRIBUTED PARAMETERS

Lumped-parameter, continuous-time, dynamic systems are described by ordinary differential equations, and distributed-parameter, continuous-time, dynamic systems by partial differential equations.

TIME-VARYING OR TIME-INVARIANT SYSTEMS

Time-varying dynamic systems are described by differential (or difference) equations having one or more coefficients as functions of time. Time-invariant (constant-parameter) dynamic systems are described by differential (or difference) equations having only constant coefficients.

DETERMINISTIC OR STOCHASTIC SYSTEMS

Deterministic systems have fixed (nonrandom) parameters and inputs, and stochastic systems have randomness in one or more parameters or inputs.

of an orbiting satellite. Onto this primary input which establishes the system operating point we superimpose a time-varying input of sufficiently small amplitude. Whether the operating point is static or dynamic, linearized dynamic equations about this operating point can be solved to provide a first approximation of the system response.

EXAMPLE 1.1

For the translational mechanical system with nonlinear spring characteristic $h(y)$ shown in Figure 1.9, the second-order nonlinear dynamic equation for the vertical

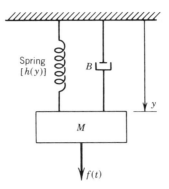

Figure 1.9 The translational mechanical system of Example 1.1.

position y of the mass M is[2]

$$M \frac{d^2 y}{dt^2} + B \frac{dy}{dt} + h(y) = Mg + f(t) \tag{1.1}$$

where B is the viscous damping constant, Mg is the gravitational force, and $f(t)$ is a small variational external force. Let the nonlinear spring characteristic be defined by

$$h(y) = K_1(y - l) + K_2(y - l)^2 \tag{1.2}$$

where l is the length of the spring when no force is applied, and K_1 and K_2 are coefficients as indicated. We want to describe the static equilibrium position y_s due to the constant gravitational force Mg in terms of Mg, l, K_1, and K_2 and to obtain the second-order differential equation for the incremental output variable $\delta y(t)$ defined by $\delta y(t) = y(t) - y_s$.

To determine y_s due to Mg acting alone, we delete $f(t)$ in (1.1) and set the time derivatives of the constant $y = y_s$ to zero to obtain $h(y_s) = Mg$. Using (1.2) with the physical restriction that $y_s \geq l$, we obtain

$$y_s = l - \frac{K_1}{2K_2} + \sqrt{\left(\frac{K_1}{2K_2}\right)^2 + \frac{Mg}{K_2}} \tag{1.3}$$

We next expand $h(y)$ in (1.2) in a Taylor series about $y = y_s$ and retain only first-order terms as an approximation to yield a linearized variational equation for $\delta y(t)$ from (1.1) as

$$M \frac{d^2(\delta y)}{dt^2} + B \frac{d(\delta y)}{dt} + [K_1 + 2K_2(y_s - l)] \delta y = f(t) \tag{1.4}$$

---

[2] We describe in Section 2.1 the details of writing differential equations for this and much more general mechanical and electrical systems.

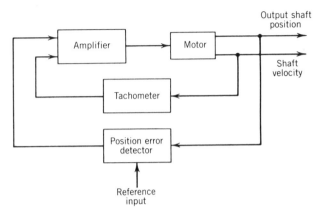

Figure 1.10 A schematic block diagram of a simplified positional servomechanism.

where y_s is given in (1.3). Higher-order terms involving $(\delta y)^2$, $(\delta y)^3$, and so forth are neglected because $\delta y(t)$ can be considered a small output variation when $f(t)$ is a sufficiently small variation.

1.5 AN OVERVIEW OF FEEDBACK CONTROL CONCEPTS

Modeling, analysis, and design considerations important in the study of linear control can be previewed by referring to the simplified positional servomechanism of Figure 1.10. Such closed-loop systems can occur, for example, in sensor-equipped robots, medical instrumentation, automotive design, aircraft tracking systems, and in many other applications involving a rotational mechanical system with feedback to control the position of an output shaft.

1.5.1 Modeling and Analysis Considerations

Models of component subsystems are obtained by forming the defining differential equations, usually from force or voltage relationships, and using measurement data to determine parameters and inputs as accurately as possible. The loading effect of each subsystem on interconnected subsystems is an important system modeling consideration. A closed-loop system that is properly designed is less sensitive than an open-loop system in terms of both system parameter variations and unwanted disturbances. Mathematical models of plants and sensors are described in Chapter 2, and output disturbance and parameter sensitivity problems are discussed in Chapter 3.

Time response characteristics are important for the analysis of closed-loop systems. Figure 1.11 shows typical time plots of the output shaft position of the simplified positional servomechanism in Figure 1.10 for three feedback strategies.

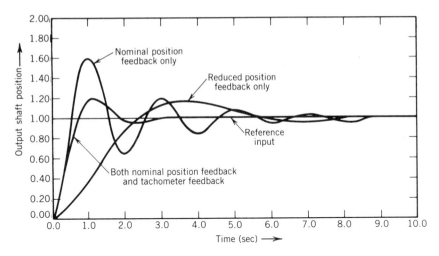

Figure 1.11 Typical time response plots for output shaft position.

The time response has an excessive overshoot when nominal position feedback only is used. A reduced amount of position feedback yields an acceptable overshoot but results in a much slower rise time. By including both a nominal position feedback and tachometer feedback, one can produce the desired time response, but this is achieved at the cost of the added complexity of an additional sensor. A complete description of time response characteristics is provided in Chapter 4.

Closely related to time response characteristics is the stability behavior of closed-loop systems. *Stability* refers to that tendency of a time response that has been altered because of the presence of disturbance inputs to remain sufficiently near the unperturbed time response. For dynamic systems described by linear ordinary differential equations with constant coefficients, the stability behavior does not depend on the system inputs but may be determined entirely by examining the *s*-plane root locations from a system's characteristic equation. These roots are also the poles of the closed-loop transfer function, provided there are no pole-zero cancellations. Analysis procedures enable us to determine the *s*-plane root locations for given closed-loop system parameters, but the design goal is to determine the controller configuration and design parameters to provide the desired system response. Chapter 5 describes techniques for determining the locus of roots of the characteristic equation as parameters are varied, and Chapter 6 provides the corresponding frequency-domain stability analysis.

1.5.2 Design Considerations

Designs of closed-loop systems such as the positional servomechanism of Figure 1.10 are generally aimed initially at achieving the desired power amplification and then making modifications to give the desired time response. In target tracking systems, for example, the closed-loop system must first be able to respond quickly, even though excessive overshoots may result. Sensors make available plant variables such as position, velocity, and acceleration for feedback control. As indicated in Figure 1.11, better control strategies can be formulated when additional variables describing the plant state are available. Sometimes series compensation networks are used to reconstruct variables that are not directly measurable.

Whether plant variables are measured directly by sensors or reconstructed by appropriate series networks, feedback controllers must then be determined to satisfy given design criteria. For linear continuous-time plants described by differential equations having constant coefficients (time-invariant plants), the classical transfer function approach may be used for design in the s-plane. It is desired to fashion a linear feedback controller with gains adjusted suitably to locate the roots of the characteristic equation for the closed-loop system within prescribed regions of the left-half s-plane. These regions correspond to acceptable time response or frequency response characteristics—for example, overshoot, settling time, and bandwidth requirements. In contrast to the classical methods, the state variable formulation may be used to design a closed-loop system optimal according to some given performance measure—for example, the integrated weighted sum of the squares of the plant input(s) and selected states. The state variable approach is also applicable for time-varying or nonlinear plants with more general performance measures. Classical and state variable design procedures are discussed in Chapters 7 and 8.

SUMMARY

Feedback and its uses have been described for a number of natural and human-devised systems. Applications are limited in this book to linear continuous-time plants or as an approximate solution for those mildly nonlinear systems that can be linearized about static or dynamic operating conditions. Brief descriptions have been given for the modeling, sensitivity, time response, stability, and design of a simplified positional servomechanism. With this overview of modeling, analysis, and design problems as background, we proceed to Chapter 2 for details of mathematical modeling both from the transfer function and state variable viewpoints.

REFERENCES

1. O. Mayr. *The Origins of Feedback Control*. Cambridge, Mass.: M.I.T. Press, 1970.

2. A. G. J. MacFarlane, ed. *Frequency-Response Methods in Control Systems*. New York: IEEE Press, 1979.

3. H. W. Bode. *Network Analysis and Feedback Amplifier Design*. Princeton, N.J.: Van Nostrand, 1945.

4. H. S. Black. "Stabilized Feedback Amplifier." *Bell System Technical Journal*, Vol. 13 (January 1934), pp. 1–18.
5. H. Nyquist. "Regeneration Theory." *Bell System Technical Journal*, Vol. 11 (January 1932), pp. 126–147.
6. W. R. Evans. *Control System Dynamics*. New York: McGraw-Hill, 1954.
7. H. S. Black. "Inventing the Negative Feedback Amplifier." *IEEE Spectrum*, Vol. 14, No. 12 (December 1977), pp. 54–60.
8. W. R. Ahrendt and C. J. Savant, Jr. *Servomechanism Practice*. New York: McGraw-Hill, 1960.
9. H. Chestnut and R. W. Mayer. *Servomechanisms and Regulating System Design*, Vols. I and II. New York: Wiley, 1951–1955.
10. J. E. Gibson. "From Control Engineering to Control Science." *IEEE Spectrum*, Vol. 2, No. 5 (May 1965), pp. 69–71.
11. G. Axelby. "The Gap—Form and Future." *IEEE Transactions on Automatic Control*, AC-9, No. 2 (April 1964), pp. 125–126.
12. M. Athans, M. L. Dertouzos, R. N. Spann, and S. J. Mason. *Systems, Networks, and Computation: Multivariable Methods*. New York: McGraw-Hill, 1974.
13. P. Silvester and N. Rumin. "In Defense of Canned Programs." *IEEE Transactions on Education*, E-20, No. 1 (February 1977), pp. 68–70.
14. J. L. Melsa and S. K. Jones. *Computer Programs for Computational Assistance in the Study of Linear Control Theory*, 2nd ed. New York: McGraw-Hill, 1973.
15. N. K. Sinha. "The Impact of the Electronic Pocket Calculator on Electrical Engineering Education." *IEEE Transactions on Education*, E-20, No. 1 (February 1977), pp. 6–9.
16. J. M. Smith. *Scientific Analysis on the Pocket Calculator*. New York: Wiley, 1975.
17. W. W. Lattin and P. M. Russo, eds. "Special Issue on Microprocessor Applications." *Proceedings of the IEEE*, Vol. 66, No. 2 (February 1978).
18. J. R. Rice. *Numerical Methods, Software and Analysis*. New York: McGraw-Hill, 1983.
19. T. L. Booth. "The Penetrating Technologies: An Analysis by the Experts." *IEEE Spectrum*, Vol. 21, No. 1 (January 1984), pp. 36–37.
20. R. J. Lauber. "Software for Industrial Process Control." *Computer*, Vol. 17, No. 2 (February 1984), pp. 6–8.

PROBLEMS

($\S$1.1)[3] **1.1** Classify each of the following feedback applications as either a natural system with inherent (regulatory process) feedback, a natu-

---

[3] This designation is placed next to each problem to indicate the section(s) to which it refers.

ral system with a human-devised controller, a human-devised system with a human-devised controller, or a human-devised system with natural feedback.

a. Tension control windup processes in textile mills.
b. Predator-prey ecosystem.
c. Optimal power distribution via power factor control.
d. Air conditioning via thermostatic control.
e. Medical instrumentation for monitoring and control.
f. Disease arrestment via body self-limiting process.
g. Wildlife management and control.
h. Missile guidance and control.

(§§1.1, 1.2) **1.2** Construct schematic block diagrams of the following closed-loop systems, indicating in each case the plant, sensor(s), feedback controller, intentional inputs, and possible disturbance inputs.

a. Skin temperature control through perspiration cooling.
b. Predator-prey ecosystem.
c. Sensor-equipped robot for product assembling.
d. Rudder control of a ship.
e. Chemical process control for mixing concentrates.

(§1.2) **1.3** Demonstrate the differences in open-loop and closed-loop lighting systems by constructing schematic block diagrams of (a) lights activated within the home at certain times of day by timers, and (b) outside mercury vapor lamps that turn on automatically with the onset of darkness.

(§1.2) **1.4** The space-saving combinational washer/dryer ("combo") shown in Figure 1.12 washes clothes and tumble dries them in the same horizontal axis tub. As the wet clothes begin the high-speed spin cycle to remove excess water prior to tumble-action drying, a load imbalance sometimes occurs and causes the machine to "walk" across the floor. To prevent this machine movement, a reed switch is activated as vibrations increase to some preset level, the machine motor is shut off, and the high-speed spin action is restarted with a new (and one hopes improved) clothes load distribution. This start-up process is repeated until the desired tub spin speed is achieved without excessive machine movement. Draw a schematic block diagram that illustrates this closed-loop operation.

(§1.2) **1.5** Halliburton Services has developed an "Automated Mud System" to help in controlling lost circulation in drilling for oil and gas. The system controls the weight of drilling mud by sensing density and adding barite as needed. Provide a schematic block diagram showing closed-loop operations.

(§1.2) **1.6** Control engineering methods have been applied in the medical field to orthotic systems to aid weakened or paralyzed limbs and to prosthetic systems to replace missing limbs. Construct a schematic

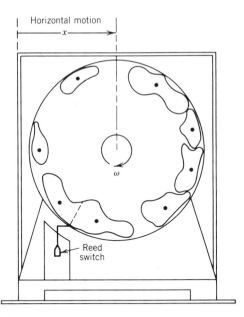

| Horizontal motion |

Figure 1.12 Simplified sketch of the "combo" washer/dryer described in Problem 1.4.

block diagram of an artificial arm being used to grasp an object on a table, and compare this operation with that of the natural arm. Identify the plant, sensor(s), and controller in each of these cases.

(§1.2) **1.7** An experimental feedback device to aid heart-attack victims has been developed and tested successfully by a team of medical researchers at Johns Hopkins University. Placed just beneath the skin in the abdomen, the device detects the ventricular fibrillation that signals the onset of a heart attack and then shocks the heart back to its normal beating. Provide a schematic block diagram showing these closed-loop operations.

(§1.2) **1.8** Construct a schematic closed-loop block diagram for the electronic engine control system shown in Figure 1.6. Show the interconnection of the plant (engine), sensors, and a multiprocessor controller for this multivariable system. Identify on your diagram the inputs and outputs of the plant and of all sensors.

(§1.2) **1.9** An unmanned cart (robot) moves among desks at the headquarter offices for New York's Citibank to deliver and receive mail. Suppose the cart leaves a dispatch area hourly and proceeds along a designated path, stopping at programmed intervals.

   a. Show a schematic block diagram of the cart's operation. Is this an open-loop or a closed-loop system?

   b. Modify your diagram in Part (a) to include a device that senses

Figure 1.13 Motorized robots with fork lifts help facilitate inventory delivery at the GMAD-Orion Plant. The plant's Automated Guide Vehicle System (AGVS) uses 22 of these vehicles, which are directed through the 77-acre Orion Plant by 19,000 feet of wire buried in the floors. The AGVS machines load and unload parts automatically in response to signals sent by two control computers. Sensors detect obstructions in the vehicle's path, stopping it until the pathway is cleared.

Source: Courtesy of General Motors Corporation.

obstructions along the route, allowing the cart to stop at least ten centimeters from any such object and to remain stopped until the pathway is cleared. Is this modified operation of the cart an open-loop or closed-loop system?

c. Make a further modification to your diagram in Part (b) to include a device that operates a light on the cart itself (and also signals the dispatch area of the cart's location) when the cart is at least ten minutes behind schedule.

d. Construct a schematic block diagram of General Motors Corporation's motorized robot in Figure 1.13 and compare its operation with the cart in Part (b).

(§1.2) **1.10** Psychologists report that the most important factor favoring the home sports team, aside from player abilities and coaching strategy, is loudly expressed home crowd support during the game. This home team edge is demonstrated most often in fast-moving sports such as collegiate or professional basketball. Show a schematic closed-loop block diagram of this phenomenon that spurs the home team onward for a decided advantage when all other factors are equal.

(§1.2) **1.11** Construct a schematic closed-loop block diagram indicating the learning progress of a student in a course. The student sets a desired goal of mastering the material sufficiently and maintains some performance level during the course by participating in class discussions, studying the textbook and reference materials, solving assigned homework problems and related exercises, and discussing the subject material with the professor, other students in the course, and others familiar with the course concepts. Marks received on course examinations and homework help the student to assess his or her understanding of the material and to modify study efforts accordingly. Indicate in your diagram the plant, sensor(s), controller, and inputs.

(§1.2) **1.12** Provide a schematic block diagram showing all functional blocks of a feedback positioning system that continuously aligns solar panels perpendicularly to the summer sun's rays for maximum solar heating in an energy-saving home. Include both a solar energy storage scheme and a supplementary source to make available the energy needed during nights and cloudy days. Provide a mechanism for directing the solar panels back toward the east for the beginning of a new day. Contrast this feedback positioning system with an open-loop system based on using as input the sun's location as a known function of time. What data could be compared to determine the better design (closed-loop or open-loop)?

(§1.4) **1.13** Classify the systems described below as either linear or nonlinear, continuous time or discrete time, time varying or time invariant, and deterministic or stochastic. Let $\alpha(t)$ be a time-varying coefficient and $n(t)$ a noisy input signal.

$$\text{a.} \qquad \frac{d^2y}{dt^2} + 3\left(\frac{dy}{dt}\right)^2 + 2y = 10 \sin t$$

$$\text{b.} \qquad \frac{dy}{dt} + y = n(t)$$

$$\text{c.} \qquad y(t_{k+2}) + 3y(t_{k+1}) + 2y(t_k) = 3$$

$$\text{d.} \qquad \frac{d^2y}{dt^2} + 3\alpha(t)\frac{dy}{dt} + y^2 + y = 0$$

(§1.4) **1.14** Consider the simple pendulum illustrated in Figure 1.14. The nonlinear ordinary differential equation describing its motion is given by

$$ML^2 \frac{d^2\theta}{dt^2} + B \frac{d\theta}{dt} + MgL \sin \theta = f(t)L$$

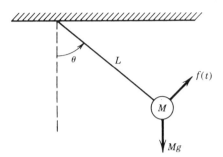

Figure 1.14 The simple pendulum described in Problem 1.14.

Determine the corresponding linearized equation about the static operating point $\theta_s = 0$.

(§§1.2, **1.15** Which of the following statements are true and which are not true?
1.5) Justify your answers by discussing each concept briefly.

    a. The open-loop control function yields the same system response as the corresponding (closed-loop) feedback law, provided there are no unknown system parameters or unknown external disturbances.

    b. The positional servomechanism of Figure 1.10 can be stable for step inputs and unstable for ramp inputs.

    c. Increasing the number of independent system variables being fed back from perfect sensor outputs improves control design possibilities.

2

Mathematical Models

The development of mathematical models for typical dynamic plants and associated sensors is an important first step in the process leading to an understanding of feedback control systems. It makes sense to quantify these relationships accurately before considering the interconnection of plant and sensor models to form closed-loop system models. Just as a track coach assesses the relative strengths of individual runners in selecting relay team members, so must we recognize the value of careful component modeling in our study which includes the more complicated behavior of interconnected feedback systems.

A primary model feature is the general form chosen for its description. Models for continuous-time dynamic plants may be described by differential equations that can be expressed in terms of state variables or, for linear plants, by transfer-function block diagrams. These model descriptions often involve several internal state variables as well as plant inputs and outputs. We are able to exercise complete control over a plant model only if the input affects every state variable. Moreover, we are able to observe completely the behavior of the plant model only if the effect of every state variable is represented in the model output. In brief, the control input message must be received at every state variable station from which continuous status reports are being received at the plant model output. Criteria for plant model controllability and observability can be established by using either state variables or transfer functions. Our goal is to relate and interpret these two forms of representations in this chapter for the modeling problem and in subsequent chapters for analysis and design.

2.1 MODELS OF SOME CONTROL SYSTEM COMPONENTS

Models that describe the dynamic behavior of some common control system components (plants and sensors) are developed in this section. Initially, the defining differential equation of a component is derived, usually from voltage or force relationships, and the Laplace transform equivalent is obtained. Detailed transfer-function block diagrams are then presented to illustrate the relationship between all physical variables. Using linearization concepts from Section 1.4 as a base, linear dynamic models are developed for typical electrical and mechanical

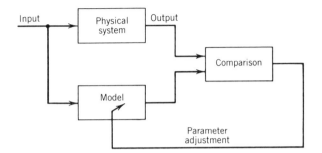

Figure 2.1 A model development or system identification procedure.

systems, including direct-current (dc) motors as electromechanical devices. Other models presented are gear trains, potentiometer error detectors, and tachometers. These component equations are used to form complete modeling details for the positional servomechanism introduced in Section 1.5. Interconnecting these devices to form a closed-loop system helps to highlight the importance of accurate component models in a practical control system design. Finally, a summary of common modeling relationships is provided in Table 2.1 near the end of this section.

2.1.1 Model Development

Constructing a dynamic model for a given physical system first requires the selection of a model structure and then some form of parameter estimation to determine acceptable model parameter values. Cause and effect relationships based on fundamental physical laws governing the process are used to determine a possible model structure or configuration. This configuration can be modified later if it fails to yield response data that agree closely enough with that of the physical system being modeled. Figure 2.1 shows a schematic block diagram of a typical model development or system identification procedure. Since only finite amounts of (usually random) data are available for identifying the system, parameter estimates contain random errors that decrease as additional data are obtained. Once the model structure and its parameters have been identified, a final step of model validation should be performed by using newly generated data from the physical system and the model [1–4].

The mathematical model development of this chapter is based on applying such fundamental physical laws as Kirchhoff's circuital laws and Newton's laws of motion to determine model structures and parameter values. Time responses of these models are described in Chapter 4 and, therefore, any identification of dynamic systems from input and output data must be determined subsequently. Modeling refinements, perhaps even resulting in a modification of the model configuration itself, might be justified upon obtaining new evidence in the form of unacceptable differences between physical system and model response data for identical input conditions.

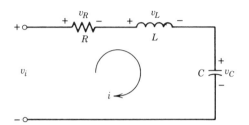

Figure 2.2 The series RLC network of Example 2.1.

2.1.2 Electrical Systems

The nature of mathematical models for electrical networks or systems may be determined by applying Kirchhoff's circuital laws on which loop and node analysis is based [5,6]. Integro-differential equations are obtained when these laws are used for an interconnection of resistors, inductors, capacitors, and either external or controlled sources. A transfer function between input and output may be obtained by taking Laplace transforms throughout and setting each initial condition to zero. The resulting algebraic equations may then be solved for an output/input ratio, which is the desired transfer function.[1] A detailed transfer-function block diagram provides a convenient format for expressing the relationships among the input, all intermediate variables, and the output.

EXAMPLE 2.1

Consider the series RLC network shown in Figure 2.2. Familiar voltage-current relationships for these individual circuit elements are

$$v_R = Ri \tag{2.1}$$

$$v_L = L\frac{di}{dt} \tag{2.2}$$

$$v_C = \frac{1}{C}\int_0^t i\,dt + v_c(0) \tag{2.3}$$

---

[1]A *transfer function* is defined as the ratio of the Laplace transform of the output to the Laplace transform of the input with all initial conditions set to zero. We show in Chapter 4 how to reinsert nonzero initial conditions to yield appropriate time responses.

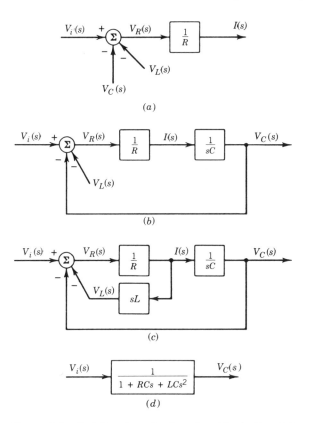

Figure 2.3 Detailed transfer-function block diagram development for the network of Figure 2.2.

Taking Laplace transforms with zero initial conditions, we obtain

$$V_R(s) = RI(s) \tag{2.4}$$

$$V_L(s) = sLI(s) \tag{2.5}$$

$$V_C(s) = \frac{1}{sC}I(s) \tag{2.6}$$

Using Kirchhoff's voltage law by summing the voltages around the loop yields

$$V_i(s) = V_R(s) + V_L(s) + V_C(s) \tag{2.7}$$

Equations (2.4) through (2.7) can be combined to form the detailed transfer-function block diagram of Figure 2.3. Arbitrarily solving (2.7) for $V_R(s)$ provides the format for the first step in developing a block diagram; that is

$$V_R(s) = V_i(s) - V_L(s) - V_C(s) \tag{2.8}$$

Using (2.8) and $I(s) = V_R(s)/R$ from (2.4), we obtain Figure 2.3a. If we regard the capacitor voltage as the system output, we utilize (2.6) to obtain $V_C(s)$ by passing $I(s)$ through a block with transfer function $1/sC$, as shown in Figure 2.3b. Equation (2.5) enables us to determine $V_L(s)$ from $I(s)$ to provide the completed block diagram of Figure 2.3c. The equivalent transfer function, obtained by solving (2.4) through (2.7) simultaneously, is shown in Figure 2.3d. Finally, we may solve (2.7) for either $V_L(s)$ or $V_C(s)$, instead of $V_R(s)$, and then use (2.5) or (2.6) to provide alternate detailed block diagrams for the series RLC network.

2.1.3 Mechanical Systems

The Newtonian laws of motion may be used to determine dynamic models for translational and rotational mechanical systems [7–9]. Translational mechanical system equations are formed by equating the product of each mass and its acceleration to the algebraic sum of all viscous damping forces, spring forces, and external forces acting on it. Moreover, we form rotational mechanical system equations by setting the product of each moment of inertia and its angular acceleration equal to the algebraic sum of all viscous damping torques, spring torques, and external torques acting on it. We present examples of both types of mechanical systems here and consider dc motors involving electrical and rotational mechanical couplings later.

EXAMPLE 2.2

A detailed block diagram is to be developed for the translational mechanical system of Figure 2.4a. A free-body diagram showing all forces acting on the mass M is given in Figure 2.4b. If we assume the mass to be moving downward, viscous damping produces an opposing upward force proportional to the velocity dy/dt as $B(dy/dt)$, and the spring yields an upward force proportional to the position difference $(y - l)$ as $K(y - l)$, where l is the unstretched length of the spring. Downward forces are the result of gravity (Mg) and an external force $f(t)$. We sum these forces algebraically with the downward direction considered positive because y is increasing in that direction and, according to Newton's law, set that sum equal to the product of the mass and its acceleration d^2y/dt^2 to yield

$$M \frac{d^2y}{dt^2} = -B \frac{dy}{dt} - K(y - l) + Mg + f(t) \qquad (2.9)$$

Rearranging (2.9) by placing terms involving y and its derivatives on the left

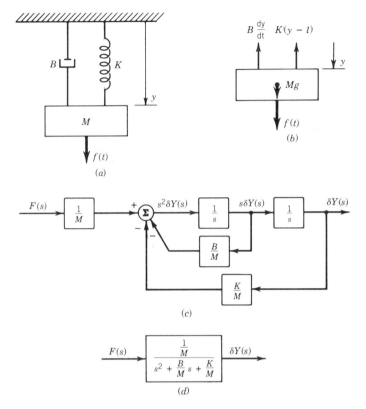

Figure 2.4 Development of a detailed block diagram for Example 2.2.

gives

$$M \frac{d^2y}{dt^2} + B \frac{dy}{dt} + K(y - l) = Mg + f(t) \tag{2.10}$$

Neglecting $f(t)$ on the right side of (2.10) momentarily, we solve for the static operating point y_s due to the gravitational force only; that is,

$$M \frac{d^2y_s}{dt^2} + B \frac{dy_s}{dt} + K(y_s - l) = Mg \tag{2.11}$$

Since y_s is a constant, its derivatives are zero and (2.11) becomes

$$K(y_s - l) = Mg \tag{2.12}$$

Therefore, the static equilibrium position y_s is

$$y_s = l + Mg/K \tag{2.13}$$

We form y as the sum of the constant y_s resulting from Mg and a variation δy resulting from $f(t)$; that is,

$$y = y_s + \delta y \tag{2.14}$$

Using (2.14) in (2.10) with y_s from (2.13) yields the linear differential equation for the variation δy as

$$M\frac{d^2(\delta y)}{dt^2} + B\frac{d(\delta y)}{dt} + K\,\delta y = f(t) \tag{2.15}$$

Solving (2.15) for $d^2(\delta y)/dt^2$ and taking Laplace transforms throughout with all initial conditions set to zero gives

$$s^2\,\delta Y(s) = -\frac{B}{M}s\,\delta Y(s) - \frac{K}{M}\delta Y(s) + \frac{1}{M}F(s) \tag{2.16}$$

where $\delta Y(s)$ and $F(s)$ are the Laplace transforms of $\delta y(t)$ and $f(t)$, respectively.

We use (2.16) to construct the detailed block diagram in Figure 2.4c. First, $s^2\,\delta Y(s)$ is obtained as the output of the summer, having inputs $-(B/M)s\,\delta Y(s)$, $-(K/M)\,\delta Y(s)$, and $(1/M)F(s)$. We next form $s\,\delta Y(s)$ by passing $s^2\,\delta Y(s)$ through a single integration block, that is, one with transfer function $1/s$. We continue by passing $s\,\delta Y(s)$ through a second integration block to yield $\delta Y(s)$. Feeding $s\,\delta Y(s)$ and $\delta Y(s)$ back through appropriate gains (B/M and K/M, respectively) provides the inputs needed in (2.16) for the summer, as shown in Figure 2.4c. Finally, solving for $\delta Y(s)/F(s)$ from (2.16) gives the system transfer function in Figure 2.4d.

EXAMPLE 2.3

The rotational mechanical system of Figure 2.5a has the free-body diagram given in Figure 2.5b. We form the algebraic sum of the viscous damping torque ($B\,d\theta/dt$), the spring torque ($K\theta$), and the externally applied torque $T(t)$ with the direction of increasing shaft angle position θ as positive and set this sum equal to the product of the moment of inertia (J) and shaft acceleration ($d^2\theta/dt^2$) to give

$$J\frac{d^2\theta}{dt^2} = -B\frac{d\theta}{dt} - K\theta + T(t) \tag{2.17}$$

Observe that $\theta_s = 0$ is the static operating condition and thus θ itself is a time variation analogous to δy in Example 2.2. Dividing (2.17) by J and taking

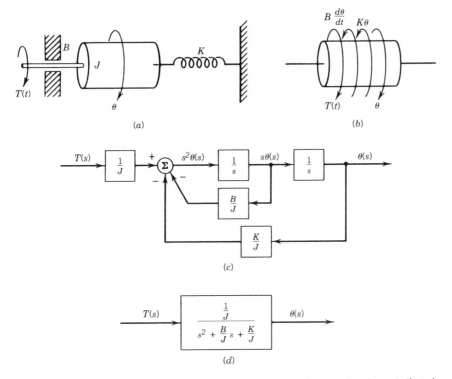

Figure 2.5 Detailed block diagram development for a simple rotational mechanical system.

Laplace transforms throughout with all initial conditions set to zero gives

$$s^2\theta(s) = \frac{B}{J}s\theta(s) - \frac{K}{J}\theta(s) + \frac{1}{J}T(s) \qquad (2.18)$$

The detailed block diagram of Figure 2.5c is constructed from (2.18) in the same way that Figure 2.4c was obtained from (2.16). We form $s\theta(s)$ by passing $s^2\theta(s)$ through an integration block and then form $\theta(s)$ by passing $s\theta(s)$ through a second integration block, as shown in Figure 2.5c. Both $s\theta(s)$ and $\theta(s)$ are fed back through the appropriate gains (B/J and K/J, respectively) and the outputs are summed (negatively) with $(1/J)T(s)$, as indicated in (2.18), to yield a detailed block diagram with an externally applied torque. The equivalent transfer function shown in Figure 2.5d is obtained by solving (2.18) for $\theta(s)/T(s)$. These equations are needed later in this section to describe the rotational mechanical parts of dynamic models for dc motors.

The Lagrangian formulation provides a unified approach for the dynamic description of a broad class of physical systems [10–12]. Systems composed of several different types of interrelated subsystems—for example, electrical, mechanical, hydraulic, and thermal systems—can be modeled in a dynamic framework by using the Lagrangian formulation. Although a nontrivial discussion of this approach with illustrative examples would take us far afield at this point, we point the interested student to the references for further information on this topic.

2.1.4 Field-Controlled dc Motor

Direct-current (dc) and alternating-current (ac) motors, such as that shown in Figure 2.6, are control system components having inherent electrical and mechanical coupling. These motors are often used as power elements in positioning shafts to follow desired reference angles [14]. As a particular case, a field-controlled dc motor is described in the schematic diagram of Figure 2.7a. We may obtain dynamic equations for this dc motor by applying Kirchhoff's voltage law to the electrical field circuit, identifying the electrical/mechanical coupling relationship between field current and torque, and then by using Newton's law of motion for the rotational mechanical system. The defining equations are

$$v_f = R_f i_f + L_f \frac{di_f}{dt} \tag{2.19}$$

$$T(t) = K_T i_f \tag{2.20}$$

$$J \frac{d^2\theta_m}{dt^2} + B \frac{d\theta_m}{dt} = T(t) \tag{2.21}$$

where v_f, i_f, R_f, and L_f are the voltage, current, resistance, and inductance of the field circuit, $T(t)$ is the torque, K_T is a torque constant proportional to the constant armature current I_a, θ_m is the motor shaft angle position, and J and B are effective values of rotational inertia and viscous friction. For the simple case indicated here (no gear train),[2] both J and B are the sums of motor and load components; that is,

$$J = J_m + J_L \tag{2.22}$$

$$B = B_m + B_L \tag{2.23}$$

The torque $T(t)$ in (2.20) is a drive torque that provides the coupling between electrical and mechanical parts of the motor. This drive torque is proportional to the product of the field current and armature current; that is,

$$T(t) = K' I_a i_f \tag{2.24}$$

---

[2] Corresponding values of J and B for the case where a gear train connects the load shaft to the motor shaft are developed later in this section.

Figure 2.6 A variable speed motor used to drive ac and dc generator assemblies over a controlled speed range.

Source: Courtesy of Feedback Incorporated.

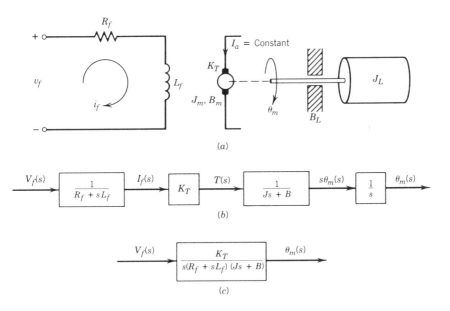

Figure 2.7 Schematic and detailed block diagrams for a field-controlled dc motor.

which may be simplified to give the expression in (2.20) by defining the torque constant K_T as $K'I_a$. The torque $T(t)$ in (2.21) is a reaction torque resulting from the motor and load. Since the drive torque and reaction torque are equal, they are both designated by the same variable $T(t)$.

To obtain a detailed block diagram of the field-controlled dc motor, we take Laplace transforms throughout in (2.19), (2.20), and (2.21) with all initial conditions set to zero to yield

$$V_f(s) = R_f I_f(s) + sL_f I_f(s) \tag{2.25}$$

$$T(s) = K_T I_f(s) \tag{2.26}$$

$$Js^2\theta_m(s) + Bs\theta_m(s) = T(s) \tag{2.27}$$

Solving (2.25) for $I_f(s)/V_f(s)$, (2.26) for $T(s)/I_f(s)$, and (2.27) for $s\theta_m(s)/T(s)$ gives

$$\frac{I_f(s)}{V_f(s)} = \frac{1}{R_f + sL_f} \tag{2.28}$$

$$\frac{T(s)}{I_f(s)} = K_T \tag{2.29}$$

$$\frac{s\theta_m(s)}{T(s)} = \frac{1}{Js + B} \tag{2.30}$$

We cascade blocks containing these three transfer functions to give the detailed block diagram of Figure 2.7b. Passing $s\theta_m(s)$ through a single integration block yields $\theta_m(s)$ to complete the diagram as shown. Solving (2.28), (2.29), and (2.30) for $\theta_m(s)/V_f(s)$ gives the transfer function for the single block in Figure 2.7c as

$$\frac{\theta_m(s)}{V_f(s)} = \frac{K_T}{s(R_f + sL_f)(Js + B)} \tag{2.31}$$

Equation (2.31) may also be expressed in the form

$$\frac{\theta_m(s)}{V_f(s)} = \frac{K}{s(1 + \tau_m s)(1 + \tau_e s)} \tag{2.32}$$

where $K = K_T/BR_f$, $\tau_m = J/B$, and $\tau_e = L_f/R_f$. The parameters τ_m and τ_e represent mechanical and electrical time constants of the field-controlled dc motor.

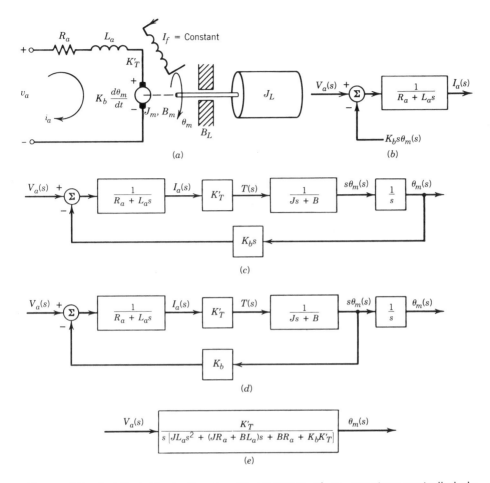

Figure 2.8 Detailed block diagram development of an armature-controlled dc motor.

2.1.5 Armature-Controlled dc Motor

The armature-controlled dc motor differs from the field-controlled motor in that the field current (I_f) is held constant and the control voltage is applied to the armature circuit. Figure 2.8a shows a schematic diagram of this motor. We use Kirchhoff's voltage law, an electrical/mechanical coupling relationship, and Newton's law of motion to obtain the defining dynamic equations as

$$v_a(t) = R_a i_a + L_a \frac{di_a}{dt} + K_b \frac{d\theta_m}{dt} \tag{2.33}$$

$$T(t) = K'_T i_a \tag{2.34}$$

$$J \frac{d^2\theta_m}{dt^2} + B \frac{d\theta_m}{dt} = T(t) \tag{2.35}$$

where the torque constant K'_T is proportional to I_f and both the drive torque in (2.34) and reaction torque in (2.35) are represented by $T(t)$. The term $K_b\, d\theta_m/dt$, appearing in (2.33), is a back electromotive force (emf) and K_b is the back emf constant.

Taking Laplace transforms throughout in (2.33), (2.34), and (2.35) with all initial conditions set to zero gives

$$V_a(s) = R_a I_a(s) + L_a s I_a(s) + K_b s \theta_m(s) \tag{2.36}$$

$$T(s) = K'_T I_a(s) \tag{2.37}$$

$$Js^2\theta_m(s) + Bs\theta_m(s) = T(s) \tag{2.38}$$

Equation (2.36) can be solved for $I_a(s)$ to give

$$I_a(s) = \frac{1}{R_a + L_a s}\left[V_a(s) - K_b s\theta_m(s)\right] \tag{2.39}$$

which is expressed in the block diagram of Figure 2.8b. Observe that an even more elemental block diagram could be formed for (2.36) by using the procedure of Example 2.1 to write

$$I_a(s) = \frac{1}{R_a}\left[V_a(s) - K_b s\theta_m(s) - L_a s I_a(s)\right] \tag{2.40}$$

However, (2.39) is appropriate for our purposes here, since only the voltage across the series combination of resistor and inductor needs to be indicated.

Solving (2.37) for $T(s)/I_a(s)$ and (2.38) for $s\theta_m(s)/T(s)$ gives

$$\frac{T(s)}{I_a(s)} = K'_T \tag{2.41}$$

$$\frac{s\theta_m(s)}{T(s)} = \frac{1}{Js + B} \tag{2.42}$$

which yields the block diagram of Figure 2.8c. Notice that $\theta_m(s)$ has been fed back through a block with transfer function $K_b s$ and subtracted from $V_a(s)$ to form the completed block diagram. An equivalent detailed block diagram that feeds back $s\theta_m(s)$—that is, the shaft rotational velocity or motor speed $(\dot\theta_m)$—through a gain K_b is shown in Figure 2.8d. Finally, the motor transfer function $\theta_m(s)/V_a(s)$ may be determined directly from (2.36), (2.37), and (2.38), as shown in Figure 2.8e. This third-order transfer function does not separate into factors having mechanical and electrical time constants exclusively, as in (2.32) for the field-controlled dc motor.

The armature-controlled dc motor is a human-devised system having inherent feedback resulting from a back electromotive force (emf) proportional to the

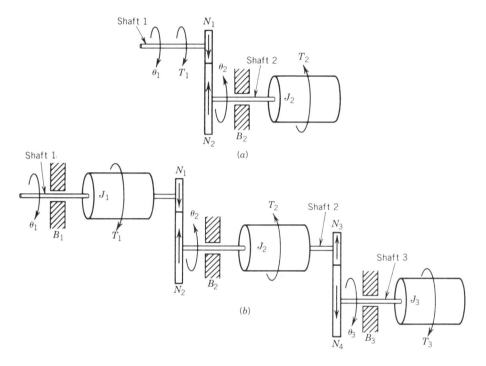

Figure 2.9   Schematic diagrams of two ideal gear trains.

motor speed. Additional control can be achieved by feeding back other plant (motor) variables as well from appropriate sensors. We note that in the SI system of units the back emf constant K_b (in volts per radian/second) is numerically equal to the torque constant K_T' (in Newton-meters/amp). We may show this equivalence by neglecting losses and setting the electrical power supplied to the motor armature equal to the mechanical power delivered by the motor to the shaft. Further details are indicated in Problem 2.12.

2.1.6 Gear Trains

An important mechanical transformation often needed in control systems design can be performed by gear trains. Proper torque levels are adjusted with gear trains in mechanical systems, obeying relationships identical to those of transformers in electrical networks. Although only the ideal linear case is considered here, gear trains in practice exhibit backlash and other undesirable characteristics. Figure 2.9 shows a schematic diagram of two ideal gear trains.

It is assumed that meshing gears have the same pitch. Therefore, the proportionality constant between the number of teeth on a particular gear (N_i) and the radius of that gear (r_i) is the same for both gears—that is, $r_1 = kN_1$ and $r_2 = kN_2$ —which yields $r_1/r_2 = N_1/N_2$. Another assumption is that the arc distance traversed on each of the two meshing gears is the same—that is, $r_1\theta_1 = r_2\theta_2$. A third assumption is that the "work in" on the first shaft is equal to the "work out" on the second shaft. Since work is the product of torque (T_i) and the angle through which it acts (θ_i), this assumption yields $T_1\theta_1 = T_2\theta_2$. Combining these results gives

$$\frac{T_1}{T_2} = \frac{\theta_2}{\theta_1} = \frac{N_1}{N_2} \tag{2.43}$$

Referring to Figure 2.9a, the torque T_2 for the second shaft satisfies

$$T_2 = J_2\frac{d^2\theta_2}{dt^2} + B_2\frac{d\theta_2}{dt} \tag{2.44}$$

Using $T_2 = (N_2/N_1)T_1$ and $\theta_2 = (N_1/N_2)\theta_1$ from (2.43) gives

$$\left(\frac{N_2}{N_1}\right)T_1 = J_2\frac{d^2}{dt^2}\left[\left(\frac{N_1}{N_2}\right)\theta_1\right] + B_2\frac{d}{dt}\left[\left(\frac{N_1}{N_2}\right)\theta_1\right] \tag{2.45}$$

which may be expressed as

$$T_1 = J\frac{d^2\theta_1}{dt^2} + B\frac{d\theta_1}{dt} \tag{2.46}$$

where J and B are effective values of these parameters defined as $(N_1/N_2)^2 J_2$ and $(N_1/N_2)^2 B_2$, respectively. For the case where several gear meshings are involved in an extended train of gears, the reflection of inertias and damping through more than one set of gears may be necessary. For example, the torque equation for Figure 2.9b with respect to Shaft 1 is given by (2.46) with effective values of J and B determined as

$$J = J_1 + \left(\frac{N_1}{N_2}\right)^2 J_2 + \left(\frac{N_1}{N_2} \cdot \frac{N_3}{N_4}\right)^2 J_3 \tag{2.47}$$

$$B = B_1 + \left(\frac{N_1}{N_2}\right)^2 B_2 + \left(\frac{N_1}{N_2} \cdot \frac{N_3}{N_4}\right)^2 B_3 \tag{2.48}$$

A convenient way to recall this transformation across a pair of gears is to remember that the square of the ratio of the driver gear to the driven gear, that is, (driver/driven)^2, applies both to rotational inertia and viscous damping.

2.1.7 Position and Velocity Sensors

Sensing devices are needed to provide measurements of the plant variables for use in the feedback control law. Two common sensors are the potentiometer error detector in Figure 2.10 and the dc tachometer in Figure 2.11.

A constant dc voltage supply of V volts is attached across the terminals of two n-turn potentiometers, as shown in Figure 2.10a. The resulting voltage v_p between

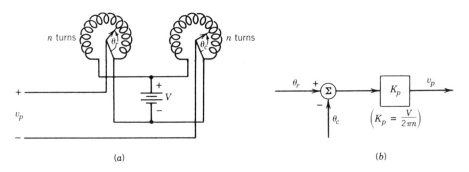

Figure 2.10 An *n*-turn potentiometer error detector.

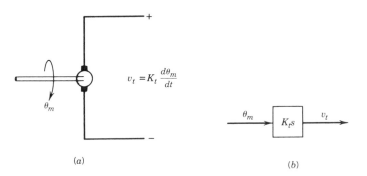

Figure 2.11 A dc tachometer.

the two wiper arms of the potentiometers is proportional to the difference $\theta_r - \theta_c$. This proportionality constant K_p is equal to $V/2\pi n$. The corresponding block diagram is shown in Figure 2.10*b*.

A dc tachometer measures the rotational velocity of a shaft and provides an output voltage proportional to this shaft speed. Figure 2.11*a* shows a schematic diagram of a dc tachometer. The corresponding block diagram of the tachometer operation is shown in Figure 2.11*b*.

2.1.8 Positional Servomechanism

We illustrate the interconnection of the plant and sensor components of this section by developing the detailed block diagram of the dc positional servomechanism of Figure 1.10. The modular system of Figure 2.12*a* is an example of instructional equipment for this purpose. We consider a field-controlled dc motor with both position and velocity sensors, as shown in Figure 2.12*b*. The dc tachometer is applied directly to the motor shaft, whereas the potentiometer is attached through a simple gear train as shown. Notice that both motor and load inertias and damping are present. The detailed block diagram in Figure 2.12*c* may be determined from (2.19), (2.20), (2.21), (2.43), (2.47), (2.48),

(a)

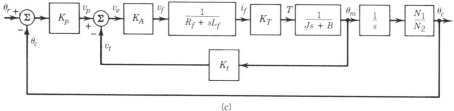

(b)

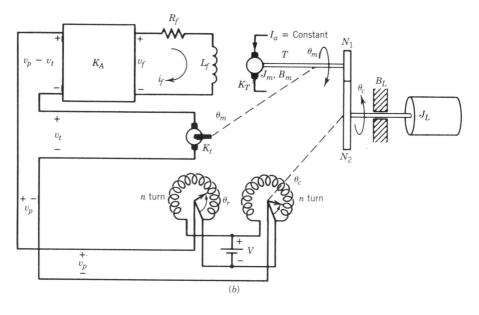

(c)

$$\theta_r \rightarrow \boxed{\dfrac{K_p K_A K_T (N_1/N_2)}{s[(R_f + sL_f)(Js + B) + K_A K_T K_t] + K_p K_A K_T (N_1/N_2)}} \rightarrow \theta_c$$

(d)

Figure 2.12 A positional servomechanism with a field-controlled dc motor and both shaft position and velocity feedback.

Source: (a) Courtesy of Feedback Incorporated.

Figure 2.10, and Figure 2.11. Solving the component equations simultaneously yields the closed-loop system transfer function in Figure 2.12d.

Table 2.1 provides brief mathematical descriptions of some additional control system components [12–14], and Figure 2.13 shows yet other components and systems for sensing, testing, and actuation.

2.2 BLOCK DIAGRAM REDUCTIONS

The dynamic models of the last section were presented in detailed block (or simulation) diagrams that showed the relationships among all the physical variables. Transfer functions for plants and systems were obtained by solving the Laplace-transformed equations for the appropriate output/input ratios. We describe in this section three methods for obtaining plant or system transfer functions from detailed block diagrams: (1) solution of the algebraic equations, (2) application of block diagram transformation rules, and (3) Mason's signal flow technique.

2.2.1 Algebraic Approach

The simplest conceptual procedure for reducing detailed block diagrams to determine the system (or plant) transfer function is the algebraic solution of the component equations. These algebraic equations may be formed from the detailed simulation diagram most easily by defining the summer outputs as the intermediate variables. If the output of the system does not come from a summer itself, then an additional equation for the output is needed. Substitution methods are commonly used to solve such equations for the required system transfer function.

2.2.2 Transformation Rules

Some elementary block diagram transformations are provided in Table 2.2 (Page 48). These rules include: (1) combining cascade blocks, (2) combining parallel blocks, (3) moving pickoff points past blocks, (4) moving summing points past blocks, and (5) eliminating single feedback loops. The key requirement for each rule is to maintain the same output/input expression between the original diagram and the equivalent diagram for all forward and feedback paths. For example, the original diagram for Rule 4 indicates that the input $R(s)$ is passed through the transfer function G and added to (or subtracted from) $Y(s)$ to form the output $E(s)$. Thus, the output/input expression for $E(s)/R(s)$ is G and for $E(s)/Y(s)$ is ± 1. These expressions must be the same for the equivalent diagram where the summing point has been moved to the left of the block containing G. In the equivalent diagram, $E(s)/R(s)$ is again G and $E(s)/Y(s)$ is $(1/G)(\pm 1)(G) = \pm 1$ in agreement with the corresponding expressions for the original diagram.

TABLE 2.1 Additional Control System Components

Component	Description

1. Two-Phase AC Servomotor

$$\frac{\theta_m(s)}{V(s)} = \frac{K_m}{s(\tau_m s + 1)}$$

$$\tau_m = \frac{J_m}{B_m + T_s/\omega_s}$$

where T_s is the stall torque and ω_s is the free shaft speed at rated voltage

2. Synchro Error Detector

3. Amplidyne

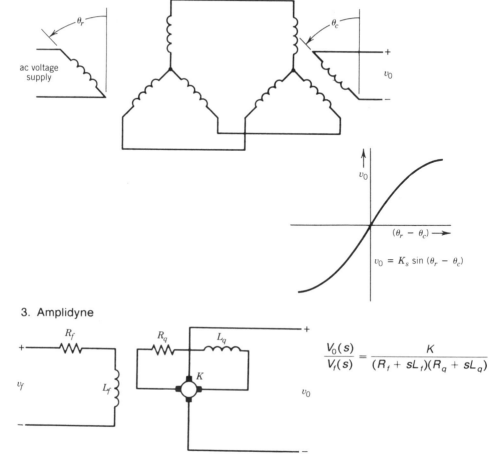

$$v_0 = K_s \sin(\theta_r - \theta_c)$$

$$\frac{V_0(s)}{V_f(s)} = \frac{K}{(R_f + sL_f)(R_q + sL_q)}$$

TABLE 2.1 (*Continued*)

Component	Description

4. Hydraulic Actuator

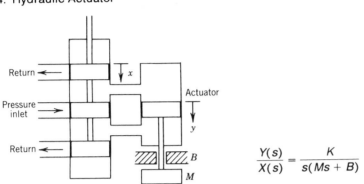

$$\frac{Y(s)}{X(s)} = \frac{K}{s(Ms + B)}$$

5. Pneumatic System

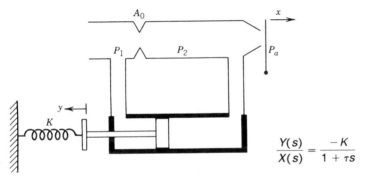

$$\frac{Y(s)}{X(s)} = \frac{-K}{1 + \tau s}$$

where τ is usually regarded as negligible

6. Thermal Heating System

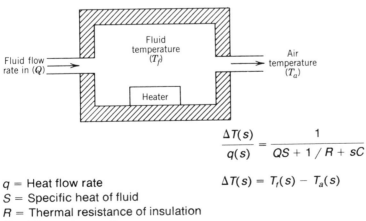

$$\frac{\Delta T(s)}{q(s)} = \frac{1}{QS + 1/R + sC}$$

q = Heat flow rate $\qquad \Delta T(s) = T_f(s) - T_a(s)$
S = Specific heat of fluid
R = Thermal resistance of insulation
C = Thermal capacitance

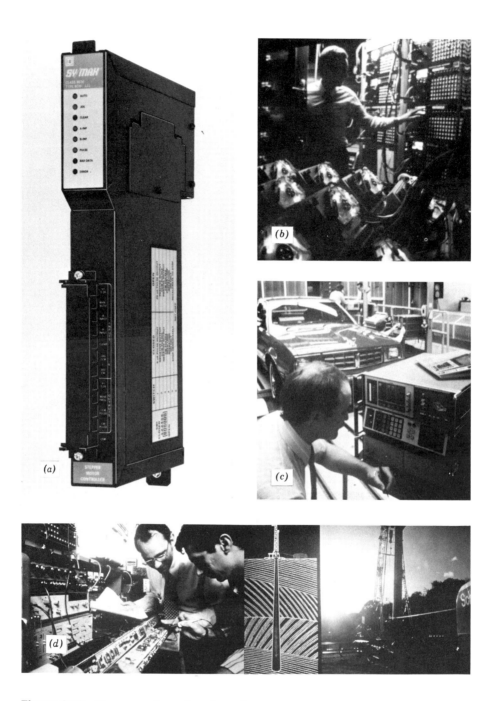

Figure 2.13 Components and systems for sensing, testing, and actuation include (*a*) the SY / MAX Programmable Controller Stepping Motor Controller (SMC) Module, which provides precise data (pulses) to the driver amplifier (translator) of a stepping motor, (*b*) ring laser gyros which employ laser beams to measure angular motion without moving parts for navigation and guidance applications, (*c*) an HP structural analyzer for use in conducting vibrations tests on an automobile prototype, and (*d*) a well-logging sensor for use in mapping the physical properties of formations deep within the earth.

Source: Courtesy of (*a*) Square D Company, (*b*) Honeywell Inc., (*c*) Hewlett-Packard Company, and (*d*) Schlumberger Well Services.

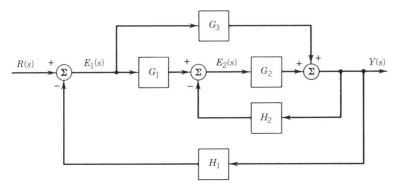

Figure 2.14 Block diagram for reduction in Example 2.4.

These transformation rules of Table 2.2 are especially useful for making minor modifications in detailed block diagrams. As noted earlier, the model for the armature-controlled dc motor can feed back either $\theta_m(s)$ through $K_b s$ or $s\theta_m(s)$ through K_b to combine with $V_a(s)$ in Figure 2.8. This option represents a direct application of Rule 3. For other cases, the sequential application of these transformation rules can be used to determine the system transfer function.

EXAMPLE 2.4

Consider the detailed block diagram of Figure 2.14. Writing equations for the summer outputs yields

$$E_1(s) = R(s) - H_1 Y(s) \tag{2.49}$$

$$E_2(s) = G_1 E_1(s) - H_2 Y(s) \tag{2.50}$$

$$Y(s) = G_3 E_1(s) + G_2 E_2(s) \tag{2.51}$$

which describe the system completely with variables $E_1(s)$, $E_2(s)$, and $Y(s)$. Substituting (2.50) into (2.51) gives

$$Y(s) = G_3 E_1(s) + G_2 [G_1 E_1(s) - H_2 Y(s)] \tag{2.52}$$

Finally, substituting (2.49) into (2.52) and simplifying yields the desired transfer function

$$\frac{Y(s)}{R(s)} = \frac{G_1 G_2 + G_3}{1 + G_2 H_2 + G_1 G_2 H_1 + G_3 H_1} \tag{2.53}$$

EXAMPLE 2.5

Rule 5 in Table 2.2 may be derived by using the defining algebraic equations

$$E(s) = R(s) \mp HY(s) \tag{2.54}$$

$$Y(s) = GE(s) \tag{2.55}$$

Substituting (2.54) into (2.55) and solving for $Y(s)/R(s)$ yields the desired result.

TABLE 2.2   Block Diagram Transformation Rules

Rules	Original Diagrams	Equivalent Diagrams
1. Combining cascade blocks		
2. Combining parallel blocks		
3. Moving pickoff points past blocks		
4. Moving summing points past blocks		
5. Eliminating single feedback loops		

EXAMPLE 2.6

It is required to apply the block diagram transformation rules of Table 2.2 to determine the system transfer function $Y(s)/R(s)$ in Figure 2.14. Initially, we use Rule 4 to move the summing point on the extreme right to the left of the block containing G_2, as indicated in Figure 2.15a. The results of this block diagram

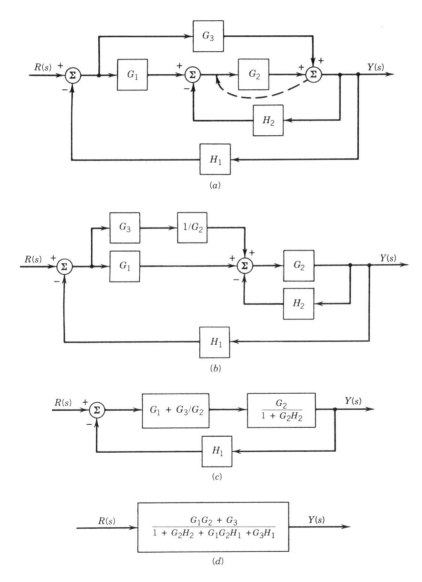

(a)

(b)

(c)

(d)

Figure 2.15 Sequential application of the transformation rules for reducing the block diagram in Figure 2.14.

transformation are shown in Figure 2.15b, where a new block containing $1/G_2$ has been inserted in cascade with the G_3 block. We next use Rule 1 to combine the cascade blocks G_3 and $1/G_2$, Rule 2 to combine this result (G_3/G_2) in parallel with G_1, and Rule 5 to eliminate the single feedback loop having G_2 in the forward path and H_2 in the feedback path. These results are shown in Figure 2.15c. Finally, Rules 1 and 5 are again used to obtain the equivalent transfer function block in Figure 2.15d.

An alternate procedure is described in Figure 2.16. Using Rule 4, the middle summer is moved to the right of the block containing G_2, as shown in Figure

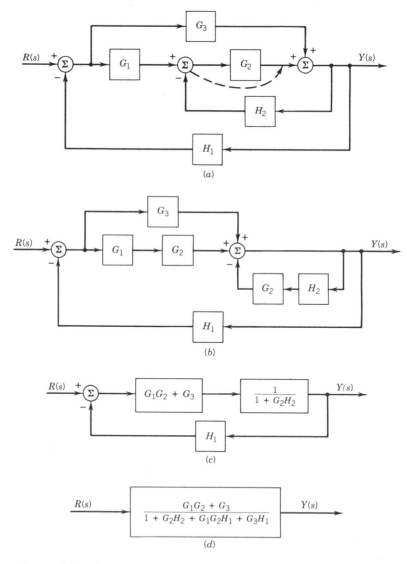

Figure 2.16 An alternate application of the transformation rules for reducing the block diagram in Figure 2.14.

2.16*a*. Blocks containing G_2 now appear following the G_1 and H_2 blocks. The resulting configuration in Figure 2.16*b* is reduced further by using Rule 1 to form cascade combinations both of G_1 and G_2 and of G_2 and H_2. Rule 2 combines G_3 with the G_1G_2 cascade combination just found, and Rule 5 eliminates the single feedback loop having unity in the forward path and G_2H_2 in the feedback path. These results are shown in Figure 2.16*c*, which is easily reduced by applying Rules 1 and 5 to give the desired solution in Figure 2.16*d*.

These two attempts at using block diagram transformation rules to determine $Y(s)/R(s)$ help to identify significant problem areas. First, it is apparent that one should have a definite plan in mind when applying the rules to yield a reduced configuration. Aimlessly moving summers or pickoff points past blocks may achieve very little without some reason for each such step. Second, it is obvious that even quite large systems with many blocks and loops may be handled rather easily when the loops are disjoint—that is, when no pickoff or summing points occur internal to these loops. Difficulty arises when this condition is not satisfied. Third, the approach to the solution is not unique, since various strategies may be employed to arrive at the same reduced transfer function. It may be difficult to prove, except in the simplest cases, that any particular strategy is definitely the best.

2.2.3 Mason's Signal Flow Technique

The application of Mason's formula to the signal flow graph corresponding to a given detailed block diagram is undoubtedly the simplest operational procedure for obtaining the system transfer function [15]. A signal flow graph is composed of various loops and one or more paths leading from an input to an output. Nodes representing system variables are interconnected by branches or unidirectional paths. This format is described in Table 2.3 with definitions of various kinds of nodes, paths, and loops.

Mason's formula is composed of three types of terms. First, we identify all forward paths and designate the path gains as P_k for $k = 1, 2, \dots$. Second, we form a parameter Δ which denotes the interactions among the various loops. The parameter Δ is defined as

$$\Delta = 1 - [\text{Sum of all single loop gains}] + [\text{Sum of gain products}$$
$$\text{of all combinations of two nontouching single loops}]$$
$$- [\text{Sum of gain products of all combinations of three}$$
$$\text{nontouching single loops}] + \cdots \qquad (2.56)$$

Third, we determine the interactions between single loops and forward paths.

TABLE 2.3 Signal Flow Graph Definitions

Term	Definition
Node	Nodes on a signal flow graph represent system variables.
Branch	Branches are unidirectional paths that connect the nodes. An arrow is assigned to indicate the direction of cause and effect.
Input node	An input node has only outgoing branches.
Output node	An output node has only incoming branches.
Path	A path is a continuous connection of branches with arrows in the same direction.
Loop	A loop is a path that starts and ends on the same node with all other nodes in the loop touched only once.
Common node	A common node is a node that is contained in two or more loops.
Nontouching loops	Nontouching loops are loops that have no common nodes.
Forward path	A forward path starts at an input node, ends at an output node, and touches no node more than once. A forward path may traverse one or more feedback branches in proceeding from input to output nodes.
Gains	Gains for paths and loops are defined as the products of branch gains for the paths or loops.

Specifically, we form a parameter Δ_k (for $k = 1, 2, \ldots$) using the definition in (2.56) but excluding terms including single loops (and combinations thereof) that touch the k th forward path. From the definitions in Table 2.3, we note that the single loops that are included in forming Δ_k have no common nodes with the forward path having gain P_k. These three types of parameters, that is, P_k, Δ, and Δ_k, are combined to form Mason's formula for the transfer function between an input $R(s)$ and an output $Y(s)$ as

$$\frac{Y(s)}{R(s)} = \frac{\sum P_k \Delta_k}{\Delta} \tag{2.57}$$

We form the signal flow graph by replacing each block in the detailed block diagram with a unidirectional branch between its input and output; we indicate the gain of this branch on the graph. One word of caution is in order when forming the signal flow graph from a detailed block diagram: If a pickoff point immediately precedes a summing point, then a unity gain branch must be inserted between these points in the signal flow graph. This inserted branch is necessary to maintain nontouching loop properties.

EXAMPLE 2.7

Let it be required that we find $Y(s)/R(s)$ for the block diagram in Figure 2.14 by using Mason's signal flow technique. The signal flow graph corresponding to this detailed block diagram is given in Figure 2.17a. Unity gain insertions preceding the summing points are not needed for this case.

Forming the signal loops indicated in Figure 2.17b yields the loop gains

$$L_1 = -G_2 H_2$$

$$L_2 = -G_1 G_2 H_1$$

$$L_3 = -G_3 H_1 \qquad (2.58)$$

Since these three single loops have at least one common node, there is no

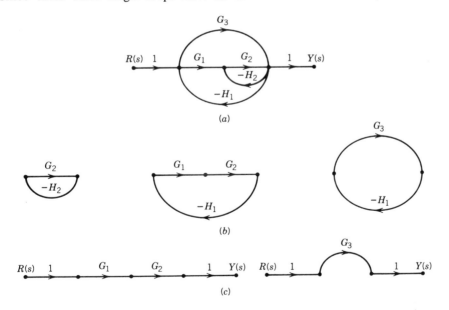

(a)

(b)

(c)

Figure 2.17 Application of Mason's signal flow technique to the block diagram in Figure 2.14.

combination of two nontouching loops. Therefore, Δ may be formed as

$$\Delta = 1 - [L_1 + L_2 + L_3]$$

$$= 1 + G_2 H_2 + G_1 G_2 H_1 + G_3 H_1 \tag{2.59}$$

Forward paths indicated in Figure 2.17c have gains given by

$$P_1 = G_1 G_2$$

$$P_2 = G_3 \tag{2.60}$$

Both paths have at least one node in common with every single loop. Therefore, $\Delta_1 = \Delta_2 = 1$, and $Y(s)/R(s)$ may be expressed directly from Mason s formula as

$$\frac{Y(s)}{R(s)} = \frac{P_1 \Delta_1 + P_2 \Delta_2}{\Delta}$$

$$= \frac{G_1 G_2(1) + G_3(1)}{1 + G_2 H_2 + G_1 G_2 H_1 + G_3 H_1} \tag{2.61}$$

which agrees with the result in (2.53).

EXAMPLE 2.8

Let the detailed block diagram of Figure 2.14 be modified by adding a feedback path having gain $-H_3$, as shown in Figure 2.18a. Pickoff points precede summing points in two places. The corresponding signal flow graph of Figure 2.18b includes the insertion of unity gain branches at these places as described earlier. Single loops have gains given by

$$L_1 = -G_1 G_2 H_1$$

$$L_2 = -G_1 G_2 H_3$$

$$L_3 = -G_2 H_2$$

$$L_4 = -G_3 H_1 \tag{2.62}$$

However, the inserted unity gain branches in the signal flow graph permit us to identify a combination of two nontouching loops—that is, the loop having gain L_2 has no nodes in common with the loop having gain L_4. Obviously, there are

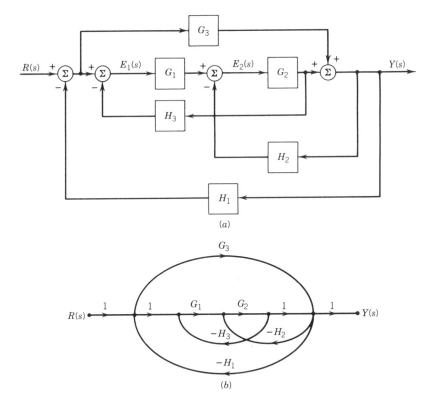

Figure 2.18 Detailed block diagram and signal flow graph for Example 2.8.

no combinations of three nontouching loops. Thus, Δ may be formed as

$$\Delta = 1 - [L_1 + L_2 + L_3 + L_4] + [(L_2)(L_4)]$$

$$= 1 + G_1G_2H_1 + G_1G_2H_3 + G_2H_2 + G_3H_1 + G_1G_2G_3H_1H_3 \quad (2.63)$$

As in Example 2.7, there are only two forward paths in the signal flow graph of Figure 2.18*b*. Their gains are given by

$$P_1 = G_1G_2$$

$$P_2 = G_3 \quad (2.64)$$

Note that the forward path having gain P_2 in (2.64) does not touch the loop having gain L_2, whereas the forward path having gain P_1 touches all loops in at least one point. Therefore,

$$\Delta_1 = 1$$

$$\Delta_2 = 1 - L_2 = 1 + G_1G_2H_3 \quad (2.65)$$

The system transfer function can be determined from Mason's formula as

$$\frac{Y(s)}{R(s)} = \frac{P_1 \Delta_1 + P_2 \Delta_2}{\Delta}$$

$$= \frac{G_1 G_2 (1) + G_3 (1 + G_1 G_2 H_3)}{1 + G_1 G_2 H_1 + G_1 G_2 H_3 + G_2 H_2 + G_3 H_1 + G_1 G_2 G_3 H_1 H_3} \quad (2.66)$$

To avoid mistakes in determining all loops and forward paths in large-scale systems, we recommend the use of a digital computer or programmable calculator having sufficient storage. Even when the detailed block diagram contains many interlocking feedback loops or forward paths, we can enter the interconnections of all branches, including directions, between all nodes of the signal flow graph into a computer program. A structured search algorithm can identify all forward paths and single feedback loops efficiently and can determine which are non-touching in combinations of two, three, four, and so forth.

In summary, the algebraic approach is straightforward, yet cumbersome when many equations are involved. The block diagram transformation approach is useful for obtaining equivalent diagrams that represent only minor modifications in the block diagram structure. Finally, it should be evident from these examples that Mason's signal flow technique is preferable for determining $Y(s)/R(s)$, either directly for simple systems or with the use of digital computers or programmable calculators for large-scale systems having many feedback loops and forward paths.

2.3 STATE VARIABLE REPRESENTATION

The concept of state, and its representation in terms of state variables for modeling purposes, permits a complete description of the internal behavior of a system [16–28]. All modes of a dynamic system are included in its state variable representation. In contrast, the classical transfer function representation describes only input-output (or external) relationships. If an unstable mode exists in a physical system, its inclusion with all other modes is ensured in the state variable representation. Yet, with the transfer function approach, a cancellation of numerator and denominator terms corresponding to the unstable mode obscures its effect in the output of the system transfer function model. In brief, the state variable representation preserves internal structure, whereas the transfer function representation focuses only on external relationships between the model input and its output.

The *state* of a system is defined as the minimum number of initial conditions that must be specified at some initial time t_0 to determine completely the dynamic behavior of the system for $t \geq t_0$ when given its input $r(t)$ for $t \geq t_0$. As an alternative to providing the system input itself for all $t < t_0$, the system state at $t = t_0$ describes the effects of $r(t)$ on the system during this time interval, exhibiting these cumulative effects at $t = t_0$. We realize that many different time functions $r(t)$ for $t < t_0$ could have been applied to a given system to arrive at the same state at $t = t_0$. Regardless of which of these past inputs were actually applied, we can determine the future dynamic behavior of the system uniquely by specifying only the system state at t_0 and its input $r(t)$ for $t \geq t_0$. The future state of the dynamic system obviously changes as time evolves beyond t_0 in response to its input and initial conditions. Nevertheless, the knowledge of the state at any time t, together with its future input, enables us to specify uniquely the state of the system at any future time. Since the system state varies as a function of time, we refer to its components as state variables.

The number and choice of state variables depend on the level of detail desired in the dynamic model selected to describe a given physical system. In general, an nth-order model has n state variables with n corresponding first-order differential

Figure 2.19 A steel jacket, which forms the base of a drilling and production platform in the Gulf of Mexico, slides off a launch barge in (a). The anchored thirty-story jacket receives the platform deck to support a drilling rig in (b).

Source: Courtesy of Kerr-McGee Corporation.

equations expressed in terms of these state variables. For example, we may choose to form a simplified model of the steel jacket sliding off a launch barge in Figure 2.19 by considering only the position of a point mass located at the center of gravity and the velocity of this point mass in a vertical plane. Thus, we might select a fourth-order model having positions and velocities in the two dimensions of the plane as the four state variables. On the other hand, we may form a more detailed model that not only includes six state variables for positions and velocities in three dimensions but also six other state variables for angular positions and angular velocities of motion about the center of gravity. Other dynamic system modeling applications, such as the tractor rollover protective structure test shown in Figure 2.20, might require selecting a model from several possible candidates with varying levels of detail. In addition to the fourth-order and twelfth-order models with structures similar to those just described for the steel jacket example, we might form a third model that includes body and head motion dynamics for the tractor driver. Such detailed models have been useful in vehicle safety testing and design in the automotive industry.

To illustrate how we select a set of state variables, we consider an nth-order linear plant model described by the differential equation

$$\frac{d^n y}{dt^n} + \alpha_{n-1}\frac{d^{n-1}y}{dt^{n-1}} + \cdots + \alpha_1 \frac{dy}{dt} + \alpha_0 y = u(t) \tag{2.67}$$

where $y(t)$ is the plant output and $u(t)$ is its input. A useful set of state variables, referred to as *phase variables*, is defined as $x_1 = y$, $x_2 = \dot{y}$, $x_3 = \ddot{y}, \ldots$, $x_n = y^{(n-1)}$, where $\dot{y}$ is dy/dt, $\ddot{y}$ is $d^2y/dt^2, \ldots$, and $y^{(n-1)}$ is $d^{n-1}y/dt^{n-1}$. We express the state variable equations by setting $\dot{x}_k = x_{k+1}$ for $k = 1, 2, \ldots, n-1$ and then solving (2.67) for $d^n y/dt^n$ and replacing y and its derivatives by the corresponding state variables to give

$$\dot{x}_1 = x_2$$

$$\dot{x}_2 = x_3$$

$$\vdots$$

$$\dot{x}_{n-1} = x_n$$

$$\dot{x}_n = -\alpha_0 x_1 - \alpha_1 x_2 - \cdots - \alpha_{n-1} x_n + u(t) \tag{2.68}$$

It is necessary for us to specify y and its first $n - 1$ derivatives at $t = t_0$ as initial conditions, together with $u(t)$ for all $t \geq t_0$, to completely define the solution of (2.67) for all $t \geq t_0$. Therefore, we see that the choice of state variables in (2.68) is a valid one. The state variables used to describe a linear plant model are not unique; we can select other state variables and find a transformation relating them to the original set.

Figure 2.20 Rollover testing, shown here in action scenes from left to right, is performed on large tractors as an aid in the design of their protective structures.

Source: Courtesy of Caterpillar Tractor Co.

Any particular set of state variables is composed of linearly independent components. The state variables $x_1, x_2, \ldots, x_n$ are linearly independent if the following equation holds for all x_j only if every k_j is zero:

$$k_1 x_1 + k_2 x_2 + \cdots + k_j x_j + \cdots + k_n x_n = 0 \tag{2.69}$$

Each state variable must be included to form the complete linearly independent set.

State variable equations, whether linear or nonlinear, can be expressed directly in the time domain by using a compact vector-matrix notation. For example, linear time-varying plant equations may be expressed as

$$\dot{\mathbf{x}}(t) = A(t)\mathbf{x}(t) + B(t)\mathbf{u}(t)$$

$$\mathbf{y}(t) = C(t)\mathbf{x}(t) + D(t)\mathbf{u}(t) \tag{2.70}$$

where $\mathbf{x}(t)$ is an n-dimensional column vector having components $x_1(t), x_2(t), \ldots, x_n(t)$ and representing the plant state; $\mathbf{u}(t)$ is an m-dimensional vector of plant inputs; $\mathbf{y}(t)$ is an r-dimensional vector of plant outputs; and $A(t)$, $B(t)$, $C(t)$, and $D(t)$ are time-varying matrices of dimensions n by n, n by m, r by n, and r by m, respectively. Linear time-invariant plants obey (2.70) under the condition that the coefficient matrices A, B, C, and D are constant with time. For single-input, single-output linear plants ($m = r = 1$), whether time varying or time invariant, the vector-matrix equations corresponding to (2.70) have the form

$$\dot{\mathbf{x}}(t) = A\mathbf{x}(t) + \mathbf{b}u(t)$$

$$y(t) = \mathbf{c}^T\mathbf{x}(t) + du(t) \tag{2.71}$$

If $y(t)$ in (2.71) corresponds to the first component of the r-vector $\mathbf{y}(t)$ in (2.70), that is, $y_1(t)$, then we may obtain the column vector $\mathbf{c}$ in (2.71) from the elements of the first row of the matrix C in (2.70). Although the resulting equations in (2.71) require that the vector $\mathbf{c}$ be transposed to yield its inner product with $\mathbf{x}$, we choose to adopt this notion so that the vector $\mathbf{c}$ and all other vectors are defined as column vectors. Finally, a general state variable format for nonlinear time-varying plants is given by

$$\dot{\mathbf{x}}(t) = \mathbf{f}[\mathbf{x}(t), \mathbf{u}(t), t]$$

$$\mathbf{y}(t) = \mathbf{h}[\mathbf{x}(t), \mathbf{u}(t), t] \tag{2.72}$$

where $\mathbf{f}[\cdot]$ and $\mathbf{h}[\cdot]$ are nonlinear vector functions of $\mathbf{x}(t)$, $\mathbf{u}(t)$, and t.

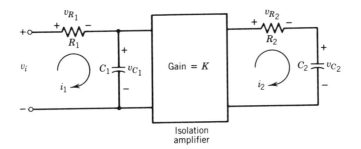

Figure 2.21   A double-ladder RC network with an isolation amplifier separating the two RC sections.

EXAMPLE 2.9

Consider the double-ladder RC network with the isolation amplifier shown in Figure 2.21. The defining equations are

$$v_i = R_1 i_1 + v_{C_1} \tag{2.73}$$

$$K v_{C_1} = R_2 i_2 + v_{C_2} \tag{2.74}$$

$$i_1 = C_1 \frac{dv_{C_1}}{dt} = C_1 \dot{v}_{C_1} \tag{2.75}$$

$$i_2 = C_2 \frac{dv_{C_2}}{dt} = C_2 \dot{v}_{C_2} \tag{2.76}$$

We select the state variables x_1 as v_{C_1} and x_2 as v_{C_2}. Substituting (2.75) into (2.73) and (2.76) into (2.74) and simplifying gives the component equations

$$\dot{x}_1 = -\frac{1}{R_1 C_1} x_1 + \frac{1}{R_1 C_1} v_i$$

$$\dot{x}_2 = \frac{K}{R_2 C_2} x_1 - \frac{1}{R_2 C_2} x_2$$

$$y = x_2 \tag{2.77}$$

Using the vector-matrix notation of (2.70) and (2.71), we identify

$$
A = \begin{pmatrix} \dfrac{-1}{R_1 C_1} & 0 \\ \dfrac{K}{R_2 C_2} & \dfrac{-1}{R_2 C_2} \end{pmatrix} \qquad B = \mathbf{b} = \begin{pmatrix} \dfrac{1}{R_1 C_1} \\ 0 \end{pmatrix}
$$

$$
C = \mathbf{c}^T = (0 \quad 1) \qquad d = 0 \tag{2.78}
$$

Let the network in Figure 2.21 be modified by permitting the amplifier gain K to be a nonlinear time-varying function—that is, $K(v_{C_1}, t)$. We express the resulting state variable equations in the vector format of (2.72) as

$$
\begin{pmatrix} \dot{x}_1 \\ \dot{x}_2 \end{pmatrix} = \begin{pmatrix} f_1[\mathbf{x}(t), \mathbf{u}(t), t] \\ f_2[\mathbf{x}(t), \mathbf{u}(t), t] \end{pmatrix} = \begin{pmatrix} -\dfrac{1}{R_1 C_1} x_1 + \dfrac{1}{R_1 C_1} u \\ \dfrac{K(x_1, t)}{R_2 C_2} x_1 - \dfrac{1}{R_2 C_2} x_2 \end{pmatrix}
$$

$$
y = \mathbf{c}^T \mathbf{x} = (0 \quad 1)\mathbf{x} = x_2 \tag{2.79}
$$

where $u = v_i$.

EXAMPLE 2.10

Consider again the field-controlled dc motor of Figure 2.7. As provided earlier in (2.19), (2.20), and (2.21), the defining equations are

$$
v_f = R_f i_f + L_f \frac{di_f}{dt} \tag{2.19}
$$

$$
T(t) = K_T i_f \tag{2.20}
$$

$$
J \frac{d^2 \theta_m}{dt^2} + B \frac{d\theta_m}{dt} = T(t) \tag{2.21}
$$

Selecting as state variables $x_1 = \theta_m$, $x_2 = \dot{\theta}_m$, and $x_3 = i_f$, we may solve (2.21) for $d^2\theta_m/dt^2$, that is, $\dot{x}_2$, and (2.19) for di_f/dt, that is, $\dot{x}_3$, to yield the state variable equations as

$$
\dot{x}_1 = x_2
$$

$$
\dot{x}_2 = -\frac{B}{J} x_2 + \frac{K_T}{J} x_3
$$

$$
\dot{x}_3 = -\frac{R_f}{L_f} x_3 + \frac{1}{L_f} v_f(t)
$$

$$
y = x_1 \tag{2.80}
$$

To illustrate state equations for multiple-input plants, we consider adding the term $(1/J)T_d(t)$ on the right side. Furthermore, suppose we specify θ_m, $\dot{\theta}_m$, and the linear combination $\theta_m + 2\dot{\theta}_m + 0.3i_f$ as the three outputs y_1, y_2, and y_3, respectively. Using vector-matrix notation, we may express the state variable and output equations as

$$\dot{\mathbf{x}} = \begin{pmatrix} 0 & 1 & 0 \\ 0 & \dfrac{-B}{J} & \dfrac{K_T}{J} \\ 0 & 0 & \dfrac{-R_f}{L_f} \end{pmatrix} \mathbf{x} + \begin{pmatrix} 0 & 0 \\ 0 & \dfrac{1}{J} \\ \dfrac{1}{L_f} & 0 \end{pmatrix} \mathbf{u}(t)$$

$$\mathbf{y} = \begin{pmatrix} y_1 \\ y_2 \\ y_3 \end{pmatrix} = \begin{pmatrix} 1 & 0 & 0 \\ 0 & 1 & 0 \\ 1 & 2 & 0.3 \end{pmatrix} \begin{pmatrix} x_1 \\ x_2 \\ x_3 \end{pmatrix} \tag{2.81}$$

where

$$\mathbf{u}(t) = \begin{pmatrix} v_f(t) \\ T_d(t) \end{pmatrix}$$

From (2.70), we identify the 3 by 3 matrix C in (2.81). Moreover, we omit the 3 by 2 matrix D in (2.81) because it contains all zeros.

Another convenient choice of state variables is the phase variable set $x_1 = \theta_m$, $x_2 = \dot{\theta}_m$, and $x_3 = \ddot{\theta}_m$. We take a time derivative of the equation in (2.21) and then substitute (2.19), (2.20), (2.21), and the derivative of the equation in (2.20) into this resulting equation to give

$$JL_f \frac{d^3\theta_m}{dt^3} + (BL_f + JR_f)\frac{d^2\theta_m}{dt^2} + BR_f\frac{d\theta_m}{dt} = K_T v_f \tag{2.82}$$

Setting $\dot{x}_1 = x_2$, $\dot{x}_2 = x_3$, and solving (2.82) for $d^3\theta_m/dt^3$, that is, for $\dot{x}_3$, we may write the state variable equations in vector-matrix notation as

$$\begin{pmatrix} \dot{x}_1 \\ \dot{x}_2 \\ \dot{x}_3 \end{pmatrix} = \begin{pmatrix} 0 & 1 & 0 \\ 0 & 0 & 1 \\ 0 & \dfrac{-BR_f}{JL_f} & \dfrac{-[BL_f + JR_f]}{JL_f} \end{pmatrix} \begin{pmatrix} x_1 \\ x_2 \\ x_3 \end{pmatrix} + \begin{pmatrix} 0 \\ 0 \\ \dfrac{K_T}{JL_f} \end{pmatrix} v_f$$

$$y = (1 \quad 0 \quad 0)\begin{pmatrix} x_1 \\ x_2 \\ x_3 \end{pmatrix} = x_1 \tag{2.83}$$

We observe that (2.83) treats θ_m as the plant output y. Alternately, we may choose to form a multiple-output $y(t)$ from θ_m, $\dot{\theta}_m$, and $\ddot{\theta}_m$ as

$$\mathbf{y} = \begin{pmatrix} y_1 \\ y_2 \\ y_3 \end{pmatrix} = \begin{pmatrix} 1 & 0 & 0 \\ 0 & 1 & 0 \\ 0 & 0 & 1 \end{pmatrix} \begin{pmatrix} x_1 \\ x_2 \\ x_3 \end{pmatrix} = I\mathbf{x} = \mathbf{x} \qquad (2.84)$$

where I is the identity matrix.

We have used a field-controlled dc motor in this example to illustrate vector-matrix notation for multivariable systems and to show the selection of two sets of state variables for the same linear plant. We develop alternate sets of state variables from a plant transfer function via simulation diagrams in the next section.

2.4 STATE VARIABLES FROM TRANSFER FUNCTIONS

The concept of choosing different state variables to describe a given model is expanded in this section to provide a systematic state variable selection procedure for single-input, single-output linear plants. The resulting state variable models may or may not represent all modes of the original physical plant; these models describe the behavior associated with only those modes remaining in the transfer function after possible cancellations of common numerator and denominator factors. Four state variable forms for a general third-order example are developed and related to corresponding computer simulation diagrams. State variables are defined as the outputs of integrators in each case, and the resulting state variable equations are expressed in vector-matrix notation. General nth-order equations and simulation diagrams for these state variable forms are provided in a table, together with the advantages and limitations in each case.

2.4.1 Direct Phase Variable Form

The direct phase variable form is presented first because it is particularly convenient for use with plant transfer functions that are not in factored form. Moreover, as shown below, the resulting set of state variables are in a phase variable format. Let the third-order plant transfer function be given as

$$\frac{Y(s)}{U(s)} = \frac{\beta_3 s^3 + \beta_2 s^2 + \beta_1 s + \beta_0}{s^3 + \alpha_2 s^2 + \alpha_1 s + \alpha_0} \qquad (2.85)$$

where each of the coefficients α_i and β_i are real numbers. This transfer function can easily be decomposed into the two blocks shown in Figure 2.22a with $W(s)$

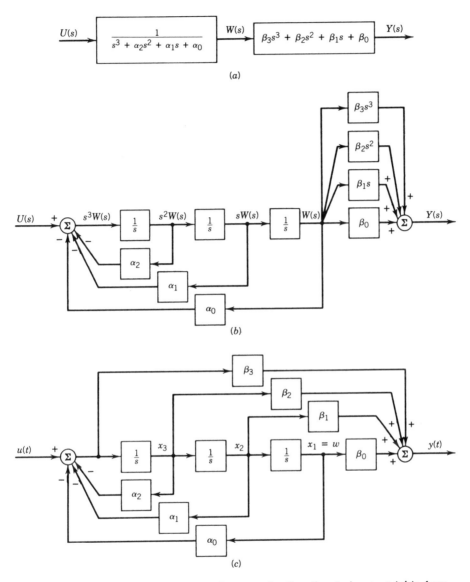

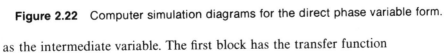

Figure 2.22 Computer simulation diagrams for the direct phase variable form.

as the intermediate variable. The first block has the transfer function

$$\frac{W(s)}{U(s)} = \frac{1}{s^3 + \alpha_2 s^2 + \alpha_1 s + \alpha_0} \qquad (2.86)$$

We rearrange (2.86) to give

$$s^3 W(s) = -\alpha_2 s^2 W(s) - \alpha_1 s W(s) - \alpha_0 W(s) + U(s) \qquad (2.87)$$

from which the first part of the simulation diagram in Figure 2.22*b* may be formed. The second part, relating $W(s)$ to $Y(s)$, is easily obtained by noting that $W(s)$ can be fed forward through four parallel blocks and their outputs summed to form $Y(s)$. (See Rule 2 of Table 2.2.) Finally, Figure 2.22*c* is obtained by moving pickoff points from $W(s)$ to other locations as noted.[3]

We select $x_1 = w$, $x_2 = \dot{w}$, and $x_3 = \ddot{w}$ and express this direct phase variable form of state variable representation as[4]

$$\dot{x}_1 = x_2$$

$$\dot{x}_2 = x_3$$

$$\dot{x}_3 = -\alpha_0 x_1 - \alpha_1 x_2 - \alpha_2 x_3 + u(t)$$

$$y = \beta_0 x_1 + \beta_1 x_2 + \beta_2 x_3 + \beta_3 \dot{x}_3$$

$$= (\beta_0 - \beta_3 \alpha_0) x_1 + (\beta_1 - \beta_3 \alpha_1) x_2$$

$$+ (\beta_2 - \beta_3 \alpha_2) x_3 + \beta_3 u(t) \tag{2.88}$$

where $\dot{x}_3$ in (2.88) was obtained by taking inverse Laplace transforms throughout in (2.87) and identifying x_1, x_2, and x_3. In vector-matrix notation, (2.88) becomes

$$\dot{\mathbf{x}} = \begin{pmatrix} 0 & 1 & 0 \\ 0 & 0 & 1 \\ -\alpha_0 & -\alpha_1 & -\alpha_2 \end{pmatrix} \mathbf{x} + \begin{pmatrix} 0 \\ 0 \\ 1 \end{pmatrix} u(t)$$

$$y = \begin{pmatrix} \beta_0 - \beta_3 \alpha_0 \\ \beta_1 - \beta_3 \alpha_1 \\ \beta_2 - \beta_3 \alpha_2 \end{pmatrix}^T \mathbf{x} + \beta_3 u(t) \tag{2.89}$$

The expression for y in (2.89) is obviously simpler when $\beta_3 = 0$.

This direct phase variable form of state variable representation is characterized by forming the states as phase variables using only the denominator terms in (2.85) and then feeding these states to the output through appropriate gains defined by numerator coefficients. The summation of these weighted state variables forms the output y, whereas the input u is fed only into the nth state equation. The advantages are that this form can be determined directly without

---

[3] For clarity in state variable descriptions, we sometimes use time-domain notation to denote signals appearing on transfer-function block diagrams.
[4] The direct phase variable form has also been designated as the controller canonical form [27] and as one of two controllability canonical forms [28]. State controllability is described in Section 2.5.

factoring the plant transfer function and that the state variables appear in the desirable phase variable form. A disadvantage is that common factors in the numerator and the denominator may easily remain unnoticed, resulting in problems when attempting to observe the plant state by using only its output.

2.4.2 Direct Feed-Forward Form[5]

To develop an alternate direct form in which the input is fed into each of the state variable equations, we first decompose the transfer function in (2.85) into the two blocks given in Figure 2.23a with $V(s)$ as the intermediate variable. The transfer function relating $Y(s)$ to $V(s)$ is equivalent to

$$sY(s) = \frac{1}{s^2}V(s) - \frac{\alpha_0}{s^2}Y(s) - \frac{\alpha_1}{s}Y(s) - \alpha_2 Y(s) \tag{2.90}$$

which corresponds to the simulation diagram of Figure 2.23b. Simple block diagram transformations, similar to those used in obtaining Figure 2.22c, result in Figures 2.23c and 2.23d. The corresponding vector-matrix state variable equations are

$$\dot{\mathbf{x}} = \begin{pmatrix} -\alpha_2 & 1 & 0 \\ -\alpha_1 & 0 & 1 \\ -\alpha_0 & 0 & 0 \end{pmatrix} \mathbf{x} + \begin{pmatrix} \beta_2 - \beta_3\alpha_2 \\ \beta_1 - \beta_3\alpha_1 \\ \beta_0 - \beta_3\alpha_0 \end{pmatrix} u$$

$$y = \begin{pmatrix} 1 \\ 0 \\ 0 \end{pmatrix}^T \mathbf{x} + \beta_3 u \tag{2.91}$$

We see that the vector multiplying u in the $\dot{\mathbf{x}}$ equation in (2.91) is simpler when $\beta_3 = 0$.

The direct phase variable form described earlier has the input entering only the nth state variable equation as a term in $\dot{x}_n$, and the output is developed from the weighted sum of the state variables. In contrast, the direct feed-forward form developed here feeds the input through gains defined by the transfer-function numerator coefficients to the state variables, and the output is the sum of the single state variable x_1 and the product $\beta_3 u$. In the first form, the vector $\mathbf{c}$ is composed of numerator and denominator coefficients, and the vector $\mathbf{b}$ has all zero elements except for the nth entry, which is unity. In this second form, the vector $\mathbf{b}$ is composed of numerator and denominator coefficients, and the vector $\mathbf{c}$ has all zero elements except for unity for the first element. In both forms, the A

---

[5] The direct feed-forward form has also been designated as the observer canonical form [27] and as one of two observability canonical forms [28]. The observability concept is described in Section 2.5.

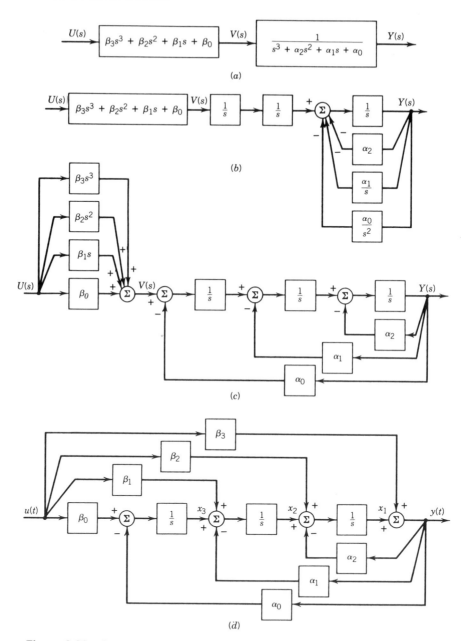

Figure 2.23 Computer simulation diagrams for the direct feed-forward form.

matrices are composed of denominator coefficients in the prescribed positions shown in (2.89) and (2.91).

The advantages and disadvantages here are essentially the same as those for the direct phase variable form with the exception that these state variable equations do not appear in phase variable form. An added advantage, however, is that the output y is defined more simply as the sum of the first state variable x_1 and the product $\beta_3 u$. A disadvantage of this form is that unnoticed common terms in the numerator and denominator present difficulties in controlling the plant with a single input.

2.4.3 Uncoupled Form

The time solution to the state variable equations can be found most easily if we select state variables that yield an uncoupled set of differential equations; the time solution is described more completely in Chapter 4. To obtain this uncoupled set of state variable equations, let the plant transfer function in (2.85) be written in a partial fraction expansion as

$$\frac{Y(s)}{U(s)} = \frac{\delta_1}{s + \gamma_1} + \frac{\delta_2}{s + \gamma_2} + \frac{\delta_3}{s + \gamma_3} + \beta_3 \tag{2.92}$$

where $-\gamma_1$, $-\gamma_2$, and $-\gamma_3$ are real and distinct roots of the denominator.

The development of the simulation diagram for (2.92) is shown in Figure 2.24. The four terms in (2.92) are placed in four parallel paths in Figure 2.24a with $U(s)$ as the input and with the outputs summed to form $Y(s)$. Either of the direct forms described earlier can be used to decompose these component transfer functions further and to define state variables. In fact, the only differences are the locations of the gains δ_1, δ_2, and δ_3 within the parallel paths. From (2.89), the direct state variable form yields

$$\dot{\mathbf{x}} = \begin{pmatrix} -\gamma_1 & 0 & 0 \\ 0 & -\gamma_2 & 0 \\ 0 & 0 & -\gamma_3 \end{pmatrix} \mathbf{x} + \begin{pmatrix} 1 \\ 1 \\ 1 \end{pmatrix} u$$

$$y = \begin{pmatrix} \delta_1 \\ \delta_2 \\ \delta_3 \end{pmatrix}^T \mathbf{x} + \beta_3 u \tag{2.93}$$

Figure 2.24b shows the computer simulation diagram corresponding to (2.93). From (2.91), the direct feed-forward form for decomposing the component transfer functions in Figure 2.24a yields

$$\dot{\mathbf{x}} = \begin{pmatrix} -\gamma_1 & 0 & 0 \\ 0 & -\gamma_2 & 0 \\ 0 & 0 & -\gamma_3 \end{pmatrix} \mathbf{x} + \begin{pmatrix} \delta_1 \\ \delta_2 \\ \delta_3 \end{pmatrix} u$$

$$y = \begin{pmatrix} 1 \\ 1 \\ 1 \end{pmatrix}^T \mathbf{x} + \beta_3 u \tag{2.94}$$

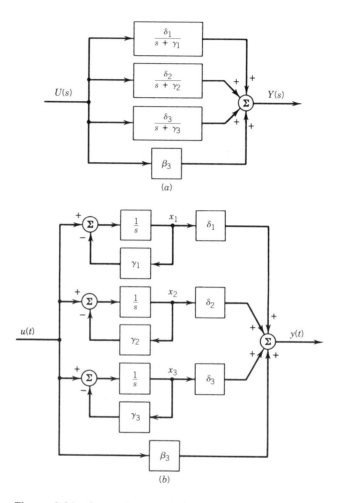

Figure 2.24 Computer simulation diagrams for the uncoupled form.

The student can easily construct the computer simulation diagram corresponding to (2.94).

The characteristic features of the uncoupled form are that the matrix A is diagonal with distinct roots of the plant transfer function denominator ($-\gamma_1$, $-\gamma_2$, and $-\gamma_3$) appearing on the main diagonal and that the vectors $\mathbf{b}$ and $\mathbf{c}$ are composed of elements whose products in corresponding positions, that is, $b_1 c_1$, $b_2 c_2$, and $b_3 c_3$, are equivalent to δ_1, δ_2, and δ_3, respectively. The advantage of this form is that each state variable describes the effect of only one mode of behavior, which makes it easier to solve for $\mathbf{x}(t)$ in (2.93) or (2.94). Disadvantages are that the transfer function denominator must be factored to yield $(s + \gamma_1)$, $(s + \gamma_2)$, and $(s + \gamma_3)$, and some additional effort is required to solve for δ_1, δ_2, and δ_3 in the partial fraction expansion.

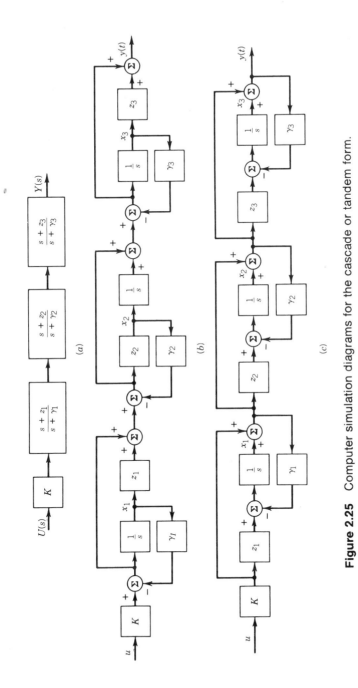

Figure 2.25   Computer simulation diagrams for the cascade or tandem form.

TABLE 2.4 Summary of State Variable Forms and Computer Simulation Diagrams

Form

1.　　　　　Direct phase variable form

$$\frac{Y(s)}{U(s)} = \frac{\beta_n s^n + \beta_{n-1} s^{n-1} + \cdots + \beta_1 s + \beta_0}{s^n + \alpha_{n-1} s^{n-1} + \cdots + \alpha_1 s + \alpha_0}$$

Computer Simulation Diagram

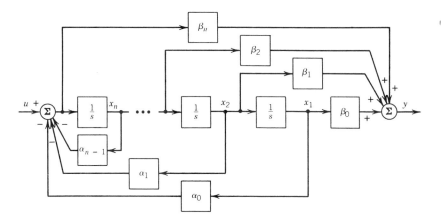

Form

2.　　　　　Direct feed-forward form

$$\frac{Y(s)}{U(s)} = \frac{\beta_n s^n + \beta_{n-1} s^{n-1} + \cdots + \beta_1 s + \beta_0}{s^n + \alpha_{n-1} s^{n-1} + \cdots + \alpha_1 s + \alpha_0}$$

Computer Simulation Diagram

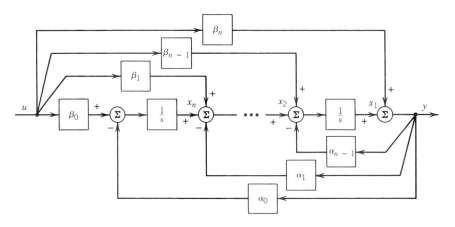

TABLE 2.4 (*Continued*)

State Variable Equations	*Advantages and Limitations*

$$\dot{x} = \begin{pmatrix} 0 & 1 & \cdots & 0 \\ 0 & 0 & & 0 \\ & \vdots & & \\ 0 & 0 & & 1 \\ -\alpha_0 & -\alpha_1 & \cdots & -\alpha_{n-1} \end{pmatrix} x + \begin{pmatrix} 0 \\ 0 \\ \vdots \\ 0 \\ 1 \end{pmatrix} u$$

$$y = \begin{pmatrix} \beta_0 - \beta_n\alpha_0 \\ \beta_1 - \beta_n\alpha_1 \\ \vdots \\ \beta_{n-1} - \beta_n\alpha_{n-1} \end{pmatrix}^T x + \beta_n u$$

Expressed directly without factoring; phase variable format; input fed only into *n*th state equation; output formed from weighted sum of states; always state controllable; common numerator and denominator terms not easily identified.

$$\dot{x} = \begin{pmatrix} -\alpha_{n-1} & 1 & \cdots & 0 \\ -\alpha_{n-2} & 0 & \cdots & 0 \\ \vdots & & & \vdots \\ -\alpha_1 & 0 & \cdots & 1 \\ -\alpha_0 & 0 & \cdots & 0 \end{pmatrix} x + \begin{pmatrix} \beta_{n-1} - \beta_n\alpha_{n-1} \\ \beta_{n-2} - \beta_n\alpha_{n-2} \\ \vdots \\ \beta_1 - \beta_n\alpha_1 \\ \beta_0 - \beta_n\alpha_0 \end{pmatrix} u$$

$$y = \begin{pmatrix} 1 \\ 0 \\ 0 \\ \vdots \\ 0 \end{pmatrix}^T x + \beta_n u$$

Expressed directly without factoring; input fed into states with numerator coefficients as gains; output formed from first state and weighted input; always observable; common numerator and denominator terms not easily identified.

Continued

TABLE 2.4 (*Continued*)

Form

3. Uncoupled form
$$\frac{Y(s)}{U(s)} = \frac{\delta_1}{s + \gamma_1} + \cdots + \frac{\delta_n}{s + \gamma_n} + \beta_n$$

Computer Simulation Diagram

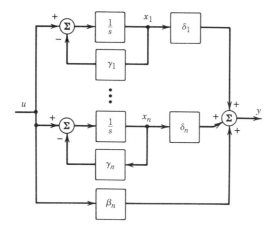

Form

4. Cascade or tandem form
$$\frac{Y(s)}{U(s)} = K\left(\frac{s + z_1}{s + \gamma_1}\right) \cdots \left(\frac{s + z_n}{s + \gamma_n}\right)$$

Computer Simulation Diagram

$$u \rightarrow \boxed{K} \rightarrow \boxed{\frac{s + z_1}{s + \delta_1}} - \cdots - \rightarrow \boxed{\frac{s + z_n}{s + \delta_n}} \rightarrow y$$

2.4.4 Cascade or Tandem Form

Let both the numerator and denominator of (2.85) be represented in factored form to yield

$$\frac{Y(s)}{U(s)} = \frac{K(s + z_1)(s + z_2)(s + z_3)}{(s + \gamma_1)(s + \gamma_2)(s + \gamma_3)} \tag{2.95}$$

TABLE 2.4 (*Continued*)

$$
\dot{\mathbf{x}} = \begin{pmatrix} -\gamma_1 & 0 & \cdots & 0 \\ 0 & -\gamma_2 & & 0 \\ \vdots & & & \vdots \\ 0 & \cdots & & -\gamma_n \end{pmatrix} \mathbf{x} + \begin{pmatrix} 1 \\ 1 \\ \vdots \\ 1 \end{pmatrix} u
$$

Must be in factored form; partial fraction expansion required; each state variable models only one mode of behavior for distinct eigenvalues; time response easily determined.

$$
y = \begin{pmatrix} \delta_1 \\ \delta_2 \\ \vdots \\ \delta_n \end{pmatrix}^T \mathbf{x} + \beta_n u
$$

Different for the various subsystem formats selected.

Must be expressed in factored form; subsystems are simulated from product terms and fed into next subsystem; may correspond to physical subsystems; difficult to write state variable equations.

By arbitrarily grouping first terms in the numerator and denominator together, second terms together, and third terms together, we obtain

$$
\frac{Y(s)}{U(s)} = K \left(\frac{s + z_1}{s + \gamma_1} \right) \cdot \left(\frac{s + z_2}{s + \gamma_2} \right) \cdot \left(\frac{s + z_3}{s + \gamma_3} \right) \tag{2.96}
$$

The cascade, or tandem, simulation diagram of Figure 2.25*b* consists of forming the connected subsystem blocks shown in Figure 2.25*a* from (2.96). The diagrams for individual subsystems are based on using the direct phase variable form of Figure 2.22 to yield

$$\dot{x}_1 = -\gamma_1 x_1 + Ku$$

$$\dot{x}_2 = (-\gamma_1 + z_1)x_1 - \gamma_2 x_2 + Ku$$

$$\dot{x}_3 = (-\gamma_1 + z_1)x_1 + (-\gamma_2 + z_2)x_2 - \gamma_3 x_3 + Ku$$

$$y = (-\gamma_1 + z_1)x_1 + (-\gamma_2 + z_2)x_2 + (-\gamma_3 + z_3)x_3 + Ku \qquad (2.97)$$

Using the direct feed-forward form for individual subsystems as an alternate procedure yields the diagram shown in Figure 2.25c. Further variations may be obtained by a different selection of subsystem blocks from (2.96) or by a mixture of the two simulation procedures for subsystem blocks.

The advantage of the cascade form is that this form often corresponds to the physical interconnection of blocks in large-scale dynamic systems. Consequently, the resulting state equations usually describe physical state variables. A disadvantage is the difficulty encountered in writing the state variable equations, because the outputs of earlier subsystem blocks feed a combination of their state variables and the input into later subsystems. The resulting substitutions, as in (2.97), are likely to lead to mistakes in writing the state variable equations for higher-order systems unless extreme caution is taken in their formation.

The four forms developed in this section for third-order plants are summarized for general nth-order plants in Table 2.4.

2.5 CONTROLLABILITY AND OBSERVABILITY

A linear plant is *state controllable* when the plant input u can be used to transfer the plant from any initial state to any arbitrary state in a finite time.[6] If not all of the state variables can be so controlled by the plant input, then the plant is said to be *not completely state controllable*, or, more simply, *state uncontrollable*. A linear plant is *observable* if the initial state $\mathbf{x}(t_0)$ can be determined uniquely when given the output $y(t)$ for $t_0 \leq t \leq t_1$ for any $t_1 > t_0$. The plant is said to be *unobservable* if not all of the initial state variables can be determined uniquely. We describe in this section how to determine the state controllability and observability of a linear plant (or system) from both the state variable and transfer function viewpoints.

2.5.1 State Variable Relationships

Let state variable equations in vector-matrix notation for a single-input, single-output linear plant be given by

$$\dot{\mathbf{x}} = A\mathbf{x} + \mathbf{b}u$$

$$y = \mathbf{c}^T \mathbf{x} + du \qquad (2.98)$$

---

[6] We describe criteria for state controllability in this section. Analogous results for output controllability are discussed in Problem 2.27.

Theorem 2.1

The plant described by (2.98) is said to be state controllable if and only if the controllability matrix M defined by

$$M \triangleq (\mathbf{b}, A\mathbf{b}, A^2\mathbf{b}, \dots, A^{(n-1)}\mathbf{b}) \tag{2.99}$$

has a nonzero determinant.

Theorem 2.2

The plant described by (2.98) is said to be observable if and only if the observability matrix Q defined by

$$Q \triangleq \left(\mathbf{c}, A^T\mathbf{c}, (A^T)^2\mathbf{c}, \dots, (A^T)^{n-1}\mathbf{c}\right) \tag{2.100}$$

has a nonzero determinant.

Since $\mathbf{b}, A\mathbf{b}, A^2\mathbf{b}, \dots, A^{(n-1)}\mathbf{b}$ are each n-dimensional column vectors, we see that the matrix M in (2.99) is an n by n matrix. Similarly, $\mathbf{c}, A^T\mathbf{c}, (A^T)^2\mathbf{c}, \dots,$ and $(A^T)^{n-1}\mathbf{c}$ are also n-dimensional column vectors that form the n by n matrix Q in (2.100). We now use these theorems to examine the state controllability and observability of state variable models obtained from a plant transfer function.

EXAMPLE 2.11

Consider the linear plant with transfer function

$$\frac{Y(s)}{U(s)} = \frac{s + z_1}{s^2 + 3s + 2} \tag{2.101}$$

Determine those values of the parameter z_1 for which the plant is both state controllable and observable by using (1) the direct phase variable form, and (2) the direct feed-forward form of state variable representation.

These two forms are obtained easily from the general case in Table 2.4. For the direct phase variable form,

$$A = \begin{pmatrix} 0 & 1 \\ -2 & -3 \end{pmatrix} \qquad \mathbf{b} = \begin{pmatrix} 0 \\ 1 \end{pmatrix} \qquad \mathbf{c} = \begin{pmatrix} z_1 \\ 1 \end{pmatrix} \tag{2.102}$$

Calculating the determinants of the matrices in (2.99) and (2.100) gives

$$\det(\mathbf{b}, A\mathbf{b}) = \det \begin{pmatrix} 0 & 1 \\ 1 & -3 \end{pmatrix} = -1 \neq 0$$

$$\det(\mathbf{c}, A^T\mathbf{c}) = \det \begin{pmatrix} z_1 & -2 \\ 1 & z_1 - 3 \end{pmatrix} = z_1^2 - 3z_1 + 2 \tag{2.103}$$

Therefore, the plant described by (2.102) is state controllable according to Theorem 2.1, since $\det(\mathbf{b}, A\mathbf{b}) \neq 0$. Using Theorem 2.2, the plant is observable, provided z_1 is not equal to $+1$ or $+2$. Observe that common factors appear in the numerator and denominator of the plant transfer function when z_1 is equal to either $+1$ or $+2$; we discuss this point more fully in the following subsection.

For Part (2) of the problem, the direct feed-forward form yields

$$A = \begin{pmatrix} -3 & 1 \\ -2 & 0 \end{pmatrix} \qquad \mathbf{b} = \begin{pmatrix} 1 \\ z_1 \end{pmatrix} \qquad \mathbf{c} = \begin{pmatrix} 1 \\ 0 \end{pmatrix} \qquad (2.104)$$

Calculating the required determinants gives

$$\det(\mathbf{b}, A\mathbf{b}) = \det \begin{pmatrix} 1 & z_1 - 3 \\ z_1 & -2 \end{pmatrix} = -\left(z_1^2 - 3z_1 + 2\right)$$

$$\det(\mathbf{c}, A^T\mathbf{c}) = \det \begin{pmatrix} 1 & -3 \\ 0 & 1 \end{pmatrix} = +1 \neq 0 \qquad (2.105)$$

By Theorems 2.1 and 2.2, the plant in (2.104) is observable for all z_1 but state controllable if and only if z_1 is neither $+1$ nor $+2$.

We have demonstrated in this example that state variable models corresponding to a given plant transfer function may be either state uncontrollable or unobservable for certain parameter values. Moreover, when a transfer function is given for a plant that is not both state controllable and observable, we can form one state variable model that is uncontrollable (but observable) and another that is unobservable (yet state controllable). It should be evident that no such arbitrariness occurs when the physical plant description is given in some particular state variable form initially. We emphasize once again that a state variable model for the original physical plant contains all modes of behavior but the plant transfer function has some modes missing when common factors cancel in the numerator and denominator. Although we can form state variable models from plant transfer functions, as in this example, and examine their state controllability and observability, we are no longer considering the original physical plant. As described in the following subsection, we can deduce the state controllability and observability of the original physical plant from its transfer function if and only if there have been no cancellations of common numerator and denominator factors. Whether such cancellations have occurred in its formation is generally not known when we are given only the plant transfer function.

2.5.2 Transfer Function Relationships

Two additional state controllability and observability theorems based on the plant transfer function are presented here. One is concerned with common numerator and denominator factors in the plant transfer function, and the other is based on examining the partial fraction expansion corresponding to the uncoupled state variable form.

Theorem 2.3

A linear plant is both state controllable and observable if and only if there are no common factors in the numerator and denominator of the plant transfer function.

Theorem 2.4

A linear plant having distinct denominator factors in its transfer function is both state controllable and observable if and only if, when its transfer function is expressed in a partial fraction expansion, the coefficients associated with individual denominator factors are all nonzero.

We see that Theorem 2.4 is more restrictive than Theorem 2.3, since Theorem 2.4 requires distinct denominator factors but Theorem 2.3 applies even when repeated factors appear in common in the numerator and denominator of the plant transfer function. We may use either theorem when these common numerator and denominator factors are distinct.

EXAMPLE 2.12

Consider again the transfer function in (2.101). The denominator factors are $(s + 1)$ and $(s + 2)$, and the numerator is simply $(s + z_1)$. By Theorem 2.3, the plant is both state controllable and observable if and only if z_1 is not $+1$ or $+2$.

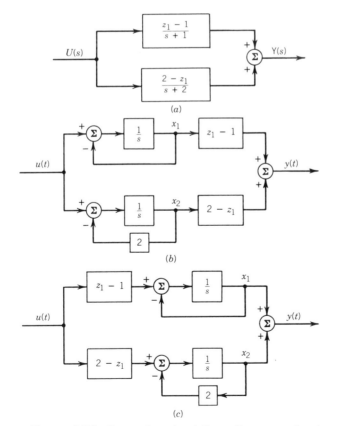

Figure 2.26 Computer simulation diagrams showing state controllability and observability for Example 2.12.

Let (2.101) be expressed in a partial fraction expansion as

$$\frac{Y(s)}{U(s)} = \frac{z_1 - 1}{s + 1} + \frac{2 - z_1}{s + 2} \tag{2.106}$$

Therefore, by Theorem 2.4, the plant again can be categorized as both state controllable and observable if and only if z_1 is neither $+1$ nor $+2$. The expansion in (2.106) corresponds to the uncoupled state variable form in Figure 2.26b, which yields

$$A = \begin{pmatrix} -1 & 0 \\ 0 & -2 \end{pmatrix} \qquad \mathbf{b} = \begin{pmatrix} 1 \\ 1 \end{pmatrix} \qquad \mathbf{c} = \begin{pmatrix} z_1 - 1 \\ 2 - z_1 \end{pmatrix} \tag{2.107}$$

The input u obviously controls both states, because it is fed into each of the two parallel paths shown in Figure 2.26b. However, if z_1 is $+1$ the state x_1 is not observed in the output. As for the direct phase variable form, the plant in (2.101) is state controllable for any z_1 but observable only if z_1 is neither $+1$ nor $+2$.

An alternate version of the uncoupled form, obtained by interchanging corresponding elements in $\mathbf{b}$ and $\mathbf{c}$, is indicated in Figure 2.26c. Thus, the plant corresponding to this figure is observable for all z_1 but state controllable if and only if z_1 is neither $+1$ or $+2$. The student should verify that Theorems 2.1 and 2.2 yield results consistent with those obtained by inspection here for these uncoupled forms.

EXAMPLE 2.13

We apply Theorems 2.1 through 2.4 to a third-order plant described by the direct phase variable form of Table 2.4. Consider the simulation diagram of Figure 2.27a with phase variables defined as the outputs of integrators as shown. In vector-matrix notation, the state variable equations are

$$\dot{\mathbf{x}} = \begin{pmatrix} 0 & 1 & 0 \\ 0 & 0 & 1 \\ 8 & 10 & 1 \end{pmatrix} \mathbf{x} + \begin{pmatrix} 0 \\ 0 \\ 1 \end{pmatrix} u$$

$$y = \begin{pmatrix} -12 \\ 3 \\ 0 \end{pmatrix}^T \mathbf{x} \tag{2.108}$$

Identifying A and $\mathbf{b}$ in (2.108), we determine the state controllability matrix in (2.99) of Theorem 2.1 as

$$M = (\mathbf{b}, A\mathbf{b}, A^2\mathbf{b}) = \begin{pmatrix} 0 & 0 & 1 \\ 0 & 1 & 1 \\ 1 & 1 & 11 \end{pmatrix} \tag{2.109}$$

which has a determinant of -1. This determinant is nonzero, so the plant in (2.108) is state controllable according to Theorem 2.1.

We next identify the vector $\mathbf{c}$ in (2.108) and, together with the matrix A already found, form the observability matrix of Theorem 2.2 as

$$Q = \left(\mathbf{c}, A^T\mathbf{c}, (A^T)^2\mathbf{c} \right) = \begin{pmatrix} -12 & 0 & 24 \\ 3 & -12 & 30 \\ 0 & 3 & -9 \end{pmatrix} \tag{2.110}$$

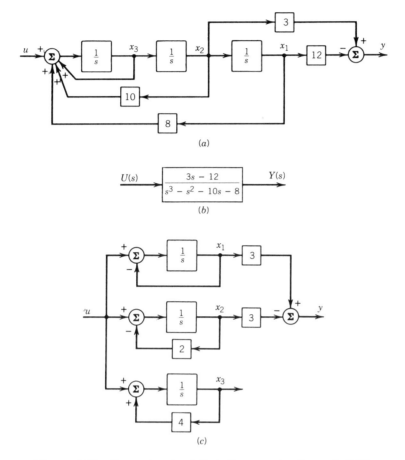

Figure 2.27 Computer simulation diagrams for Example 2.13.

which has a determinant of zero. According to Theorem 2.2, the plant in (2.108) is not observable.

Using Figure 2.22c and identifying corresponding coefficients in the third-order plant transfer function in (2.85), we may immediately express the plant transfer function for $Y(s)/U(s)$, as shown in Figure 2.27b, as

$$\frac{Y(s)}{U(s)} = \frac{3s - 12}{s^3 - s^2 - 10s - 8} \tag{2.111}$$

Factoring (2.111) yields

$$\frac{Y(s)}{U(s)} = \frac{3(s - 4)}{(s + 1)(s + 2)(s - 4)} \tag{2.112}$$

According to Theorem 2.3, the $(s - 4)$ factor common to both numerator and denominator ensures that the plant is not both state controllable and observable. This result is in agreement with our previous finding from Theorem 2.2 that the plant is unobservable. Moreover, we have now identified which particular mode is unobservable—that is, the mode associated with the denominator factor $(s - 4)$.

Expressing $Y(s)/U(s)$ in (2.112) in a partial fraction expansion yields

$$\frac{Y(s)}{U(s)} = \frac{\delta_1}{s+1} + \frac{\delta_2}{s+2} + \frac{\delta_3}{s-4} \tag{2.113}$$

where $\delta_1 = 3$, $\delta_2 = -3$, and $\delta_3 = 0$. By Theorem 2.4, the plant in (2.108) is not both state controllable and observable. The corresponding computer simulation diagram (uncoupled form) is shown in Figure 2.27c. Although the transfer function in (2.101) for Examples 2.11 and 2.12 could be expressed in the simulation diagram forms of either Figure 2.26b or 2.26c, the option has been specified for the plant in this example. Because the state variable equations are given in (2.108) and because we have shown by Theorem 2.2 that the plant is unobservable, we realize that the form corresponding to Figure 2.26b must be used for this case.

We have demonstrated the application of the four theorems on state controllability and observability in Example 2.13. Theorems 2.1 and 2.2 were used to determine that the given plant was state controllable but not observable. Theorems 2.3 and 2.4 confirmed that the plant was not both state controllable and observable. The unobservable mode for this third-order plant was an unstable one, corresponding to the denominator factor $(s - 4)$. Therefore, the time response for the state variables of the linear plant in (2.108) contains an unbounded term, that is, a term involving e^{4t}, and yet this unbounded time response is not observed at the plant output.

2.6 TRANSFER MATRICES FROM STATE VARIABLE FORMS

The problem considered in this section is the inverse of the one examined in Section 2.4; that is, here we determine the plant transfer function for single-input, single-output plants when a state variable representation is given. General transfer matrices are obtained for multivariable cases initially and then specialized to form transfer functions for single-input, single-output cases. Although it is true that the block diagram reduction procedures of Section 2.3 can be applied to a computer simulation diagram to obtain the transfer function or transfer matrix, the emphasis here is on the development of matrix relationships by performing matrix operations directly on the state equations.

The state variable description of a linear multivariable plant was given in (2.70) for the time-varying case. The corresponding time-invariant vector-matrix notation yields

$$\dot{\mathbf{x}} = A\mathbf{x} + B\mathbf{u}$$

$$\mathbf{y} = C\mathbf{x} + D\mathbf{u} \tag{2.114}$$

where the input vector $\mathbf{u}$ has m components and the output $\mathbf{y}$ has r components. As before in (2.70), the coefficient matrices are compatible in dimension with these vectors—that is, A is n by n, B is n by m, C is r by n, and D is r by m.

By taking Laplace transforms throughout in (2.114) and by setting all initial conditions to zero, we obtain

$$s\mathbf{X}(s) = A\mathbf{X}(s) + B\mathbf{U}(s)$$

$$\mathbf{Y}(s) = C\mathbf{X}(s) + D\mathbf{U}(s) \tag{2.115}$$

Solving the first equation in (2.115) for $\mathbf{X}(s)$ and substituting into the second equation yields

$$\mathbf{Y}(s) = \left[C(sI - A)^{-1}B + D\right]\mathbf{U}(s) = G(s)\mathbf{U}(s) \tag{2.116}$$

where $G(s)$ is the plant transfer matrix of dimension r by m given by

$$G(s) = C(sI - A)^{-1}B + D \tag{2.117}$$

For the single-input, single-output case ($m = r = 1$), the plant transfer matrix in (2.117) becomes a plant transfer function given by

$$G(s) = \frac{Y(s)}{U(s)} = \mathbf{c}^T(sI - A)^{-1}\mathbf{b} + d \tag{2.118}$$

EXAMPLE 2.14

We want to determine $Y(s)/U(s)$ by using (2.118) for the linear plant described by

$$\dot{x}_1 = x_2$$
$$\dot{x}_2 = x_3$$
$$\dot{x}_3 = -6x_1 - 11x_2 - 6x_3 + u(t)$$
$$y = x_1 + 4x_2 + u(t) \tag{2.119}$$

Solving for $(sI - A)^{-1}$ yields

$$(sI - A)^{-1}$$

$$= \left[\begin{pmatrix} s & 0 & 0 \\ 0 & s & 0 \\ 0 & 0 & s \end{pmatrix} - \begin{pmatrix} 0 & 1 & 0 \\ 0 & 0 & 1 \\ -6 & -11 & -6 \end{pmatrix} \right]^{-1} = \begin{pmatrix} s & -1 & 0 \\ 0 & s & -1 \\ 6 & 11 & s+6 \end{pmatrix}^{-1}$$

$$= \frac{\begin{pmatrix} s(s+6)+11 & s+6 & 1 \\ -6 & s(s+6) & s \\ -6s & -(11s+6) & s^2 \end{pmatrix}}{\det \begin{pmatrix} s & -1 & 0 \\ 0 & s & -1 \\ 6 & 11 & s+6 \end{pmatrix}}$$

$$
= \begin{pmatrix}
\dfrac{s^2 + 6s + 11}{s^3 + 6s^2 + 11s + 6} & \dfrac{s + 6}{s^3 + 6s^2 + 11s + 6} & \dfrac{1}{s^3 + 6s^2 + 11s + 6} \\[3mm]
\dfrac{-6}{s^3 + 6s^2 + 11s + 6} & \dfrac{s(s + 6)}{s^3 + 6s^2 + 11s + 6} & \dfrac{s}{s^3 + 6s^2 + 11s + 6} \\[3mm]
\dfrac{-6s}{s^3 + 6s^2 + 11s + 6} & \dfrac{-(11s + 6)}{s^3 + 6s^2 + 11s + 6} & \dfrac{s^2}{s^3 + 6s^2 + 11s + 6}
\end{pmatrix}
$$

$$(2.120)$$

where the matrix inverse was determined by dividing the transpose of the matrix of cofactors by the determinant of the matrix. Forming $Y(s)/U(s)$ from (2.118) gives

$$
\frac{Y(s)}{U(s)} = \begin{pmatrix} 1 \\ 4 \\ 0 \end{pmatrix}^T (sI - A)^{-1} \begin{pmatrix} 0 \\ 0 \\ 1 \end{pmatrix} + 1
$$

$$
= \frac{1 + 4s}{s^3 + 6s^2 + 11s + 6} + 1
$$

$$
= \frac{s^3 + 6s^2 + 15s + 7}{s^3 + 6s^2 + 11s + 6} \qquad (2.121)
$$

Alternatively, the general direct phase variable form in Table 2.4 may be utilized to obtain the result in (2.121) immediately. Moreover, the component state variable equations in (2.119) may be used directly by taking Laplace transforms of each equation and solving for the required $Y(s)/U(s)$. Finally, block diagram transformation rules or Mason's signal flow technique can be applied to the computer simulation diagram corresponding to (2.119). Although these alternate approaches require fewer computational steps for this particular example, the generality of the vector-matrix solution in (2.118) is useful in the derivation of new relationships for general, linear, time-invariant plants.

2.7 LINEAR TRANSFORMATIONS

As illustrated earlier in this chapter, a linear time-invariant plant may be described by any of several state variable forms. We can convert from one form to another by using a nonsingular linear transformation. A particular state variable form corresponds to the selection of coordinate axes describing an n-dimensional state space, where each state variable represents one of the dimension coordinates. At any time t, the state is represented by a single point in the state space defined by these n coordinate axes. As time evolves for all $t \geq t_0$, the state moves along some path in this state space. Converting from one state variable form to another is simply a transformation of coordinates—a combined rotating and stretching (or compressing) of coordinate axes. One practical utility of this linear transformation concept is that we may derive a closed-loop system design which requires, for example, that phase variables be used in some given feedback control law by performing an appropriate transformation on any other set of plant state variables.

Mathematically, it is necessary to convert from one state variable form **x** to a different form **x*** by using the linear transformation

$$\mathbf{x}^* = P^{-1}\mathbf{x} \qquad (2.122)$$

where P is a constant, nonsingular, n by n matrix. The use of a nonsingular transformation is necessary to preserve complete information about the state **x** in the new form **x***. Obviously, we may also express (2.122) as

$$\mathbf{x} = P\mathbf{x}^* \qquad (2.123)$$

2.7.1 Vector-Matrix Equations

Let the **x** form be described by

$$\dot{\mathbf{x}} = A\mathbf{x} + \mathbf{b}u$$
$$y = \mathbf{c}^T\mathbf{x} + du \qquad (2.124)$$

and the **x*** form by

$$\dot{\mathbf{x}}^* = A^*\mathbf{x}^* + \mathbf{b}^*u$$
$$y = \mathbf{c}^{*T}\mathbf{x}^* + d^*u \qquad (2.125)$$

Using (2.123) in (2.124) yields

$$P\dot{\mathbf{x}}^* = A(P\mathbf{x}^*) + \mathbf{b}u$$
$$y = \mathbf{c}^T(P\mathbf{x}^*) + du \qquad (2.126)$$

from which

$$\dot{\mathbf{x}}^* = P^{-1}AP\mathbf{x}^* + P^{-1}\mathbf{b}u$$
$$y = \mathbf{c}^TP\mathbf{x}^* + du \qquad (2.127)$$

Equating (2.125) and (2.127) and rearranging gives

$$PA^* = AP$$
$$P\mathbf{b}^* = \mathbf{b}$$
$$\mathbf{c}^* = P^T\mathbf{c} \qquad (2.128)$$

and $d^* = d$. Since P is an n by n matrix, the first equation in (2.128) yields n^2 component equations, and the second and third yield n such component equations each for a total of $n^2 + 2n$ equations. Not all of these equations are independent of each other. However, when the state variable models for both **x** and **x*** are state controllable and observable, only n^2 of these equations are linearly independent and, therefore, a unique solution for P is obtained.

EXAMPLE 2.15

Determine the matrix P that transforms the state for a second-order plant from the direct phase variable form given by

$$\dot{x}_1 = x_2$$

$$\dot{x}_2 = -2x_1 - 3x_2 + u$$

$$y = 3x_1 + x_2 \qquad (2.129)$$

to the uncoupled form

$$\dot{x}_1^* = -x_1^* + u$$

$$\dot{x}_2^* = -2x_2^* + u$$

$$y = 2x_1^* - x_2^* \qquad (2.130)$$

The solution may be obtained by using (2.128) directly and solving for the elements of the matrix P. Using the first equation in (2.128) yields

$$\begin{pmatrix} p_{11} & p_{12} \\ p_{21} & p_{22} \end{pmatrix} \begin{pmatrix} -1 & 0 \\ 0 & -2 \end{pmatrix} = \begin{pmatrix} 0 & 1 \\ -2 & -3 \end{pmatrix} \begin{pmatrix} p_{11} & p_{12} \\ p_{21} & p_{22} \end{pmatrix} \qquad (2.131)$$

which can be simplified to yield only the two equations $p_{11} = -p_{21}$ and $p_{22} = -2p_{12}$. These must be supplemented by the remaining two vector equations in (2.128) for solution. The second equation in (2.128) gives

$$\begin{pmatrix} p_{11} & p_{12} \\ p_{21} & p_{22} \end{pmatrix} \begin{pmatrix} 1 \\ 1 \end{pmatrix} = \begin{pmatrix} 0 \\ 1 \end{pmatrix} \qquad (2.132)$$

which yields $p_{11} = -p_{12}$ and $p_{21} + p_{22} = 1$. Solving these equations with those

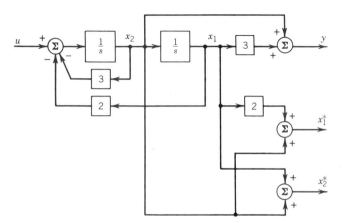

Figure 2.28 A linear transformation of state variables for Example 2.15.

obtained from (2.131), the matrix P and its inverse are

$$P = \begin{pmatrix} 1 & -1 \\ -1 & 2 \end{pmatrix}$$

$$P^{-1} = \begin{pmatrix} 2 & 1 \\ 1 & 1 \end{pmatrix} \tag{2.133}$$

Finally, we may easily verify that the third equation in (2.128) is also satisfied by the P in (2.133).

Figure 2.28 shows a simulation diagram of the original state variables in the direct phase variable form of (2.129) and the application of the transformation in (2.133) to form the new state variables $\mathbf{x}^*$. We can show by a block diagram transformation of this simulation diagram that the output combination $(2x_1^* - x_2^*)$ is identical to y, as indicated by (2.130).

2.7.2 Multivariable Plants

The procedure for linear transformations for multivariable plants is the same as for single-input, single-output plants. However, the interpretation of the results of these transformations in multivariable cases with regard to state controllability and observability is noteworthy.

Suppose a state variable set $\mathbf{x}$ is to be transformed to another set $\mathbf{x}^*$, where both sets are state controllable and observable. If we let $\mathbf{x}^* = P^{-1}\mathbf{x}$, the matrix equations for P are given by

$$PA^* = AP$$

$$PB^* = B$$

$$C^* = CP \tag{2.134}$$

The first matrix equation in (2.134) yields n^2 component equations, the second matrix equation gives nm component equations, and the third nr component equations. Of these $n^2 + nm + nr$ equations, only n^2 are linearly independent, resulting in a unique matrix P.

Similar to the single-input, single-output case, transformations to the uncoupled form are also useful in the multivariable case. Consider the general uncoupled form for multivariable plants given by

$$\dot{\mathbf{x}}^* = A^*\mathbf{x}^* + B^*\mathbf{u}$$

$$= \begin{pmatrix} \lambda_1 & 0 & \cdots & 0 \\ 0 & \lambda_2 & \cdots & 0 \\ \vdots & \vdots & & \vdots \\ 0 & 0 & \cdots & \lambda_n \end{pmatrix} \mathbf{x}^* + \begin{pmatrix} b_{11}^* & \cdots & b_{1m}^* \\ b_{21}^* & \cdots & b_{2m}^* \\ \vdots & & \vdots \\ b_{n1}^* & \cdots & b_{nm}^* \end{pmatrix} \mathbf{u}$$

$$\mathbf{y} = C^*\mathbf{x}^* + D^*\mathbf{u}$$

$$= \begin{pmatrix} c^*_{11} & \cdots & c^*_{1n} \\ c^*_{21} & \cdots & c^*_{2n} \\ \vdots & & \vdots \\ c^*_{r1} & \cdots & c^*_{rn} \end{pmatrix}\mathbf{x}^* + \begin{pmatrix} d^*_{11} & \cdots & d^*_{1m} \\ d^*_{21} & \cdots & d^*_{2m} \\ \vdots & & \vdots \\ d^*_{r1} & \cdots & d^*_{rm} \end{pmatrix}\mathbf{u} \qquad (2.135)$$

where $\lambda_1, \lambda_2, \ldots, \lambda_n$ are real and distinct. The multivariable plant in (2.135) is state controllable if and only if no row of B^* contains all zero elements. If only zeros appear in some row of B^*, then the plant inputs cannot affect the particular state variable corresponding to that row. Similarly, the plant is observable if and only if no column of the C^* matrix contains all zero elements. Otherwise, the dynamic behavior of the state variable corresponding to that column with all zero entries will not be detectable in the plant output.

SUMMARY

In this chapter we have looked at the fundamentals of mathematical modeling from both state variable and transfer function viewpoints. State variable models were formed by constructing computer simulation diagrams from transfer functions. Transfer function models were formed either by using block diagram reduction procedures on the state variable models or by performing vector-matrix operations directly. An integration of the two viewpoints was achieved in the modeling of some common control system components, in the representation of any linear system in both frameworks, and in state controllability and observability studies. Linear transformations were presented as a means of converting from one state variable form to another or from one transfer function format to any other desired format.

REFERENCES

1. D. P. Maki and M. Thompson. *Mathematical Models and Applications.* Englewood Cliffs, N.J.: Prentice-Hall, 1973.
2. A. P. Sage and J. L. Melsa. *System Identification*, Mathematics in Science and Engineering Series, Vol. 80. New York: Academic Press, 1971.
3. D. Graupe. *Identification of Systems.* Huntington, N.Y.: Robert E. Kreiger, 1976.
4. J. M. Smith. *Mathematical Modeling and Digital Simulation for Engineers and Scientists.* New York: Wiley, 1977.
5. W. H. Hayt, Jr., and J. E. Kemmerly. *Engineering Circuit Analysis*, 3rd ed. New York: McGraw-Hill, 1977.
6. M. E. Van Valkenburg. *Network Analysis*, 3rd ed. Englewood Cliffs, N.J.: Prentice-Hall, 1974.
7. E. O. Doebelin. *System Dynamics: Modeling and Response.* Columbus, Ohio: Charles E. Merrill, 1972.

8. R. Cannon. *Dynamics of Physical Systems*. New York: McGraw-Hill, 1967.

9. C. M. Close and D. K. Frederick. *Modeling and Analysis of Dynamic Systems*. Boston, Mass.: Houghton Mifflin, 1978.

10. D. A. Wells. *Lagrangian Dynamics*. Schaum's Outline Series. New York: McGraw-Hill, 1967.

11. J. Meisel. *Principles of Electromechanical Energy Conversion*. New York: McGraw-Hill, 1966.

12. J. J. D'Azzo and C. H. Houpis. *Linear Control System Analysis and Design: Conventional and Modern*, 2nd ed. New York: McGraw-Hill, 1981.

13. J. E. Gibson and F. B. Tuteur. *Control System Components*. New York: McGraw-Hill, 1958.

14. R. C. Dorf. *Modern Control Systems*, 3rd ed. Reading, Mass.: Addison-Wesley, 1980.

15. S. J. Mason and H. J. Zimmerman. *Electronic Circuits, Signals, and Systems*. Cambridge, Mass.: M.I.T. Press, 1960.

16. L. A. Zadeh and C. A. Desoer. *Linear System Theory*. New York: McGraw-Hill, 1963.

17. P. M. DeRusso, R. J. Roy, and C. M. Close. *State Variables for Engineers*. New York: Wiley, 1965.

18. R. J. Schwarz and B. Friedland. *Linear Systems*. New York: McGraw-Hill, 1965.

19. S. C. Gupta. *Transform and State Variable Methods in Linear Systems*. New York: Wiley, 1966.

20. K. Ogata. *State Space Analysis of Control Systems*. Englewood Cliffs, N.J.: Prentice-Hall, 1967.

21. H. H. Rosenbrock and C. Storey. *Mathematics of Dynamical Systems*. New York: Wiley, 1970.

22. D. M. Wiberg. *Theory and Problems of State Space and Linear Systems*, Schaum's Outline Series. New York: McGraw-Hill, 1971.

23. L. Padulo and M. A. Arbib. *System Theory*. Philadelphia, Pa.: Saunders, 1974.

24. J. R. Johnson and D. E. Johnson. *Linear Systems Analysis*. New York: Wiley, 1975.

25. J. B. Lewis, *Analysis of Linear Dynamic Systems*. Champaign, Ill.: Matrix Publishers, 1978.

26. G. M. Swisher. *Introduction to Linear System Analysis*. Champaign, Ill.: Matrix Publishers, 1978.

27. T. Kailath. *Linear Systems*. Englewood Cliffs, N.J.: Prentice-Hall, 1980.

28. D. G. Luenberger. *Introduction to Dynamic Systems*. New York: Wiley, 1979.

PROBLEMS

(§2.1) **2.1** Explain why it is desirable to utilize new data in validating a model as opposed to the original data used in generating the model.

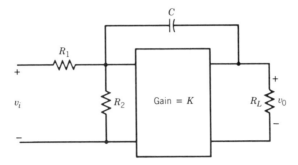

Figure 2.29 The feedback amplifier described in Problem 2.3.

(§2.1) **2.2** For the network shown in Figure 2.21, develop a detailed block diagram and determine $V_{C_2}(s)/V_i(s)$ for

    a. The network as given.
    b. The modified network obtained by removing the isolation amplifier and reconnecting it such that $v_{C_1} = R_2 i_2 + v_{C_2}$.

(§2.1) **2.3** Construct a detailed block diagram and determine $V_0(s)/V_i(s)$ for the feedback amplifier given in Figure 2.29. The amplifier has a very high gain K, infinite input impedance, and zero output impedance.

(§2.1) **2.4** For the translational mechanical system of Figure 2.30, determine static operating points (y_{is}), write dynamic equations and construct detailed block diagrams for the variations δy_i, and determine $\delta Y_1(s)/F(s)$. Let l_i be the unstretched length of spring K_i, and let d_i be the width of the block with mass M_i for $i = 1, 2$.

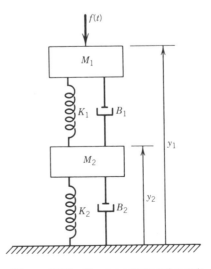

Figure 2.30 The translational mechanical system for Problem 2.4.

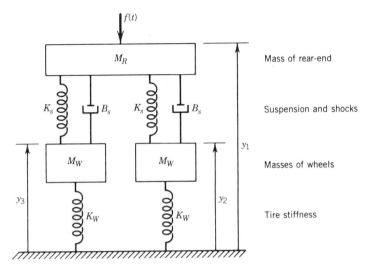

Figure 2.31 The automobile rear-end suspension system described in Problem 2.5.

(§2.1) **2.5** An automobile rear-end suspension system behaves like the translational mechanical system shown in Figure 2.31. Assume rotations about the mass centers of gravity are negligible, and solve for the static operating conditions—that is, find y_{1s}, y_{2s}, and y_{3s}. Provide a detailed block diagram for variations δy_1, δy_2, and δy_3, and then determine the transfer function $\delta Y_1(s)/F(s)$.

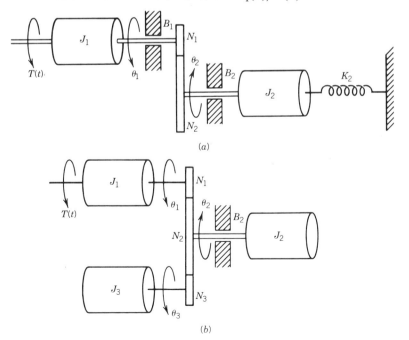

Figure 2.32 Rotational mechanical system for Problem 2.7.

(§2.1) **2.6** A space capsule, tumbling slowly in the light atmosphere of a distant planet, has an input control torque gauged to maintain a constant rotational spin of 0.1 rad/sec. Draw a schematic diagram and a detailed block diagram, and determine the plant transfer function relating the input control torque to the capsule's angular position.

(§2.1) **2.7** For the rotational mechanical systems of Figure 2.32, construct detailed block diagrams and find $\theta_1(s)/T(s)$.

(§2.1) **2.8** Explain the operation of the solenoid shown in Figure 2.33 and write the electromechanical equations describing its dynamic behavior. Draw a detailed block diagram and find the open-loop transfer function for $\delta Y(s)/V_i(s)$.

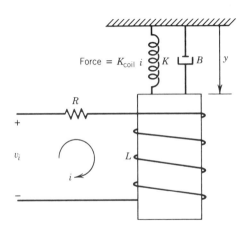

Figure 2.33 The solenoid described in Problem 2.8.

(§2.1) **2.9** Use the static linearization procedure of Section 1.4 to form a linearized equation for the electromechanical system with the movable plate capacitor shown in Figure 2.34. This linearized equation is valid for sufficiently small variations about the static operating point (y_0, q_0). The nonlinear differential equations that describe the system dynamics are

$$M\ddot{y} + B\dot{y} + K(y_0 + y) + \frac{1}{2\varepsilon A}(q_0 + q)^2 = f(t)$$

$$Rq + \frac{1}{\varepsilon A}(q_0 + q)(y_0 + y) = V + v_i$$

where y is the plate position, q is the charge $(q = di/dt)$, A is the area of the plate, ε is the dielectric constant, and V is a dc voltage

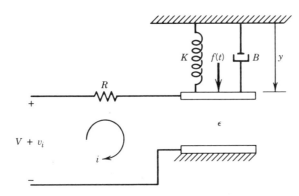

Figure 2.34 The movable plate capacitor described in Problem 2.9.

equal to $q_0 y_0 / \varepsilon A$. Construct a detailed block diagram using the linearized system equations and determine the four transfer functions relating the outputs $\delta Y(s)$ and $\delta Q(s)$ to inputs $V_i(s)$ and $F(s)$.

(§2.1) **2.10** Construct a detailed block diagram of the field-controlled dc motor with the specified parameters in Figure 2.35. Find the plant transfer

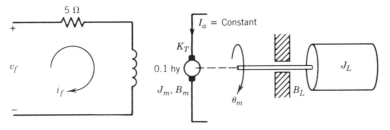

Figure 2.35 The field-controlled dc motor described in Problem 2.10.

function $\theta_m(s)/V_f(s)$. In addition to the specified R_f and L_f, other motor and load parameters are $K_T = 0.1$ newton-meters/amp, J_L and J_m are 0.2 and 0.05 newton-meters/radian per sec^2, and B_L and B_m are 0.1 and 0.02 newton-meters/radian per sec.

(§2.1) **2.11** Repeat the instructions of Problem 2.10 for the armature-controlled dc motor in Figure 2.36. Find the plant transfer function $\theta_m(s)/V_a(s)$. The values of J_L and B_L are identical to those of Problem 2.10 and K_T' is 0.2 newton-meters/amp. The value of K_b is 0.2 V/radians per second. J_m and B_m both may be considered negligible.

(§2.1) **2.12** Show that for an armature-controlled dc motor the back emf constant K_b and the torque constant K_T' are numerically equivalent

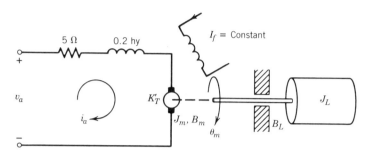

Figure 2.36 The armature-controlled dc motor described in Problem 2.11.

in SI units, but that $K_b \simeq 1.36K_T'$ in English units. [*Hint:* Neglecting losses, we equate the electrical power supplied to the motor armature $(v_b i_a)$ to the mechanical power delivered to the motor shaft $(T_D \dot{\theta}_m)$, where $v_b = K_b \dot{\theta}_m$ and $T_D = K_T' i_a$.]

(§2.1) **2.13** Develop a detailed block diagram for the positional servomechanism in Figure 2.37 that includes a field-controlled dc motor. Identify all physical variables on your block diagram, and then determine the closed-loop transfer function $\theta_c(s)/\theta_r(s)$. The motor parameters are given in Problem 2.10.

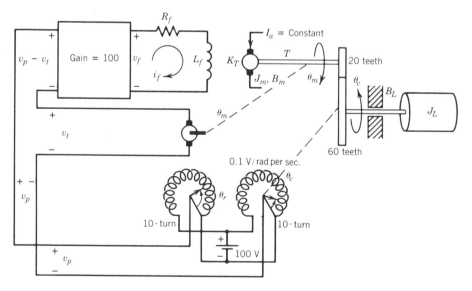

Figure 2.37 The dc positional servomechanism for Problem 2.13.

(§2.1) **2.14** The speed control system of Figure 2.38 features a generator-motor combination known as a Ward-Leonard set. Explain the physical operation of the system, construct a detailed block diagram model, and determine $\theta_c(s)/V_R(s)$.

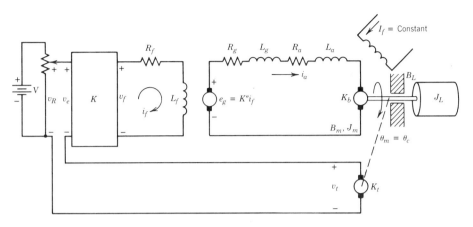

Figure 2.38   The speed control system of Problem 2.14.

(§2.2) **2.15**   Determine $Y(s)/R(s)$ for the block diagram in Figure 2.39 by using (a) the algebraic approach, (b) the transformation rules of Table 2.2, and (c) Mason's signal flow technique.

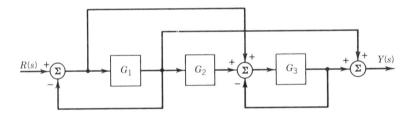

Figure 2.39   The block diagram for reduction in Problem 2.15.

(§2.2) **2.16**   Determine $Y(s)/R(s)$ for the block diagram shown in Figure 2.40.

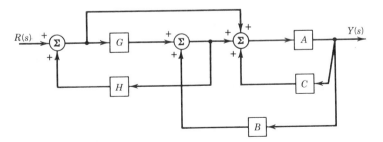

Figure 2.40   The block diagram for reduction in Problem 2.16.

(§2.2) **2.17**   Use Mason's signal flow technique to find $Y(s)/R(s)$ for the block diagram in Figure 2.41.

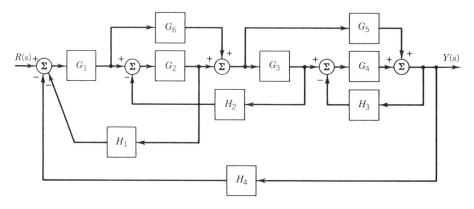

Figure 2.41 The block diagram for reduction in Problem 2.17.

(§2.2) **2.18** Use Mason's signal flow technique to determine the elements of the matrix $G(s)$ in Figure 2.42, where

$$\begin{pmatrix} Y(s) \\ V(s) \end{pmatrix} = G(s) \begin{pmatrix} R(s) \\ T(s) \end{pmatrix}$$

and

$$G(s) = \begin{pmatrix} G_{11}(s) & G_{12}(s) \\ G_{21}(s) & G_{22}(s) \end{pmatrix}$$

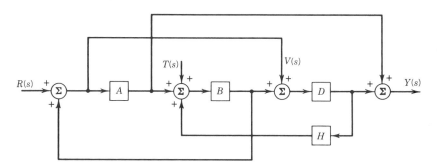

Figure 2.42 The block diagram of a multivariable system for reduction in Problem 2.18.

(§2.2) **2.19** Given the following plant transfer functions, construct detailed block diagrams from which each could have been derived. The final diagram should contain only one "letter" in each block, the same "letter" should not appear in two blocks, and no two "letters" should appear in a simple cascade arrangement without either a summer or pickoff point between them. Outputs from blocks may be either added or subtracted at summing points. It may be necessary to introduce some unity-gain branches into the diagram.

a. $\dfrac{Y(s)}{U(s)} = \dfrac{CFP}{1 - AF - GP - DFP - BC + ABCF}$

b. $\dfrac{Y(s)}{U(s)}$

$$= \dfrac{ACE - E + CE - BE - GE}{1 - AG - AB - EF - CD + AEFG + BD + DG + ABEF}$$

(§§2.1, **2.20** For the electrical network given in Figure 2.43,
2.3)

    a. Write network equations describing the dynamic behavior of the network.

    b. Express these equations in state variable vector-matrix notation using the voltages across the capacitors as state variables.

    c. Determine $V_0(s)/V_i(s)$.

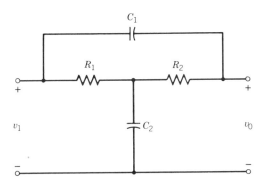

Figure 2.43 The electrical network described in Problem 2.20.

(§§2.1, **2.21** Write state variable equations in vector-matrix form for the electri-
2.3, cal network of Figure 2.44 by selecting
2.4)

    a. Voltages across capacitors as state variables.

    b. The direct phase variable form of state variables.

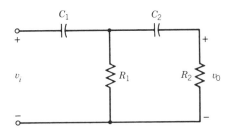

Figure 2.44 The electrical network described in Problem 2.21.

(§2.3) **2.22** Using vector-matrix notation, express the dynamic equations for incremental variations δy_1 and δy_2 in a state variable form for the translational mechanical system of Figure 2.30. Let the output vector $\delta \mathbf{y}$ be composed of δy_1 and δy_2.

(§2.4) **2.23** Consider the plant transfer function

$$G(s) = \frac{Y(s)}{U(s)} = \frac{5(s^2 + 9s + 20)}{s^3 + 6s^2 + 11s + 6} = \frac{5(s + 4)(s + 5)}{(s + 1)(s + 2)(s + 3)}$$

where $Y(s)$ is the output and $U(s)$ is the input. Construct computer simulation diagrams, select state variables as the outputs of integrators, and express the resulting state variable equations in vector-matrix notation for the

a. Direct phase variable form.
b. Direct feed-forward form.
c. Uncoupled form.
d. Any one of several possible cascade or tandem forms.

(§2.4) **2.24** Express the following linear plant with input u and output y in the direct feed-forward form. Construct an appropriate computer simulation diagram and write the resulting state variable equations in vector-matrix notation for the direct feed-forward form.

$$\dot{x}_1 = -x_1 + u(t)$$

$$\dot{x}_2 = x_3$$

$$\dot{x}_3 = -12x_2 - 7x_3 + \dot{u}(t)$$

$$y = x_1 + 2x_2$$

(§§2.4, 2.5) **2.25** Express the plant transfer function given below in state variable representation by using (a) the direct phase variable form, (b) the direct feed-forward form, and (c) a tandem or cascade form. For each form, determine whether the plant is state controllable and observable by using the matrices of Theorems 2.1 and 2.2. Show that your answers are consistent with Theorems 2.3 and 2.4 on the cancellation of common numerator and denominator factors.

$$G(s) = \frac{5s + 5}{s^2 + 7s + 6}$$

(§2.5) **2.26** Consider the two subsystems described below, where the input of Subsystem 2 is obtained as the output y_1 of Subsystem 1. Use Theorems 2.1 and 2.2 to show that the combined (cascaded) system having input u and output y is not both state controllable and observable, even though each individual subsystem is state controllable and observable. Use Theorem 2.3 to aid in explaining your result.

Subsystem 1:

$$\dot{x}_1 = 2x_2 + 4u$$

$$\dot{x}_2 = x_1 + x_2 + u$$

$$y_1 = x_2$$

Subsystem 2:

$$\dot{x}_3 = -3x_3 + y_1$$

$$y = x_2 - 5x_3$$

(§2.5) **2.27** A linear plant is said to be output controllable if the input u can be used to transfer the output y to any prescribed state in finite time. In terms of matrices, output controllability is guaranteed if and only if at least one of the following scalars is nonzero: $\mathbf{c}^T\mathbf{b}, \mathbf{c}^T A\mathbf{b}, \mathbf{c}^T A^2\mathbf{b}, \ldots, \mathbf{c}^T A^{(n-1)}\mathbf{b}, d$. Determine whether the following plants are state controllable, observable, and output controllable.

a.

$$\dot{x}_1 = x_2$$

$$\dot{x}_2 = -2x_1 - 3x_2 + u$$

$$y = x_1 + x_2$$

b.

$$\dot{x}_1 = -3x_1 + x_2 + u$$

$$\dot{x}_2 = -2x_2 + u$$

$$y = x_1$$

(§§2.2, **2.28** Determine $Y(s)/U(s)$ for the plant given below by using
2.6)

a. $\dfrac{Y(s)}{U(s)} = \mathbf{c}^T(sI - A)^{-1}\mathbf{b} + d.$

b. Any one of the block diagram reduction procedures of Section 2.2.

$$\dot{x}_1 = -3x_1 + x_2 + u$$

$$\dot{x}_2 = -2x_2 + x_3 + 5u$$

$$\dot{x}_3 = -x_1 + 2u$$

$$y = x_1 + u$$

(§§2.2, **2.29** Repeat the instructions of Problem 2.28 for the plant transfer
2.6)   function given by

$$\dot{x}_1 = -x_1 + u$$

$$\dot{x}_2 = -3x_2 + 4x_3 + u$$

$$\dot{x}_3 = -2x_2 - 5x_3 + 2u$$

$$y = x_1 + x_3 + 3u$$

(§§2.4, **2.30** Given the plant transfer function
2.7)

$$\frac{Y(s)}{U(s)} = \frac{s+4}{s^2 + 5s + 6}$$

    a. Express the plant in direct phase variable form and label this
state variable set as **x**.

    b. Express the plant in the uncoupled form and label this state
variable set as **x***.

    c. Determine the linear transformation matrix P such that $\mathbf{x}^* = P^{-1}\mathbf{x}$.

(§§2.1, **2.31** Let the state variables **x** for an RLC series circuit be defined as the
2.3,   current through the inductor i_L and the voltage across the capacitor
2.7) v_C; that is,

$$\mathbf{x} = \begin{pmatrix} i_L \\ v_C \end{pmatrix}$$

Let **x*** be defined as the voltage across the inductor v_L and the
voltage across the resistor v_R; that is,

$$\mathbf{x}^* = \begin{pmatrix} v_L \\ v_R \end{pmatrix}$$

    a. Write the state variable equations for **x** and **x***.

    b. Determine the linear transformation matrix P such that $\mathbf{x}^* = P^{-1}\mathbf{x}$.

(§§2.1, **2.32** Treat $v_1(t)$ and $v_2(t)$ as plant inputs and $i_L(t)$ and $v_C(t)$ as outputs
2.3,   in the electrical network shown in Figure 2.45.

2.4,
2.5,   a. Write the dynamic equations that describe the system behavior.

2.7)   b. Form the plant transfer matrix $G(s)$.

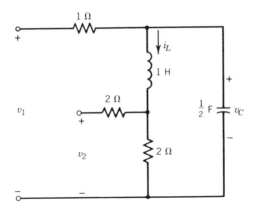

Figure 2.45 The multivariable electrical network described in Problem 2.32.

c. Express the multivariable plant equations in the direct phase variable form, the direct feed-forward form, and the uncoupled form using vector-matrix notation.

d. Determine whether the multivariable plant is state controllable, observable, or output controllable (see Problem 2.27) by using the matrix definitions. Verify by examining B^* and C^* for the uncoupled form.

(§§2.1, **2.33** Repeat the instructions of Problem 2.32 for the mechanical system
2.3,   shown in Figure 2.46. Treat $f_1(t)$ and $f_2(t)$ as inputs and y and θ
2.4,   as outputs. Let l be the length of the spring when no force is
2.5,   applied.
2.7)

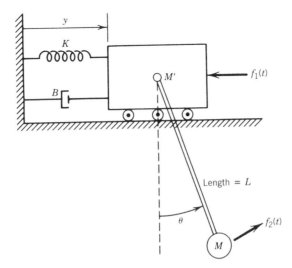

Figure 2.46 The multivariable mechanical system described in Problem 2.33.

3

Feedback Characteristics

Interconnecting plants, sensors, and controllers for closed-loop operation yields a system capable of continuously monitoring its own performance. Sensors keep track of plant state variables for use in a feedback controller that applies corrective signals to the plant based on computed errors between desired and actual plant outputs. The feedback configuration has many interesting properties: among others, it can provide an increased system bandwidth at the expense of reduced system gain; it can reduce or eliminate steady-state errors for certain polynomial test inputs; and it can provide a reduced system sensitivity in the presence of plant modeling errors and output disturbances. Robotics and other sensory applications employing the feedback mechanism as a key feature continue to expand rapidly and lead the way for high-technology thrusts.

We introduced the feedback concept in Chapter 1 with a description of open-loop and closed-loop system applications and proceeded in Chapter 2 to form mathematical models of typical system components. Continuing this modeling emphasis, we begin this chapter by discussing some of the properties evident in closed-loop system operations. We describe the characteristics of plant models in negative unity-feedback systems that yield zero steady-state errors for certain polynomial test inputs. We present sensitivity studies to show that overall system errors resulting from errors in the models of Chapter 2 are reduced when those models are placed in feedback configurations. We then introduce a traditional control law (PID control) and conclude the chapter by discussing linear state variable feedback control.

3.1 FEEDBACK EFFECTS

It is important at this early stage of our study of control systems to pinpoint key characteristics of systems having feedback [1-16]. We identify some of these characteristics or properties as

1. Gain versus bandwidth tradeoffs.
2. Reduced steady-state errors for polynomial test inputs.
3. Reduced system sensitivity to parameter and output disturbances.

102

4. Shaping of the time response.
5. Reduced stability with increased gain.

We describe in this section the relative importance of these feedback characteristics and their interrelationships in selected applications. Although the control engineer may regard some features as having major importance, other features as highly desirable, and yet others as having only minor importance, feedback characteristics are often not independent of each other. As examples, we examine systems that illustrate communication-at-a-distance, tracking capabilities, and power amplification.

3.1.1 Communication-at-a-Distance

The satellite shown in Figure 3.1 is built for long-term reliability and adaptability and, consequently, many of those operations ordinarily assigned to the satellite are performed instead by the earth station. For example, the earth station

Figure 3.1 One of three telecommunications satellites in the Tracking and Data Relay Satellite System operated by the National Aeronautics and Space Administration.

Source: Courtesy of TRW Electronics and Defense Sector.

performs the control and tracking functions of the satellite's single-access antennas as well as forming and controlling the beam of the multiple-access phased array. In this case a major consideration is to reduce sensitivity to undesirable disturbance inputs, such as spurious signals and impacting micrometeorites. Moreover, the elimination of steady-state errors for step and ramp steering command inputs is essential. It is also desirable to have an adequate closed-loop bandwidth for feedback control and for the band of frequencies encountered in these disturbance inputs. However, the transient response and the associated relative stability consideration have only minor importance, provided the system remains stable and transients are damped out quickly enough for a suitable operation.

3.1.2 Tracking Systems

Rapid tracking of maneuvering targets is illustrated in Figure 3.2. The most important feedback characteristics for these tracking system applications are the need for speed and accuracy in the shaping of the tracking system time response and the elimination of steady-state errors. It is also desirable to have a suitable closed-loop bandwidth and to have a reduced sensitivity to unwanted disturbances. Relative stability considerations are of minor importance, since the destabilizing effect of increased loop gain can actually be beneficial by resulting in a faster speed of response in following a maneuvering target.

3.1.3 Power Amplification

Consider the rudder control system of the ship shown in Figure 3.3a. The main feedback characteristic in this application is the elimination of steady-state errors resulting from step command inputs. It is also desirable from a practical standpoint to prohibit oscillations about the desired course and, therefore, the proper shaping of the time response has some importance. In addition, increasing the forward-loop gain in this power amplification example can lead to instabilities, which must be avoided in all cases. Again, a reduced sensitivity to unwanted disturbances is desirable. However, since only slowly varying commands are expected, a wide bandwidth is not needed in this case.

Table 3.1 shows a chart summarizing the results of these three applications. We emphasize once again that the characteristics identified here are not to be regarded as independent considerations. For example, a highly oscillatory transient response indicates a low relative stability. In fact, stability is always a concern in any application and must be guaranteed as an inherent part of the design procedure.

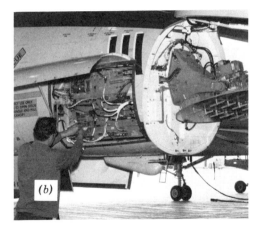

Figure 3.2 Aspects of target tracking are illustrated by (*a*) this rendezvous radar which is capable of detecting and tracking targets as small as one square meter at ranges up to 22 km in space, (*b*) the AWG-9 weapons control system which can track multiple aircraft targets at distances of more than 180 km, and (*c*) the assembling of an automatic tracking sensor of the A-6E target recognition attack multisensor.

Source: Courtesy of Hughes Aircraft Company.

Figure 3.3 Automated controls guide virtually every function of this Sea vessel (*a*) as it performs complex well stimulation treatments in the North Sea. The computer-aided control center in (*b*) provides automatic monitoring and operation control of pressures, temperatures, and chemical variables, which are transmitted to shore in real time.

Sources: Courtesy of Halliburton Services.

TABLE 3.1   Feedback Applications and Feedback Characteristics

Application \ Characteristic	Bandwidth Considerations	Steady-State Errors	System Sensitivity	Time Response Shaping	Relative Stability
Satellite attitude control (Communication-at-a-distance)	●	★	★	○	○
Radar tracking systems	●	★	●	★	○
Ship rudder control (Power amplification)	○	★	●	●	●

Legend: ★ Major Importance
          ● Desirable
          ○ Minor Importance

3.2 GAIN VERSUS BANDWIDTH TRADEOFFS

In this section we demonstrate that increasing negative feedback typically results in a larger bandwidth for the closed-loop system, but at the expense of a reduced system gain. This characteristic has been important historically in the design of linear electronic feedback amplifiers. We show that the system gain-bandwidth product is constant with respect to variations in the feedback gain β for first-order systems and approximately constant for those higher-order systems that have a single dominant pole.

Tradeoffs between system gain and system bandwidth are based on the system response to sinusoidal inputs of frequency ω. The procedure is to express the closed-loop system transfer function in terms of ω by setting $s = j\omega$. We define the system bandwidth as that value of ω at which the magnitude of the system frequency response is reduced to $1/\sqrt{2} \simeq 0.707$ of its low-frequency (or dc) gain, which is nonzero for the systems considered here.

EXAMPLE 3.1

Consider the first-order system of Figure 3.4*a*. By using Rule 5 of Table 2.2, we obtain the closed-loop transfer function as

$$\frac{Y(s)}{R(s)} = \frac{G(s)}{1 + G(s)H(s)} = \frac{\dfrac{K}{s+1}}{1 + \left(\dfrac{K}{s+1}\right)\beta} = \frac{K}{s + 1 + K\beta} \tag{3.1}$$

Setting $s = j\omega$ in (3.1) gives

$$\frac{Y(j\omega)}{R(j\omega)} = \frac{K}{j\omega + 1 + K\beta} \tag{3.2}$$

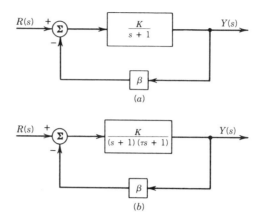

Figure 3.4 Systems for gain versus bandwidth calculations.

The low-frequency gain of the closed-loop system is obtained from (3.2) by setting ω equal zero to give

$$\text{Gain}\bigg|_{\omega=0} = \frac{Y(j\omega)}{R(j\omega)}\bigg|_{\omega=0} = \frac{K}{1 + K\beta} \tag{3.3}$$

The bandwidth (BW) satisfies

$$\left|\frac{Y(j\omega)}{R(j\omega)}\right|\bigg|_{\omega=BW} = \frac{1}{\sqrt{2}}\left|\frac{Y(j\omega)}{R(j\omega)}\right|\bigg|_{\omega=0} \tag{3.4}$$

which may be expressed as

$$\frac{K}{\sqrt{(BW)^2 + (1 + K\beta)^2}} = \frac{1}{\sqrt{2}}\left(\frac{K}{1 + K\beta}\right) \tag{3.5}$$

Solving (3.5) for the bandwidth BW gives

$$BW = 1 + K\beta \tag{3.6}$$

Using (3.3) and (3.6) yields a constant system gain-bandwidth product as

$$\left[\text{Gain}\big|_{\omega=0}\right] \cdot \left[\text{Bandwidth}\right] = K \tag{3.7}$$

While the system gain-bandwidth product depends on K, we observe that it is constant with respect to variations in the feedback gain β.

EXAMPLE 3.2

Repeating the procedures of Example 3.1 for the second-order system of Figure 3.4*b* yields a closed-loop transfer function as

$$\frac{Y(s)}{R(s)} = \frac{K}{\tau s^2 + (\tau + 1)s + 1 + K\beta} \tag{3.8}$$

Setting $s = j\omega$ gives

$$\frac{Y(j\omega)}{R(j\omega)} = \frac{K}{j\omega(\tau + 1) + 1 + K\beta - \tau\omega^2} \tag{3.9}$$

Setting ω equal zero in (3.9) yields the low-frequency gain as

$$\left. \text{Gain} \right|_{\omega=0} = \left. \frac{Y(j\omega)}{R(j\omega)} \right|_{\omega=0} = \frac{K}{1 + K\beta} \tag{3.10}$$

The system bandwidth BW is determined from

$$\left| \frac{K}{j(BW)(\tau + 1) + 1 + K\beta - \tau(BW)^2} \right| = \frac{1}{\sqrt{2}}\left(\frac{K}{1 + K\beta} \right) \tag{3.11}$$

which gives

$$\frac{K}{\sqrt{[(\tau + 1)(BW)]^2 + \left(1 + K\beta - \tau(BW)^2\right)^2}} = \frac{1}{\sqrt{2}}\left(\frac{K}{1 + K\beta} \right) \tag{3.12}$$

The desired bandwidth BW, from (3.12), satisfies

$$\tau^2(BW)^4 + \left[(\tau + 1)^2 - 2\tau(1 + K\beta)\right](BW)^2 - (1 + K\beta)^2 = 0 \tag{3.13}$$

We solve the quadratic equation in (3.13) for $(BW)^2$ for this second-order system and then form the square root to give BW as

$$BW = \sqrt{\frac{-(1 + \tau^2 - 2\tau K\beta) + \sqrt{(1 + \tau^2 - 2\tau K\beta)^2 + 4\tau^2(1 + K\beta)^2}}{2\tau^2}} \tag{3.14}$$

For $K = 1$ and $\tau = 0.1$, the system bandwidth BW in (3.14) is 0.99 for $\beta = 0$, 1.55 for $\beta = 0.5$, and 2.16 for $\beta = 1$, corresponding to low-frequency system gains; that is, $1/(1 + \beta)$, of 1, 0.67, and 0.5 to yield system gain-bandwidth products of 0.99, 1.03, and 1.08, respectively. With a maximum difference of only approximately 9% between system gain-bandwidth products for $\beta = 0$ and $\beta = 1$,

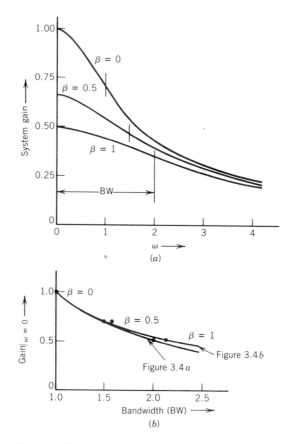

Figure 3.5 System gain and system bandwidth relationships for the systems in Figure 3.4.

we note that the closed-loop system transfer function appears to maintain a single dominant pole as β varies over the range $[0, 1]$.

Figure 3.5a shows typical plots of system gain versus frequency using several values of β for the system in Figure 3.4a. Figure 3.5b shows curves of low-frequency system gain versus system bandwidth for both systems in Figure 3.4. A gain-bandwidth tradeoff is necessary in those applications where both features are important.

3.3 STEADY-STATE ERRORS

Both transfer functions and state variables are utilized in this section for a systematic study of steady-state errors for unity-feedback systems [1–7]. From the transfer function approach, we show how to apply the final-value theorem to the

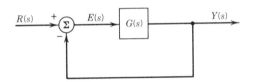

Figure 3.6 System configuration for steady-state error analysis.

transfer function of the error for stable systems to yield its steady-state value for given inputs. Moreover, closed-loop systems are characterized as Type $0, 1, 2, \ldots, m$ and give prescribed errors according to their types and inputs. Steady-state errors are also determined from state variable equations by first identifying which components of $\dot{\mathbf{x}}$ can be set equal to zero and then solving the resulting linear algebraic equations for the equilibrium states.

3.3.1 Error Constants and Types of Systems

Consider the unity-feedback system of Figure 3.6 with $E(s)$ given by

$$E(s) = \frac{R(s)}{1 + G(s)} \tag{3.15}$$

Let $G(s)$ be expressed in the special form

$$G(s) = \frac{K(1 + T_a s)(1 + T_b s) \cdots}{s^m (1 + T_1 s)(1 + T_2 s) \cdots} \tag{3.16}$$

The number m of pure integrations—that is, the exponent of s in the denominator of $G(s)$—determines the "type" of the system. If m equals 0, the system is a Type 0 system; m equals 1 denotes a Type 1 system; m equals 2 identifies the system as Type 2; and so forth.

Steady-state errors due to polynomial test inputs, such as steps, ramps, and parabolic inputs, can be determined in terms of error constants. Let the positional error constant be defined as

$$K_p \triangleq \lim_{s \to 0} G(s) \tag{3.17}$$

Furthermore, let velocity and acceleration error constants be defined as

$$K_v \triangleq \lim_{s \to 0} [sG(s)] \tag{3.18}$$

$$K_a \triangleq \lim_{s \to 0} [s^2 G(s)] \tag{3.19}$$

The final-value theorem used to obtain the steady-state error e_{ss} for stable systems can be stated briefly as

$$f_{ss} = \lim_{t \to \infty} f(t) = \lim_{s \to 0} \left[sF(s) \right] \tag{3.20}$$

where $f(t)$ is the time function of interest, f_{ss} is its final value, and $F(s)$ is the corresponding Laplace transform. This final-value theorem may be used to determine the steady-state value of $f(t)$ without explicitly solving for $f(t)$ but by using only its Laplace transform $F(s)$. Restrictions are that the Laplace transform must be defined for both $f(t)$ and its first derivative and the inverse Laplace transform of $sF(s)$ must decay to zero as t approaches infinity. If the function of interest is the error $e(t)$ for some system, we require that the system be stable.[1]

By using the final-value theorem on $E(s)$ in (3.15), we obtain

$$e_{ss} = \lim_{t \to \infty} e(t) = \lim_{s \to 0} \left[sE(s) \right]$$

$$= \lim_{s \to 0} \left[\frac{sR(s)}{1 + G(s)} \right] \tag{3.21}$$

If $R(s)$ is the Laplace transform $1/s$ of a unit step input, then e_{ss} in (3.21) becomes

$$e_{ss} = \lim_{s \to 0} \left[\frac{s(1/s)}{1 + G(s)} \right]$$

$$= \frac{1}{1 + \lim_{s \to 0} G(s)} = \frac{1}{1 + K_p} \tag{3.22}$$

For a Type 0 system of the form given in (3.16), K_p in (3.17) is K, but for Type 1 and 2 systems, K_p is ∞. Thus, e_{ss} in (3.21) is $1/(1 + K)$ for a Type 0 system and zero for Type 1 and Type 2 systems. In brief, there will be a finite offset, that is, $1/(1 + K)$, for a Type 0 system, but the output will match the step input as $t \to \infty$ for Type 1 and Type 2 systems, provided these systems are stable.

For a unit ramp input, $r(t) = t$ for $t \geq 0$ and $R(s) = 1/s^2$. Therefore,

$$e_{ss} = \lim_{s \to 0} \left[\frac{s(1/s^2)}{1 + G(s)} \right]$$

$$= \frac{1}{\lim_{s \to 0} \left[s + sG(s) \right]} = \frac{1}{\lim_{s \to 0} \left[sG(s) \right]} = \frac{1}{K_v} \tag{3.23}$$

---

[1]Specifically, we require that the system be asymptotically stable, a concept discussed more fully in Chapter 5. This restriction eliminates the possibility of steady-state sinusoidal oscillations.

TABLE 3.2   Steady-State Errors for Stable Unity-Feedback Systems

System Type	Unit Step Input $[R(s) = 1/s]$ $e_{ss} = \dfrac{1}{1 + K_p}$	Unit Ramp Input $[R(s) = 1/s^2]$ $e_{ss} = \dfrac{1}{K_v}$	Parabolic Input $[R(s) = 1/s^3]$ $e_{ss} = \dfrac{1}{K_a}$
0	$\dfrac{1}{1 + K}$	∞	∞
1	0	$\dfrac{1}{K}$	∞
2	0	0	$\dfrac{1}{K}$
$\vdots$			
$m \,[\geq 3]$	0	0	0

The velocity error constant K_v in (3.18) is zero for a Type 0 system, K for a Type 1 system, and ∞ for a Type 2 system. The steady-state errors are ∞, $1/K$, and zero, respectively. Corresponding results for a parabolic input—that is, $r(t) = \frac{1}{2}t^2$ for $t \geq 0$ and $R(s) = 1/s^3$—yield

$$e_{ss} = \frac{1}{K_a} \tag{3.24}$$

The acceleration error constant K_a in (3.19) is zero for Type 0 and Type 1 systems and K for Type 2 systems, which gives e_{ss} of ∞, ∞, and $1/K$, respectively. These cases are summarized in Table 3.2.

We have shown that the steady-state error for a unity-feedback system is zero for step inputs, provided the system is Type 1 or higher, that is, at least one pure integration appears in the plant transfer function. Moreover, e_{ss} is zero for ramp inputs if the system is Type 2 or higher, and so forth. Rather than viewing Table 3.2 as a series of columns having particular inputs and different types of systems, we may examine the rows with particular system types and let the inputs be the varying quantity. For example, a Type 1 system can follow a step input with no steady-state error and a ramp input with a finite offset $(1/K)$, but the error increases without bounds as $t \to \infty$ for parabolic inputs. Similar comparisons may be made from the other rows of the table.

EXAMPLE 3.3

Consider the positional servomechanism of Figure 3.7a. The dc motor is field controlled, and positional sensing is achieved by using an n-turn potentiometer error detector. This system is a simplified version of the one shown in Figure 2.12. A detailed block diagram is given in Figure 3.7b, with reductions shown in Figure

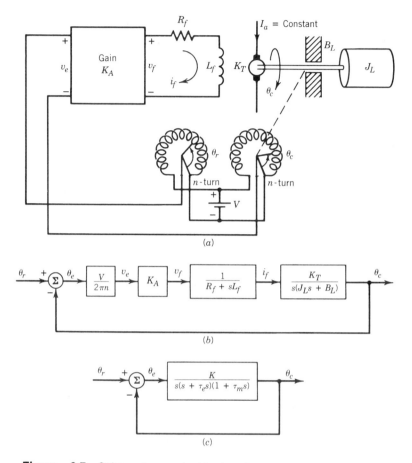

Figure 3.7 Schematic and block diagrams of a positional servomechanism.

3.7c. The plant transfer function is given by

$$G(s) = \frac{K}{s(1 + \tau_e s)(1 + \tau_m s)} \tag{3.25}$$

where

$$K = \frac{V}{2\pi n}\left(\frac{K_A K_T}{R_f B_L}\right) \tag{3.26}$$

$$\tau_e = \frac{L_f}{R_f} \tag{3.27}$$

$$\tau_m = \frac{J_L}{B_L} \tag{3.28}$$

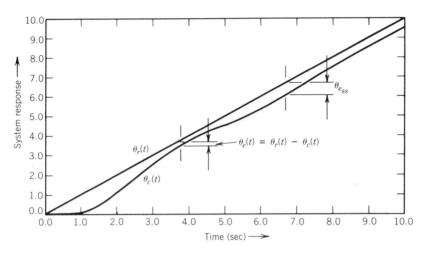

Figure 3.8 A ramp-input time response and steady-state error for the positional servomechanism of Figure 3.7.

We easily identify the system in Figure 3.7c as a Type 1 system, since an s factor appears in the denominator of the plant transfer function and the system has negative unity feedback. Consequently, the positional error constant K_p is ∞, the velocity error constant K_v is K, and the acceleration error constant K_a is zero. From Table 3.2, the steady-state errors are zero for step inputs, $1/K$ for unit ramp inputs, and infinite for parabolic inputs. A sketch of the ramp input case is shown in Figure 3.8.

Suppose $\theta_r(t)$ for $t \geq 0$ is given by

$$\theta_r(t) = 2 + 3t + 10t^2 \tag{3.29}$$

The steady-state error e_{ss} is

$$e_{ss} = 2\left(\frac{1}{1 + K_p}\right) + 3\left(\frac{1}{K_v}\right) + 20\left(\frac{1}{K_a}\right)$$

$$= 2\left(\frac{1}{1 + \infty}\right) + 3\left(\frac{1}{K}\right) + 20\left(\frac{1}{0}\right)$$

$$= 2(0) + \frac{3}{K} + 20(\infty) = \infty \tag{3.30}$$

Therefore, e_{ss} is infinite because this Type 1 system cannot follow a parabolic input. Observe that if B_L were zero in Figure 3.7b, the system would become a Type 2 system. However, we can show by the methods of Chapter 5 that this system is unstable for $B_L = 0$, and the steady-state error is infinite. This case with an s^2 term appearing in the plant transfer function denominator occurs, for example, for spacecraft control in free space where atmospheric and other rotational damping forces are negligible. The system may be stabilized by using tachometer feedback, but the resulting system would again be Type 1 and could not follow a parabolic input.

3.3.2 State Variable Approach

The state variable equations to be considered for this error analysis problem are given by

$$\dot{\mathbf{x}} = A\mathbf{x} + \mathbf{b}e$$

$$y = \mathbf{c}^T\mathbf{x}$$

$$e = r(t) - y = r(t) - \mathbf{c}^T\mathbf{x} \qquad (3.31)$$

The matrix A defines the denominator of the plant transfer function. In particular, the denominator is given by $\det(sI - A)$. If A is singular, that is, $\det(A) = 0$ and A^{-1} does not exist, then the system is Type 1 or greater. This condition can readily be determined by examining the characteristic equation

$$\det(sI - A) = s^n + \alpha_{n-1}s^{n-1} + \cdots + \alpha_2 s^2 + \alpha_1 s + \alpha_0 = 0 \qquad (3.32)$$

The matrix A is singular if and only if α_0 is zero. If $\alpha_0 \neq 0$, the system is Type 0. Moreover, if $\alpha_0 = 0$ but $\alpha_1 \neq 0$, then the system is Type 1. Similarly, if $\alpha_0 = \alpha_1 = 0$ but $\alpha_2 \neq 0$, then the system is Type 2, and so forth. In each of these cases, we assume that there are no cancellations of s^m factors (for $m = 1, 2, \ldots$) between the numerator and denominator of the plant transfer function.

The attractiveness of the state variable approach is in determining e_{ss} for stable systems directly from the vector-matrix equations. Consider the case where (3.31) describes a Type 0 stable system, and let $r(t)$ be a unit step input. Since each of the state variables approaches a constant value as $t \to \infty$, we may set $\dot{\mathbf{x}} = 0$ and solve for e_{ss} as

$$\dot{\mathbf{x}} = 0 = A\mathbf{x}_{ss} + \mathbf{b}e_{ss}$$

$$\mathbf{x}_{ss} = -A^{-1}\mathbf{b}e_{ss}$$

$$e_{ss} = r(t) - \mathbf{c}^T\mathbf{x}_{ss}$$

$$e_{ss} = 1 - \mathbf{c}^T[-A^{-1}\mathbf{b}]e_{ss}$$

$$e_{ss} = \frac{1}{1 - \mathbf{c}^T A^{-1}\mathbf{b}} \qquad (3.33)$$

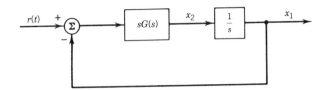

Figure 3.9 Cascaded plant arrangement of a Type 1 system for state variable steady-state error analysis.

From (3.22) and (3.33), we can easily identify

$$K_p = -\mathbf{c}^T A^{-1} \mathbf{b} \tag{3.34}$$

which agrees with $K_p = \lim_{s \to 0} G(s)$, where $G(s)$ is given by (2.118) with d set equal to zero.

Next, consider the case where (3.31) describes a Type 1 stable system. Since $\alpha_0 = 0$ for this case, A^{-1} does not exist. However, we may form a reduced-order nonsingular matrix A_r. Let the output of the single pure integration in the plant transfer function be defined as x_1, as shown in Figure 3.9, and the remaining states for $sG(s)$ be defined as $\mathbf{x}_r$. Therefore, we may write state equations for this system in Figure 3.9 as

$$\dot{x}_1 = x_2$$

$$\dot{\mathbf{x}}_r = A_r \mathbf{x}_r + \mathbf{b}_r e$$

$$e = r(t) - x_1 \tag{3.35}$$

Let $r(t)$ be either a step input or a ramp input. In either of these cases, we postulate that e_{ss} is a constant. From Table 3.2, we know that e_{ss} for Type 1 systems is actually zero for a step input and some nonzero constant for a ramp input. However, in using this state-space approach, we only assume at this point that e_{ss} does not vary with time. Thus,

$$\dot{e}_{ss} = \dot{r}_{ss}(t) - \dot{x}_{1_{ss}} = 0$$

$$x_{2_{ss}} = \dot{x}_{1_{ss}} = \dot{r}_{ss}(t) \tag{3.36}$$

Of all the state variables, only $\dot{x}_{1_{ss}}$ is potentially nonzero. In fact, $\dot{x}_{1_{ss}}$ is zero for a step input and has a value of unity for a unit ramp input. Since $\dot{\mathbf{x}}_{r_{ss}} = \mathbf{0}$,

$$\mathbf{0} = A_r \mathbf{x}_{r_{ss}} + \mathbf{b}_r e_{ss} \tag{3.37}$$

and the value of e_{ss} may be determined directly. For the Type 1 system, only $x_{2_{ss}}$ will be nonzero for a ramp input. If the direct phase variable form (see Table 2.4) is used for A_r, then the last of the component equations in (3.37) yields

$$0 = -\alpha_1 x_{2_{ss}} + K e_{ss} \tag{3.38}$$

where $\alpha_1 = 1$ and K is defined in (3.16). Therefore,

$$e_{ss} = \frac{1}{K} x_{2_{ss}} \tag{3.39}$$

which yields a zero steady-state error for a step input and $e_{ss} = 1/K$ for a unit ramp input. This result agrees with our earlier findings and satisfies the constant steady-state error assumption made initially for this Type 1 system.

EXAMPLE 3.4

Consider a unity-feedback system described by the state variable equations given by

$$\dot{x}_1 = x_2$$

$$\dot{x}_2 = x_3$$

$$\dot{x}_3 = -x_2 - 15x_3 + 10e$$

$$e = r(t) - x_1 \tag{3.40}$$

where $r(t) = 3t$. Therefore, we assume $\dot{e} = 0$ and

$$x_{2_{ss}} = \dot{x}_1 = \dot{r}(t) = 3 \tag{3.41}$$

From (3.37),

$$0 = \begin{pmatrix} 0 & 1 \\ -1 & -15 \end{pmatrix} \begin{pmatrix} x_{2_{ss}} \\ x_{3_{ss}} \end{pmatrix} + \begin{pmatrix} 0 \\ 10 \end{pmatrix} e_{ss} \tag{3.42}$$

The first component equation in (3.42) yields $x_{3_{ss}} = 0$, and the second gives

$$0 = -x_{2_{ss}} - 15x_{3_{ss}} + 10e_{ss}$$

$$e_{ss} = \frac{x_{2_{ss}}}{10} = \frac{3}{10} \tag{3.43}$$

which agrees with (3.23) and Table 3.2.

3.4 SYSTEM SENSITIVITY

Sensitivity effects in feedback systems are described by transfer functions and state variable methods in this section [17–20]. We define *system sensitivity* as the ratio of the fractional change in a system output function to the corresponding fractional change in a system parameter as the change in the system parameter approaches zero. Relationships for system sensitivity are determined for parameter variations in open-loop systems and in forward and feedback paths of closed-loop systems. The usefulness of feedback configurations in reducing the effect of output disturbances is identified and related to system parameter sensitivity. We define a sensitivity coefficient vector to denote first-order variations in parameters of the state variable equations. The relationship between system sensitivity and sensitivity coefficients is identified, and examples are presented to illustrate this interconnection of concepts.

3.4.1 Transfer Function Relationships

For any system output function F, we define the system sensitivity resulting from variations in a parameter α as

$$S_\alpha^F = \lim_{\Delta\alpha \to 0} \left[\frac{\text{Fractional change in } F}{\text{Fractional change in } \alpha} \right]$$

$$= \lim_{\Delta\alpha \to 0} \left[\frac{\Delta F/F}{\Delta\alpha/\alpha} \right] = \lim_{\Delta\alpha \to 0} \left[\frac{\Delta F}{\Delta\alpha} \left(\frac{\alpha}{F} \right) \right] = \frac{\partial F}{\partial\alpha} \left(\frac{\alpha}{F} \right) \qquad (3.44)$$

For small incremental changes in F and α, we may approximate S_α^F as

$$S_\alpha^F \cong \frac{\Delta F}{\Delta\alpha} \left(\frac{\alpha}{F} \right) \qquad (3.45)$$

Suppose we are interested in determining the effect on the closed-loop transfer function M of a variation in a system parameter α about some nominal operating value α_N. We define the system sensitivity S_α^M as

$$S_\alpha^M = \lim_{\Delta\alpha \to 0} \left[\frac{\text{Fractional change in } M}{\text{Fractional change in } \alpha} \right] = \frac{\partial M}{\partial\alpha} \left(\frac{\alpha}{M} \right) \qquad (3.46)$$

where S_α^M is generally frequency dependent, that is, a function of s, since M is generally a function of s. Assume that α is contained only in a single transfer function G within a larger system. We want to determine the sensitivity of M to α in terms of the sensitivities of M to G and of G to α. We use the chain rule to form

$$\frac{\partial M}{\partial\alpha} = \frac{\partial M}{\partial G} \left(\frac{\partial G}{\partial\alpha} \right) \qquad (3.47)$$

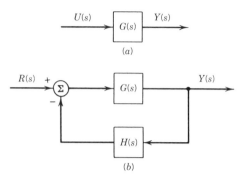

Figure 3.10 System configurations for sensitivity calculations.

from which

$$S_\alpha^M = \frac{\partial M}{\partial \alpha}\left(\frac{\alpha}{M}\right) = \left[\frac{\partial M}{\partial G}\left(\frac{\partial G}{\partial \alpha}\right)\right]\cdot\left[\frac{\alpha}{G}\left(\frac{G}{M}\right)\right] \qquad (3.48)$$

Rearranging (3.48) gives

$$S_\alpha^M = \left[\frac{\partial M}{\partial G}\left(\frac{G}{M}\right)\right]\cdot\left[\frac{\partial G}{\partial \alpha}\left(\frac{\alpha}{G}\right)\right]$$

$$= S_G^M S_\alpha^G \qquad (3.49)$$

Although S_G^M may be formed for some standard cases, S_α^G is generally computed anew for each individual problem. The sensitivity of M to variations within component transfer functions will be described in detail for the two configurations shown in Figures 3.10a and 3.10b.

3.4.2 Case 1: Variations in Open-Loop Systems

The sensitivity S_G^M for Figure 3.10a can easily be determined as

$$S_G^M = \frac{\partial M}{\partial G}\left(\frac{G}{M}\right) = \frac{\partial[G]}{\partial G}\left(\frac{G}{[G]}\right) = 1 \qquad (3.50)$$

Therefore, the sensitivity of $Y(s)/U(s)$ for variations in $G(s)$ is unity for the open-loop case, and no reduction in sensitivity is obtained for this configuration. If some parameter within G varies, then

$$S_\alpha^M = S_G^M S_\alpha^G = [1]\cdot\left[\frac{\partial G}{\partial \alpha}\left(\frac{\alpha}{G}\right)\right] = \frac{\partial G}{\partial \alpha}\left(\frac{\alpha}{G}\right) \qquad (3.51)$$

3.4.3 Case 2: Variations within the Forward Path of a Closed-Loop System

Consider the closed-loop system of Figure 3.10b. The sensitivity of M to variations in G are given by

$$S_G^M = \frac{\partial M}{\partial G}\left(\frac{G}{M}\right) = \left[\frac{\partial}{\partial G}\left(\frac{G}{1 + GH}\right)\right]\left(\frac{G}{M}\right)$$

$$= \frac{[1 + GH](1) - G[H]}{(1 + GH)^2}\left(\frac{G}{\dfrac{G}{1 + GH}}\right) = \frac{1}{1 + GH} \qquad (3.52)$$

Since $|1 + GH|$ is usually much greater than unity in a well-designed feedback system, there will be a significant reduction in sensitivity for this feedback configuration. The sensitivity of M resulting from variations of a parameter α within G may be determined by using (3.52) in (3.49).

3.4.4 Case 3: Variations within the Feedback Path

Let us determine S_H^M for the system of Figure 3.10b as

$$S_H^M = \frac{\partial M}{\partial H}\left(\frac{H}{M}\right) = \left\{\frac{\partial}{\partial H}\left[\frac{G}{1 + GH}\right]\right\}\left(\frac{H}{M}\right)$$

$$= \frac{-G^2}{(1 + GH)^2}\left[\frac{H}{\dfrac{G}{1 + GH}}\right] = \frac{-GH}{1 + GH} \qquad (3.53)$$

A reduction in sensitivity is achieved if $|1 + GH|$ is larger than $|GH|$. In a manner identical to the derivation of (3.49), we can calculate the sensitivity of M due to variations in a parameter α within H by using

$$S_\alpha^M = S_H^M S_\alpha^H \qquad (3.54)$$

Table 3.3 indicates these sensitivity results for two feedback configurations.

EXAMPLE 3.5

The problem to be considered here is to determine S_α^M and S_β^M for the closed-loop system of Figure 3.11. Notice that α appears in the forward path and β in the

TABLE 3.3 System Sensitivity Results

Configuration	Sensitivity Results
	$Y(s) = \dfrac{G_0 G_1}{1 + G_1 H_1} R(s)$ $M(s) = \dfrac{G_0 G_1}{1 + G_1 H_1}$ $S_{G_0}^M = \dfrac{\partial M}{\partial G_0}\left(\dfrac{G_0}{M}\right) = 1$ $S_{G_1}^M = \dfrac{\partial M}{\partial G_1}\left(\dfrac{G_1}{M}\right) = \dfrac{1}{1 + G_1 H_1}$ $S_{H_1}^M = \dfrac{\partial M}{\partial H_1}\left(\dfrac{H_1}{M}\right) = \dfrac{-G_1 H_1}{1 + G_1 H_1}$
	$M(s) = \dfrac{G_1 G_2}{1 + G_1 H_1 + G_1 G_2 H_2}$ $S_{G_2}^M = \dfrac{\partial M}{\partial G_2}\left(\dfrac{G_2}{M}\right) = \dfrac{1 + G_1 H_1}{1 + G_1 H_1 + G_1 G_2 H_2}$ $S_{H_2}^M = \dfrac{\partial M}{\partial H_2}\left(\dfrac{H_2}{M}\right) = \dfrac{-G_1 G_2 H_2}{1 + G_1 H_1 + G_1 G_2 H_2}$

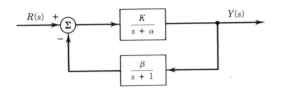

Figure 3.11 Feedback configuration for Example 3.5.

feedback path. For variations in M resulting from α, we determine S_α^M as

$$S_\alpha^M = S_G^M S_\alpha^G = \left[\frac{1}{1 + GH} \right] \cdot \left[\frac{-K}{(s + \alpha)^2} \left(\frac{\alpha}{\frac{K}{s + \alpha}} \right) \right]$$

$$= \frac{\dfrac{-\alpha}{s + \alpha}}{1 + \left(\dfrac{K}{s + \alpha} \right) \left(\dfrac{\beta}{s + 1} \right)} = \frac{-\alpha(s + 1)}{(s + \alpha)(s + 1) + K\beta} \qquad (3.55)$$

The sensitivity of M due to variations in β is

$$S_\beta^M = S_H^M S_\beta^H = \left[\frac{-GH}{1 + GH} \right] \cdot \left[\frac{1}{s + 1} \left(\frac{\beta}{\frac{\beta}{s + 1}} \right) \right]$$

$$= \frac{-K\beta}{(s + \alpha)(s + 1) + K\beta} \qquad (3.56)$$

We observe that both S_α^M and S_β^M are frequency-dependent functions.

3.4.5 Output Disturbances in Feedback Systems

Figure 3.12 shows a closed-loop system with an output disturbance $D(s)$. The effect of this disturbance on the system output $Y(s)$ is given by

$$\left. \frac{Y(s)}{D(s)} \right|_{R(s)=0} = \frac{1}{1 + G(s)H(s)} \qquad (3.57)$$

Therefore, since $|1 + GH|$ is usually designed to be much greater than unity, the

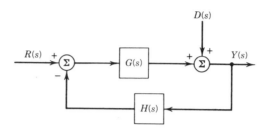

Figure 3.12　A feedback system with output disturbance.

effect of output disturbances are reduced for feedback systems. Notice that (3.57) is identical to the expression for S_G^M. Thus, reductions of both output disturbance effects and system sensitivity are obtained with a feedback configuration.

EXAMPLE 3.6

It is required to determine the effect of the output disturbance $D(s)$ in Figure 3.13. Using (3.57), we write directly

$$\frac{Y(s)}{D(s)}\bigg|_{R(s)=0} = \frac{1}{1 + G(s)H(s)} = \frac{1}{1 + \left(\dfrac{K}{s + \alpha}\right)\left(\dfrac{\beta}{s + 1}\right)}$$

$$= \frac{(s + 1)(s + \alpha)}{(s + 1)(s + \alpha) + K\beta} \tag{3.58}$$

For sufficiently large values of the product $K\beta$, (3.58) indicates that the effect of the disturbance $D(s)$ on the output $Y(s)$ can be made as small as desired.

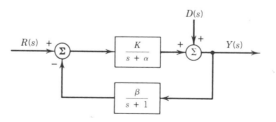

Figure 3.13　Feedback configuration for Example 3.6.

3.4.6 State Variable Relationships

Those system sensitivity concepts presented above can be related to the state variable systems representation in the form of sensitivity coefficients. Consider the general nonlinear system

$$\dot{\mathbf{x}} = \mathbf{f}(\mathbf{x}, \alpha, t) \tag{3.59}$$

where α is the system parameter whose variations we wish to investigate. Let an associated sensitivity coefficient vector $\mathbf{v}_\alpha$ be defined as

$$\mathbf{v}_\alpha \triangleq \frac{\partial \mathbf{x}}{\partial \alpha} = \begin{vmatrix} \dfrac{\partial x_1}{\partial \alpha} \\ \vdots \\ \dfrac{\partial x_n}{\partial \alpha} \end{vmatrix} \tag{3.60}$$

An approximation based on retaining only the first-order terms of a Taylor series for $\Delta \mathbf{x}$ yields

$$\Delta \mathbf{x} = \mathbf{v}_\alpha(\Delta \alpha) \tag{3.61}$$

Taking the time derivative of (3.60) gives

$$\dot{\mathbf{v}}_\alpha = \frac{d}{dt}\left(\frac{\partial \mathbf{x}}{\partial \alpha}\right) = \frac{\partial}{\partial \alpha}(\dot{\mathbf{x}}) = \frac{\partial}{\partial \alpha}[\mathbf{f}(\mathbf{x}, \alpha, t)]$$

$$= \left(\frac{\partial \mathbf{f}}{\partial \mathbf{x}}\right)\frac{\partial \mathbf{x}}{\partial \alpha} + \frac{\partial \mathbf{f}}{\partial \alpha} = \left(\frac{\partial \mathbf{f}}{\partial \mathbf{x}}\right)\mathbf{v}_\alpha + \frac{\partial \mathbf{f}}{\partial \alpha} \tag{3.62}$$

where the Jacobian $\partial \mathbf{f}/\partial \mathbf{x}$ is an n by n matrix and $\partial \mathbf{f}/\partial \alpha$ is an n-vector. Observe that the solution of (3.62) also depends on the solution of (3.59). We select some nominal value of α about which variations are to be considered and denote this nominal value by α_N. We want to determine the effect of small variations in α about α_N. To accomplish this objective, we solve (3.59) for $\mathbf{x}$ and (3.62) for $\mathbf{v}_\alpha$ with α_N inserted for α in all places. Then we may use (3.61) to solve for the first-order approximation of changes $\Delta \mathbf{x}$ in the state due to variations $\Delta \alpha$ in the parameter α about the nominal value α_N.

EXAMPLE 3.7

Consider again the system in Figure 3.11. Defining the outputs of the forward and feedback blocks as state variables, as shown in Figure 3.14, we obtain the

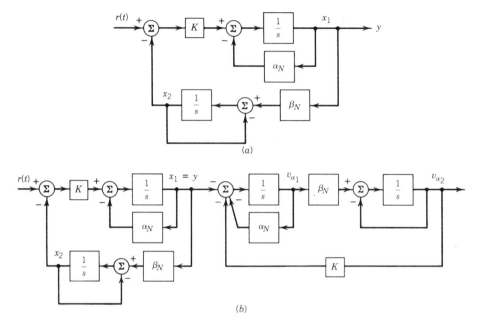

Figure 3.14 Computer simulation diagrams for the state **x** and for the combined state **x** and sensitivity coefficient $\mathbf{v}_\alpha$ evaluated at nominal conditions.

component state equations as

$$\dot{x}_1 = -\alpha x_1 - Kx_2 + Kr(t)$$

$$\dot{x}_2 = \beta x_1 - x_2$$

$$y = x_1 \tag{3.63}$$

We want to form $\partial \mathbf{f}/\partial \mathbf{x}$ and $\partial \mathbf{f}/\partial \alpha$ to use in the dynamic equation (3.62) for $\mathbf{v}_\alpha$. Suppose β is set at some known value β_N for the moment and we wish to consider only variations in α. Thus, we identify

$$\mathbf{f}(\mathbf{x}, \alpha, t) = \begin{pmatrix} f_1 \\ f_2 \end{pmatrix} = \begin{pmatrix} -\alpha x_1 - Kx_2 + Kr(t) \\ \beta_N x_1 - x_2 \end{pmatrix} \tag{3.64}$$

from which

$$\frac{\partial \mathbf{f}}{\partial \mathbf{x}} = \begin{vmatrix} \dfrac{\partial f_1}{\partial x_1} & \dfrac{\partial f_1}{\partial x_2} \\ \dfrac{\partial f_2}{\partial x_1} & \dfrac{\partial f_2}{\partial x_2} \end{vmatrix} = \begin{pmatrix} -\alpha & -K \\ \beta_N & -1 \end{pmatrix} \tag{3.65}$$

and

$$\frac{\partial \mathbf{f}}{\partial \alpha} = \begin{pmatrix} \dfrac{\partial f_1}{\partial \alpha} \\[2mm] \dfrac{\partial f_2}{\partial \alpha} \end{pmatrix} = \begin{pmatrix} -x_1 \\ 0 \end{pmatrix} \tag{3.66}$$

Therefore, evaluating the matrix and vector in (3.65) and (3.66) at some nominal value α_N and using them in (3.62) gives

$$\dot{\mathbf{v}}_\alpha = \begin{pmatrix} -\alpha_N & -K \\ \beta_N & -1 \end{pmatrix} \mathbf{v}_\alpha + \begin{pmatrix} -x_1 \\ 0 \end{pmatrix} \tag{3.67}$$

Figure 3.14*b* shows a computer simulation diagram for the sensitivity coefficient $\mathbf{v}_\alpha$ using the input obtained from the output of the state variable diagram with the nominal value α_N inserted as shown. Similar calculations may be performed to yield an equation for variations in β. The student is asked in Problem 3.21 to provide these details.

3.4.7 Steady-State Values

Time expressions corresponding to S_α^M and the solution of the differential equations for $\mathbf{v}_\alpha$ are presented in Chapter 4 of this book. As a preliminary step in examining the time response, we consider here the steady-state values of changes in the system output resulting from incremental variations in a system parameter. Both transfer function and state variable methods are presented in this context. We use the final-value theorem of Laplace transform theory for the transfer function approach, and we set $\dot{\mathbf{v}}_\alpha$ equal to zero and solve for the resulting steady-state value of $\mathbf{v}_\alpha$ in the state variable approach.

The incremental change in the system output resulting from a small variation in α can be determined from the system sensitivity S_α^M. For an incremental variation in α, we may write as an approximation

$$\frac{\dfrac{\Delta M}{M}}{\dfrac{\Delta \alpha}{\alpha}} \cong S_\alpha^M \tag{3.68}$$

Therefore, we may solve for $\Delta M(s)$ to obtain

$$\Delta M(s) \cong S_\alpha^M \left(\frac{M}{\alpha} \right) \Delta \alpha \tag{3.69}$$

For a stable system, the steady-state effect on the output y from step changes in α from some α_N to $\alpha_N + \Delta\alpha$ may be found by using the final-value theorem. First, we solve for $\Delta Y(s)$ from

$$\Delta M = \frac{\Delta Y(s)}{R(s)} \tag{3.70}$$

where it is assumed that $R(s)$ is unaffected by changes in α. As a particular case, let $R(s)$ be the Laplace transform of a unit step input; that is, $R(s) = 1/s$. Therefore, from (3.69) and (3.70)

$$\mathscr{L}\{\Delta y(t)\} = \Delta Y(s) = (\Delta M)R(s) = S_\alpha^M\left(\frac{M}{\alpha}\right)(\Delta\alpha)\left(\frac{1}{s}\right) \tag{3.71}$$

Using the final-value theorem, we have

$$\Delta y_{ss} = \lim_{s \to 0}\left[s\mathscr{L}\{\Delta y(t)\}\right] = \lim_{s \to 0}\left[s\left\{S_\alpha^M\left(\frac{M}{\alpha}\right)\left(\frac{1}{s}\right)\Delta\alpha\right\}\right] \tag{3.72}$$

For the state variable approach, we set $\dot{\mathbf{v}}_\alpha = \mathbf{0}$ and solve for the resulting $\mathbf{v}_\alpha$ to determine its steady-state value. Therefore, setting $\dot{\mathbf{v}}_\alpha = \mathbf{0}$ in (3.62), we find $\mathbf{v}_{\alpha_{ss}}$ as the solution of a set of linear algebraic equations, that is,

$$\mathbf{v}_{\alpha_{ss}} = -\left(\frac{\partial \mathbf{f}}{\partial \mathbf{x}}\right)^{-1}\left(\frac{\partial \mathbf{f}}{\partial \alpha}\right)\bigg|_{ss} \tag{3.73}$$

provided the indicated inverse of the Jacobian matrix $\partial\mathbf{f}/\partial\mathbf{x}$ exists. The steady-state value of the system output change Δy_{ss} may be determined by

$$\Delta y_{ss} = \mathbf{c}^T \Delta \mathbf{x}_{ss} = \mathbf{c}^T \mathbf{v}_{\alpha_{ss}}(\Delta\alpha) \tag{3.74}$$

where $\mathbf{c}$ is the output vector designating the weighted sum of the state variables being fed into the output, that is, $y = \mathbf{c}^T\mathbf{x}$.

In general, parameter variations $\Delta\alpha$ from the nominal parameters in the model of the open-loop plant, sensors, or controller result in a corresponding variation Δy in the system response y. Equations (3.72) and (3.74) provide these results in the steady state for the transfer function and state variable approaches, respectively. When such variations in parameters do occur, their effects can be reduced considerably by using a properly designed feedback configuration.

EXAMPLE 3.8

To illustrate these two approaches for determining steady-state values of Δy due to incremental parameter changes $\Delta\alpha$, consider once again the system in Figure 3.11 and the results of Examples 3.5 and 3.7.

For the transfer function approach, we use S_α^M from (3.55) in (3.72) to determine variations in the output y due to $\Delta\alpha$ as

$$
\Delta y_{ss} = \lim_{s \to 0} s \left[S_\alpha^M \left(\frac{M}{\alpha} \right) \left(\frac{1}{s} \right) \Delta\alpha \right]
$$

$$
= \lim_{s \to 0} s \left[\frac{-\alpha(s+1)}{(s+\alpha)(s+1)+K\beta} \left(\frac{K(s+1)}{(s+\alpha)(s+1)+K\beta} \right) \left(\frac{1}{\alpha s} \right) \Delta\alpha \right]
$$

$$
= \frac{-K[\Delta\alpha]}{(\alpha+K\beta)^2}
\tag{3.75}
$$

where α and β in (3.75) are evaluated at nominal values α_N and β_N. Similarly, for incremental changes resulting from variations in β about β_N, we have

$$
\Delta y_{ss} = \lim_{s \to 0} s \left[S_\beta^M \left(\frac{M}{\beta} \right) \left(\frac{1}{s} \right) \Delta\beta \right]
$$

$$
= \lim_{s \to 0} s \left[\frac{-K\beta}{(s+\alpha)(s+1)+K\beta} \left(\frac{K(s+1)}{(s+\alpha)(s+1)+K\beta} \right) \left(\frac{1}{\beta s} \right) \Delta\beta \right]
$$

$$
= \frac{-K^2[\Delta\beta]}{(\alpha+K\beta)^2}
\tag{3.76}
$$

Thus, by superposition, the total output variation Δy_{ss} in the steady state due to variations in both α and β about nominal values α_N and β_N is given by

$$
\Delta y_{ss} = \frac{-K}{(\alpha_N+K\beta_N)^2}(\Delta\alpha) + \frac{-K^2}{(\alpha_N+K\beta_N)}(\Delta\beta)
\tag{3.77}
$$

For the state variable approach, we first evaluate $\partial \mathbf{f}/\partial \mathbf{x}$ and $\partial \mathbf{f}/\partial \alpha$ at the nominal conditions α_N and β_N in the steady state. From (3.65) and (3.66), we have

$$
\frac{\partial \mathbf{f}}{\partial \mathbf{x}} = \begin{pmatrix} -\alpha_N & -K \\ \beta_N & -1 \end{pmatrix}
\tag{3.78}
$$

and

$$
\frac{\partial \mathbf{f}}{\partial \alpha} = \begin{pmatrix} -x_{1_{ss}} \\ 0 \end{pmatrix}
\tag{3.79}
$$

The steady-state value of x_1 may be found by setting $\dot{x} = 0$ in (3.63) to give

$$0 = -\alpha_N x_{1_{ss}} - Kx_{2_{ss}} + K$$

$$0 = \beta_N x_{1_{ss}} - x_{2_{ss}} \tag{3.80}$$

from which

$$x_{1_{ss}} = \frac{K}{\alpha_N + K\beta_N} \tag{3.81}$$

Using (3.81) in (3.73) and performing the indicated operations, we find $\mathbf{v}_{\alpha_{ss}}$ as

$$\mathbf{v}_{\alpha_{ss}} = -\begin{pmatrix} -\alpha_N & -K \\ \beta_N & -1 \end{pmatrix}^{-1} \begin{pmatrix} -x_{1_{ss}} \\ 0 \end{pmatrix} = \frac{\begin{pmatrix} -1 \\ \beta_N \end{pmatrix}}{\alpha_N + K\beta_N} x_{1_{ss}}$$

$$= \frac{\begin{pmatrix} -1 \\ \beta_N \end{pmatrix}}{\alpha_N + K\beta_N} \left(\frac{K}{\alpha_N + K\beta_N} \right) = \frac{\begin{pmatrix} -1 \\ \beta_N \end{pmatrix} K}{(\alpha_N + K\beta_N)^2} \tag{3.82}$$

Since $y = x_1$, that is, $\mathbf{c} = \begin{pmatrix} 1 \\ 0 \end{pmatrix}$, we may use (3.74) to determine the steady-state value Δy_{ss} due to variations $\Delta\alpha$, that is,

$$\Delta y_{ss} = (1 \quad 0) \frac{\begin{pmatrix} -1 \\ \beta_N \end{pmatrix} K}{(\alpha_N + K\beta_N)^2} (\Delta\alpha) = \frac{-K}{(\alpha_N + K\beta_N)^2} (\Delta\alpha) \tag{3.83}$$

which agrees with the result in (3.75) by using the transfer function approach to parameter sensitivity. Similar calculations may be performed to yield the result in (3.76) for variations in β.

We have presented sensitivity concepts from both transfer function and state variable viewpoints in this section. We identified sensitivity functions or coefficients in terms of Laplace transforms or differential equations and determined steady-state solutions.

3.5 PID CONTROL

As a prelude to analysis and design procedures in Parts II and III of this book, the feedback effects associated with two vastly different control laws are examined in this and the following sections. We describe in this section control laws based on only the plant output being available for feedback. The controller is composed of either a simple gain (proportional control), an integrator (integral control), a differentiator (derivative control), or some weighted combination of these possibilities. In contrast to this proportional-integral-derivative (PID) control law, we assume the other extreme for which all state variables are available for feedback. In Section 3.6, we describe the associated controller, which has been referred to as linear state variable feedback (SVFB) control.

3.5.1 Proportional Control

The proportional control law of Figure 3.15a has a plant input u that is linearly proportional to the error e between the system input and output; that is,

$$u = k_p e = k_p(r - y) \tag{3.84}$$

This simple feedback control has only the single parameter k_p available for adjustment to meet conflicting design requirements. Time response curves for a typical control system with an adjustable proportional control are shown in Figure 3.15b for a unit step input. Since a relatively slow speed of response is

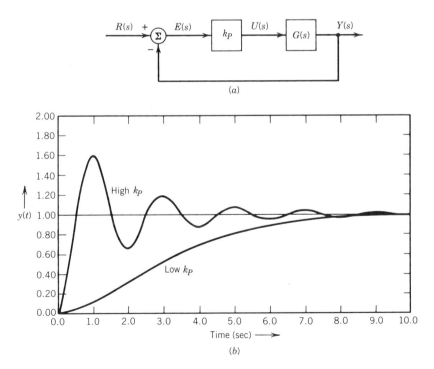

Figure 3.15 Proportional control and associated typical time response curves.

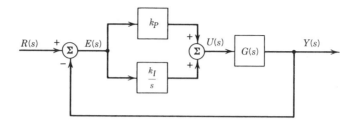

Figure 3.16 Proportional-plus-integral control.

obtained for low values of k_P, we would like to adjust k_P upwards. However, a highly oscillatory response occurs for these high values of k_P. Thus, we would like to select a large k_P to yield a fast response and to reduce steady-state errors, and yet we need a small k_P to avoid a highly oscillatory transient time response. Clearly, more flexibility in the adjustment of controller parameters is required to yield acceptable transient and steady-state responses simultaneously.

3.5.2 Integral Control

Figure 3.16 depicts proportional-plus-integral control, which resolves the steady-state error problem just described. If a steady-state error exists for the proportional control of Figure 3.15, then the integral of that error creates an actuating signal to drive the plant to the desired state, which reduces the steady-state error to zero. The proportional gain k_P can then be adjusted to improve the transient response.

We should point out that the use of integral control increases the system order by one, that is, from order n to order $n + 1$. Suppose the state variable equations for an nth-order linear plant are given by

$$\dot{\mathbf{x}} = A\mathbf{x} + \mathbf{b}u$$

$$y = \mathbf{c}^T\mathbf{x} \tag{3.85}$$

We define an additional state variable x_{n+1} as the output of the $1/s$ block in the integral controller. Therefore, the state variable equations for the complete closed-loop system with proportional-plus-integral control are

$$\dot{\mathbf{x}} = A\mathbf{x} + \mathbf{b}u$$

$$\dot{x}_{n+1} = e$$

$$y = \mathbf{c}^T\mathbf{x} \tag{3.86}$$

where

$$u = k_P e + k_I x_{n+1} \tag{3.87}$$

We form an augmented state vector of dimension $(n + 1)$ as

$$\mathbf{x}_a = \begin{pmatrix} \mathbf{x} \\ x_{n+1} \end{pmatrix} \tag{3.88}$$

Substituting (3.87) and (3.88) into (3.86), we obtain the vector-matrix equations

$$\dot{\mathbf{x}}_a = \begin{pmatrix} A & k_I \mathbf{b} \\ \mathbf{0}^T & 0 \end{pmatrix} \mathbf{x}_a + \begin{pmatrix} k_P \mathbf{b} \\ 1 \end{pmatrix} e \tag{3.89}$$

We define A_a and $\mathbf{b}_a$ as

$$A_a = \begin{pmatrix} A & k_I \mathbf{b} \\ \mathbf{0}^T & 0 \end{pmatrix} \qquad \mathbf{b}_a = \begin{pmatrix} k_P \mathbf{b} \\ 1 \end{pmatrix} \tag{3.90}$$

Since

$$e = r - y = r - \mathbf{c}^T \mathbf{x} \tag{3.91}$$

the system equations in vector-matrix notation become

$$\dot{\mathbf{x}}_a = A_a \mathbf{x}_a + \mathbf{b}_a e$$

$$= A_a \mathbf{x}_a + \mathbf{b}_a \left[r(t) - \mathbf{c}^T \mathbf{x} \right]$$

$$= \left[A_a - \mathbf{b}_a \mathbf{c}_a^T \right] \mathbf{x}_a + \mathbf{b}_a r(t) \tag{3.92}$$

where

$$\mathbf{c}_a = \begin{pmatrix} \mathbf{c} \\ 0 \end{pmatrix} \tag{3.93}$$

We demonstrate the effects of integral control for a specific system in the following example.

EXAMPLE 3.9

Suppose the plant transfer function in the system of Figure 3.16 is given by

$$G(s) = \frac{10}{s(s + 1)} \tag{3.94}$$

If only proportional control is being used, then the steady-state error due to a unit ramp input is $e_{ss} = 1/K_v$, where

$$K_v = \lim_{s \to 0} s \left[k_P G(s) \right]$$

$$= \lim_{s \to 0} s \left(\frac{10 k_P}{s(s + 1)} \right) = 10 k_P \tag{3.95}$$

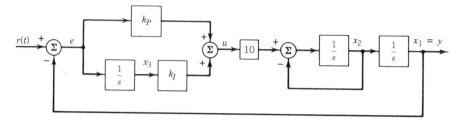

Figure 3.17 State variables for Example 3.9.

However, for proportional-plus-integral control,

$$K_v = \lim_{s \to 0} s\left[\left(k_P + \frac{k_I}{s}\right)G(s)\right]$$

$$= \lim_{s \to 0} s\left[\frac{10\left(k_P + \dfrac{k_I}{s}\right)}{s(s+1)}\right] = \infty \qquad (3.96)$$

which yields, as expected, $e_{ss} = 1/K_v = 0$, provided the closed-loop system is stable.

State variables are shown in Figure 3.17 for the plant transfer function in (3.94). In terms of these direct phase variables, we have

$$\begin{pmatrix} \dot{x}_1 \\ \dot{x}_2 \end{pmatrix} = \begin{pmatrix} 0 & 1 \\ 0 & -1 \end{pmatrix}\begin{pmatrix} x_1 \\ x_2 \end{pmatrix} + \begin{pmatrix} 0 \\ 10 \end{pmatrix}u$$

$$\dot{x}_3 = e$$

$$y = \begin{pmatrix} 1 \\ 0 \end{pmatrix}^T \begin{pmatrix} x_1 \\ x_2 \end{pmatrix} = x_1 \qquad (3.97)$$

We may express the equations in (3.89) for the augmented state vector as

$$\dot{\mathbf{x}}_a = \begin{pmatrix} 0 & 1 & 0 \\ 0 & -1 & 10k_I \\ 0 & 0 & 0 \end{pmatrix}\mathbf{x}_a + \begin{pmatrix} 0 \\ 10k_P \\ 1 \end{pmatrix}e \qquad (3.98)$$

Using (3.92) and (3.93) with the error e from (3.91), we may write the system vector-matrix equations in (3.98) as

$$\dot{\mathbf{x}}_a = \left[A_a - \mathbf{b}_a\mathbf{c}_a^T\right]\mathbf{x}_a + \mathbf{b}_a r(t)$$

$$= \left[\begin{pmatrix} 0 & 1 & 0 \\ 0 & -1 & 10k_I \\ 0 & 0 & 0 \end{pmatrix} - \begin{pmatrix} 0 \\ 10k_P \\ 1 \end{pmatrix}\begin{pmatrix} 1 \\ 0 \\ 0 \end{pmatrix}^T\right]\mathbf{x}_a + \begin{pmatrix} 0 \\ 10k_P \\ 1 \end{pmatrix}r(t)$$

$$= \begin{pmatrix} 0 & 1 & 0 \\ -10k_P & -1 & 10k_I \\ -1 & 0 & 0 \end{pmatrix}\mathbf{x}_a + \begin{pmatrix} 0 \\ 10k_P \\ 1 \end{pmatrix}r(t) \qquad (3.99)$$

Let us determine, for a unit ramp input, the steady-state error with and without integral control by state variable procedures. Using (3.97) with u from (3.84) and e from (3.91), the closed-loop system equations for the case of proportional control only are

$$\dot{x}_1 = x_2$$

$$\dot{x}_2 = -x_2 + 10k_p(r - x_1) \tag{3.100}$$

Initially, we make the assumption that e_{ss} is some constant value; this assumption (if true) must be verified before completing the problem. According to the results of Section 3.3 on steady-state errors, we may set $\dot{x}_{1_{ss}} = 1$ and $\dot{x}_{2_{ss}} = 0$, since $r - x_{1_{ss}} = e_{ss}$ and $r(t) = t$, a unit ramp input. Therefore, (3.100) becomes

$$1 = x_{2_{ss}}$$

$$0 = -x_{2_{ss}} + 10k_p e_{ss} \tag{3.101}$$

Solving (3.101) gives $e_{ss} = 1/10k_p$, which corresponds to the transfer function result $e_{ss} = 1/K_v$ from (3.95). Moreover, we see that the initial assumption (e_{ss} is a constant value) has been verified.

For the case of proportional-plus-integral control, the closed-loop system equations in (3.98) in component form are

$$\dot{x}_1 = x_2$$

$$\dot{x}_2 = -x_2 + 10k_I x_3 + 10k_p e$$

$$\dot{x}_3 = e \tag{3.102}$$

Assuming e_{ss} is a constant, we obtain $\dot{x}_{1_{ss}} = 1$ and $\dot{x}_{2_{ss}} = 0$ as before. For steady-state conditions, we use the second equation in (3.102) to obtain

$$\dot{x}_{2_{ss}} = -x_{2_{ss}} + 10k_I x_{3_{ss}} + 10k_p e_{ss} = 0 \tag{3.103}$$

Taking the time derivative in (3.103) yields

$$-\dot{x}_{2_{ss}} + 10k_I \dot{x}_{3_{ss}} + 10k_p \dot{e}_{ss} = 0 \tag{3.104}$$

from which $\dot{x}_{3_{ss}} = 0$, since both $\dot{x}_{2_{ss}}$ and $\dot{e}_{ss}$ are zero. The third equation in (3.102) states that $\dot{x}_3 = e$, which gives $e_{ss} = \dot{x}_{3_{ss}} = 0$, provided the closed-loop system is stable. This result is identical to the one obtained in (3.96). Again, we have shown that e_{ss} is a constant (zero) in accordance with the initial assumption.

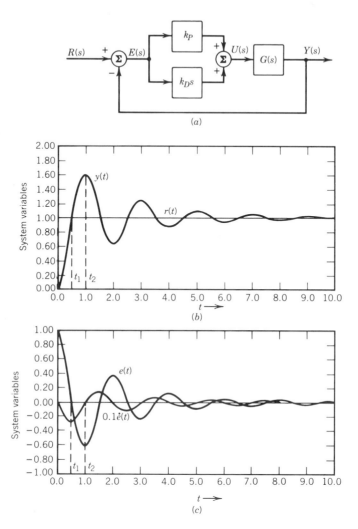

Figure 3.18 Block diagram and typical time response curves for proportional plus-derivative control.

3.5.3 Derivative Control

Figure 3.18a shows the configuration for proportional-plus-derivative control. An output time response for a step input to a typical system having only proportional control is shown in Figure 3.18b. The error $e(t)$ and its time derivative $\dot{e}(t)$ are shown in Figure 3.18c. With a maximum at $t = 0$, the error is reduced to zero at $t = t_1$ and continues to become more negative until $t = t_2$, when the output $y(t)$ in Figure 3.18b attains its maximum value. On the other hand, the error derivative $\dot{e}(t)$ is zero at $t = 0$, reaches a minimum at $t = t_1$, and increases to zero at $t = t_2$. The error derivative enables us to anticipate an overshoot, even though the error itself has been reduced to zero. By forming the plant input u as a

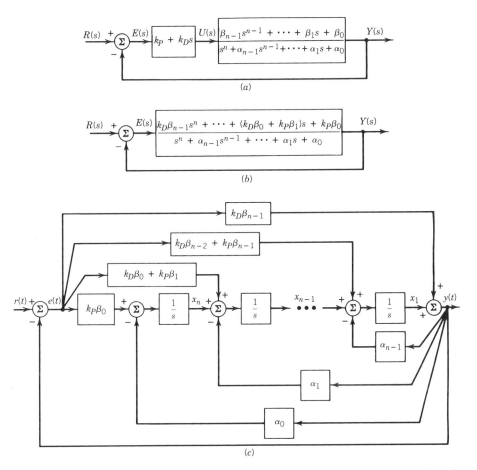

Figure 3.19 Computer simulation diagram development for proportional-plus-derivative control.

linear combination of e and $\dot{e}$, we can force the error to zero without these oscillations having excessive amplitudes.

We prefer forming state variable models for derivative control that eliminate terms involving $\dot{r}(t)$ directly. Using $u = k_P e + k_D \dot{e}$ in (3.85) gives

$$\dot{x} = Ax + bu$$

$$= Ax + b(k_P e + k_D \dot{e}) \qquad (3.105)$$

Since $e = r - c^T x$ from (3.91) and hence $\dot{e} = \dot{r} - c^T \dot{x}$, problems arise in the introduction of both $\dot{r}(t)$ and $\dot{x}$ on the right side of (3.105). To circumvent these problems, we first combine the transfer function blocks of the controller ($k_P + k_D s$) and plant, as shown in Figures 3.19a and 3.19b. Using the nth-order plant transfer function $G(s)$ defined in Table 2.4, we choose to restrict the numerator of this plant transfer function in Figure 3.19a to be no greater than $n - 1$. This restriction ensures that the combined controller-plant transfer function in Figure 3.19b has a numerator polynomial of order n or less and, consequently, terms involving $\dot{r}(t)$ directly are avoided in the system state variable equations. We note at this point that any state variable model can be used for the controller-plant configuration, including any of the four forms in Table 2.4. Figure 3.19c shows the direct feed-forward form yielding state variable equations as

$$\dot{\mathbf{x}} = A\mathbf{x} + \mathbf{b}'e$$

$$= \begin{pmatrix} -\alpha_{n-1} & 1 & \cdots & 0 \\ -\alpha_{n-2} & 0 & \cdots & 0 \\ \vdots & \vdots & & \vdots \\ -\alpha_1 & 0 & \cdots & 1 \\ -\alpha_0 & 0 & \cdots & 0 \end{pmatrix} \mathbf{x} + \begin{pmatrix} k_D\beta_{n-2} + k_P\beta_{n-1} - k_D\beta_{n-1}\alpha_{n-1} \\ k_D\beta_{n-3} + k_P\beta_{n-2} - k_D\beta_{n-1}\alpha_{n-2} \\ \vdots \\ k_D\beta_0 + k_P\beta_1 - k_D\beta_{n-1}\alpha_1 \\ k_P\beta_0 - k_D\beta_{n-1}\alpha_0 \end{pmatrix} e \quad (3.106)$$

which is the desired result.

EXAMPLE 3.10

Let the linear plant transfer function given by (3.94) be used in the feedback system of Figure 3.18a. We select the direct feed-forward form shown in Figure 3.19c to express the state variable equations for this proportional-plus-derivative control system as

$$\begin{pmatrix} \dot{x}_1 \\ \dot{x}_2 \end{pmatrix} = \begin{pmatrix} -1 & 1 \\ 0 & 0 \end{pmatrix} \begin{pmatrix} x_1 \\ x_2 \end{pmatrix} + \begin{pmatrix} 10k_D \\ 10k_P \end{pmatrix} e$$

$$y = x_1 \quad (3.107)$$

where $e = r - x_1$.

Using transform techniques, we may solve for $Y(s)$ in Figure 3.18a to give

$$Y(s) = \left(\frac{\dfrac{10(k_P + k_D s)}{s(s+1)}}{1 + \dfrac{10(k_P + k_D s)}{s(s+1)}} \right) R(s)$$

$$= \frac{10(k_P + k_D s)R(s)}{s^2 + [1 + 10k_D]s + 10k_P} \quad (3.108)$$

Often referred to simply as derivative control, proportional-plus-derivative control exhibits many of the characteristics of tachometer feedback. Both derivative control and tachometer feedback yield system transfer functions in which the

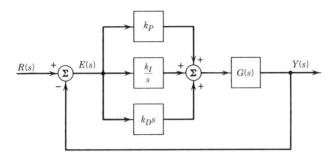

Figure 3.20 Combined PID control.

coefficient of s in the denominator increases as the gain k_D or K_t increases. Therefore, we find that excessive oscillations for step inputs are reduced by both control schemes. An important difference is that a numerator factor $(k_P + k_D s)$ appears in (3.108) for derivative control. We determine a quantitative measure of the reduction in oscillations for derivative control and tachometer feedback in Section 4.2.

3.5.4 Combined PID Control

A combination of proportional, integral, and derivative control is shown in Figure 3.20. We may express the transfer function between $E(s)$ and $Y(s)$ as

$$\frac{Y(s)}{E(s)} = (k_P + k_I/s + k_D s)G(s) \tag{3.109}$$

Therefore, the closed-loop transfer function is

$$\frac{Y(s)}{R(s)} = \frac{(k_P + k_I/s + K_D s)G(s)}{1 + (k_P + k_I/s + k_D s)G(s)} \tag{3.110}$$

The corresponding state variable equations for PID control may be obtained by modifying (3.89), which was derived for proportional-plus-integral control. Choosing the direct feed-forward form, we need to change only the augmented vector $\mathbf{b}_a$ in (3.90) to include the derivative control effects shown in the vector $\mathbf{b}'$ in (3.106). The resulting state variable equations for PID control using the direct feed-forward form are

$$\dot{\mathbf{x}}_a = A_a \mathbf{x}_a + \mathbf{b}_a e$$

$$\begin{pmatrix} \dot{\mathbf{x}} \\ \dot{x}_{n+1} \end{pmatrix} = \begin{pmatrix} A & k_I \mathbf{b} \\ \mathbf{0}^T & 0 \end{pmatrix} \begin{pmatrix} \mathbf{x} \\ x_{n+1} \end{pmatrix} + \begin{pmatrix} \mathbf{b}' \\ 1 \end{pmatrix} e$$

$$y = \mathbf{c}_a^T \mathbf{x}_a = x_1 \tag{3.111}$$

The matrix A_a and vector $\mathbf{c}_a$ remain unchanged from (3.90) and (3.93). Regardless of the procedure (transfer functions or state variables) used to describe the system having PID control, the presence of three adjustable controller parameters permits improvements in both transient and steady-state parts of the time response. We demonstrate the effectiveness and limitations of PID control further in Section 4.7.

3.6 LINEAR STATE VARIABLE FEEDBACK CONTROL

The linear state variable feedback (SVFB) control law assumes that all state variables are accessible from sensor measurements. The state variables are fed through linear gains and then combined with the system input to form the control. Because there are several feedback gains that can be varied, we can often adjust these values to meet a somewhat restrictive design requirement. In fact, the linear SVFB control law has the maximum flexibility in attempting to relocate open-loop poles to satisfy design requirements because it uses a weighted sum of all state variables. However, seldom in practice are all state variables available for feedback. Quite often, only a few of the state variables can be measured directly. Observer theory, described in Chapter 8, can be utilized to form useful approximations for the inaccessible state variables.

Consider the single-input, single-output plant in Figure 3.21a described by

$$\dot{\mathbf{x}} = A\mathbf{x} + \mathbf{b}u$$

$$y = \mathbf{c}^T\mathbf{x} \tag{3.112}$$

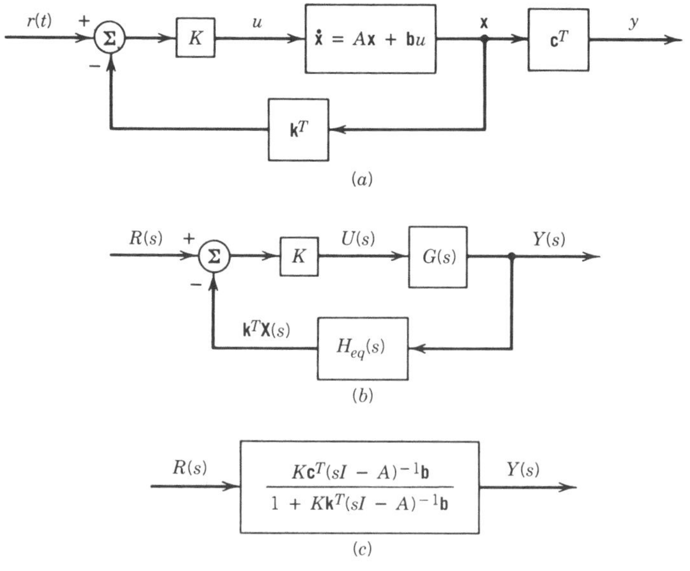

(a)

(b)

(c)

Figure 3.21 Development of the linear SVFB control algorithm.

with the linear SVFB controller given by

$$u = K\left[r(t) - \mathbf{k}^T \mathbf{x}\right] \tag{3.113}$$

where $\mathbf{k}$ is the n-vector of controller gains. Using (3.113) in (3.112) yields

$$\dot{\mathbf{x}} = A\mathbf{x} + \mathbf{b}K\left[r(t) - \mathbf{k}^T \mathbf{x}\right]$$

$$= \left[A - K\mathbf{b}\mathbf{k}^T\right]\mathbf{x} + K\mathbf{b}r(t)$$

$$= A_k \mathbf{x} + K\mathbf{b}r(t) \tag{3.114}$$

where A_k, defined as $A - K\mathbf{b}\mathbf{k}^T$, describes the closed-loop system interactions. The closed-loop poles are determined by the roots of the determinant of $(sI - A_k)$. We want to show how to select the controller parameters $\mathbf{k}$ such that a desired closed-loop pole placement is obtained.

We could express the closed-loop transfer function from (3.112), (3.114), and (2.118) as

$$\frac{Y(s)}{R(s)} = K\mathbf{c}^T(sI - A_k)^{-1}\mathbf{b} \tag{3.115}$$

However, a more enlightening approach is to utilize the equivalent block diagram in Figure 3.21b where $H_{eq}(s)$ is defined as[2]

$$H_{eq}(s) = \frac{\mathbf{k}^T \mathbf{X}(s)}{Y(s)} = \frac{\mathbf{k}^T(sI - A)^{-1}\mathbf{b}}{\mathbf{c}^T(sI - A)^{-1}\mathbf{b}} \tag{3.116}$$

From Figure 3.21b, we have

$$\frac{Y(s)}{R(s)} = \frac{KG(s)}{1 + KG(s)H_{eq}(s)}$$

$$= \frac{K\mathbf{c}^T(sI - A)^{-1}\mathbf{b}}{1 + K\mathbf{k}^T(sI - A)^{-1}\mathbf{b}} \tag{3.117}$$

which is shown in Figure 3.21c. The dependence of the closed-loop poles on $\mathbf{k}$ can now be identified more easily. Because the numerator of (3.117) is not a function of $\mathbf{k}$, the closed-loop zeros must be the same as the open-loop zeros.

---

[2] Melsa and Schultz [3] used this equivalent feedback transfer function concept extensively in their treatment of linear SVFB control.

EXAMPLE 3.11

It is required to design a linear SVFB controller such that the plant transfer function

$$G(s) = \frac{2s + 5}{s^2 + 1} \tag{3.118}$$

will have closed-loop poles at -1 and -2, that is,

$$\frac{Y(s)}{R(s)} = \frac{2s + 5}{(s + 1)(s + 2)} = \frac{2s + 5}{s^2 + 3s + 2} \tag{3.119}$$

The linear SVFB controller is indicated in Figure 3.22a for the direct phase variable form of state variables. The configuration in Figure 3.22b can be

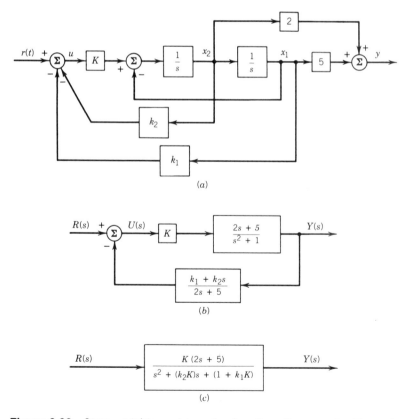

Figure 3.22   State variable and transfer function diagrams for Example 3.11.

obtained by noting that $X_2(s) = sX_1(s)$ so that

$$Y(s) = 5X_1(s) + 2X_2(s)$$

$$= (2s + 5)X_1(s) \tag{3.120}$$

Moreover,

$$\mathbf{k}^T\mathbf{X}(s) = k_1 X_1(s) + k_2 X_2(s)$$

$$= (k_1 + k_2 s)X_1(s) \tag{3.121}$$

Therefore, from (3.115)

$$H_{eq}(s) = \frac{\mathbf{k}^T\mathbf{X}(s)}{Y(s)} = \frac{k_1 + k_2 s}{2s + 5} \tag{3.122}$$

From Figure 3.22b and (3.117),

$$\frac{Y(s)}{R(s)} = \frac{KG(s)}{1 + KG(s)H_{eq}(s)}$$

$$= \frac{K(2s + 5)}{s^2 + k_2 Ks + (1 + k_1 K)} \tag{3.123}$$

Setting (3.123) equal to (3.119) yields

$$K = 1$$

$$1 + k_1 K = 2$$

$$k_2 K = 3 \tag{3.124}$$

from which $K = 1$, $k_1 = 1$, and $k_2 = 3$ as the linear SVFB controller parameters. Alternatively, we may write the closed-loop system state equations as

$$\dot{x}_1 = x_2$$

$$\dot{x}_2 = -(1 + k_1 K)x_1 - k_2 Kx_2 + Kr(t)$$

$$y = 5x_1 + 2x_2 \tag{3.125}$$

From (3.119), the corresponding equations for the desired closed-loop system response are

$$\dot{x}_1 = x_2$$

$$\dot{x}_2 = -2x_1 - 3x_2 + r(t)$$

$$y = 5x_1 + 2x_2 \tag{3.126}$$

Equating (3.125) and (3.126) yields the same equations as in (3.124) and, consequently, the same controller parameters.

SUMMARY

We have described closed-loop system behavior and representation in a quantitative manner in this chapter. As a continuation of the mathematical modeling of components in Chapter 2, we have emphasized the modeling of the complete system. We discussed feedback properties and their application to systems both generally and in specific examples. We presented gain versus bandwidth tradeoffs, steady-state error analysis, and system sensitivity concepts and described two control laws (PID and SVFB).

REFERENCES

1. B. C. Kuo. *Automatic Control Systems*, 4th ed. Englewood Cliffs, N.J.: Prentice-Hall, 1982, pp. 7–11.
2. R. C. Dorf. *Modern Control Systems*, 3rd ed. Reading, Mass.: Addison-Wesley, 1980, Chapter 3.
3. J. L. Melsa and D. G. Schultz. *Linear Control Systems*. New York: McGraw-Hill, 1969, Chapter 3.
4. J. J. D'Azzo and C. H. Houpis. *Linear Control System Analysis and Design: Conventional and Modern*, 2nd ed. New York: McGraw-Hill, 1981, Chapter 6.
5. K. Ogata. *Modern Control Engineering*. Englewood Cliffs, N.J.: Prentice-Hall, 1970, Chapter 5.
6. C. T. Chen. *Analysis and Synthesis of Linear Control Systems*. New York: Holt, Rinehart, and Winston, 1975, Chapters 6 and 7.
7. A. P. Sage. *Linear Systems Control*. Champaign, Ill.: Matrix Publishers, 1978, Chapters 4 and 5.
8. W. R. Perkins and J. B. Cruz, Jr. *Engineering of Dynamic Systems*. New York: Wiley, Sections 3.6, 6.8, 11.7, 11.8, and 13.3.
9. F. H. Raven. *Automatic Control Engineering*, 3rd ed. New York: McGraw-Hill, 1978, Chapter 4.
10. V. W. Eveleigh. *Introduction to Control Systems Design*. New York: McGraw-Hill, 1972.
11. J. G. Truxal. *Introductory System Engineering*. New York: McGraw-Hill, 1972.
12. W. L. Brogan. *Modern Control Theory*, 2nd ed. Englewood Cliffs, N.J.: Prentice-Hall, 1985.

13. J. J. DiStefano III, A. R. Stubberud, and I. J. Williams. *Feedback and Control Systems*, Schaum's Outline Series. New York: McGraw-Hill, 1967.
14. T. E. Fortmann and K. L. Hitz. *An Introduction to Linear Control Systems.* New York: Marcel Dekker, 1977.
15. T. Takahashi. *Mathematics of Automatic Control.* English Translation by Holt, Rinehart, and Winston. New York: Holt, Rinehart, and Winston, 1966.
16. H. L. Harrison and J. G. Bollinger. *Introduction to Automatic Controls*, 2nd ed. Scranton, Pa.: International Textbook Co., 1969, Chapter 8.
17. W. H. Bode. *Network Analysis and Feedback Amplifier Design.* New York: Van Nostrand, 1945.
18. R. Tomovic. *Sensitivity Analysis of Dynamic Systems.* New York: McGraw-Hill, 1963.
19. J. B. Cruz, Jr., ed. *System Sensitivity Analysis.* Stroudsburg, Pa.: Dowden, Hutchinson, and Ross, 1973.
20. P. M. Frank. *Introduction to System Sensitivity Theory.* New York: Academic Press, 1978.

PROBLEMS

(§3.1) **3.1** Show a schematic diagram and then determine which of the five characteristics of Table 3.1 are most important in each of the following closed-loop system operations.

   a. Automobile automatic speed control.
   b. Anti-aircraft artillery (with sensor).
   c. Mid-course correction of an unmanned interplanetary spacecraft.
   d. Voice-activated tape recording system.

(§3.1) **3.2** Describe different feedback systems for which each of the five characteristics of Table 3.1 are of major importance. In particular, describe one system for which the first characteristic is especially important, another for which the second is a primary consideration, and so forth.

(§3.1) **3.3** Describe three applications as alternatives to those in Table 3.1 to illustrate communication-at-a-distance, tracking operations, and power amplification. Indicate which feedback characteristics are of major importance for each application.

(§3.2) **3.4** Sketch frequency response curves of system gain versus frequency ω for the system of Example 3.1 in Figure 3.4*a*. Let $K = 1$, and sketch curves for $\beta = 0$, 0.5, and 1.

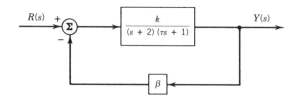

Figure 3.23   The closed-loop system described in
Problem 3.7.

(§3.2) **3.5** Repeat the instruction of Problem 3.4 for the system of Example 3.2
in Figure 3.4b with $K = 1$ and $\tau = 0.1$ for $\beta = 0$ and 1.

(§3.2) **3.6** Sketch curves of low-frequency system gain versus system bandwidth for the two systems of Figure 3.4 for feedback gains β of 0,
0.3, and 0.5. Let $K = 1$ and $\tau = 0.2$.

(§3.2) **3.7** Sketch the curve of low-frequency system gain versus system bandwidth for the system of Figure 3.23 as β varies over the values 0,
0.5, and 1. Let $K = 1$ and $\tau = 0.1$.

(§3.3) **3.8** Identify the system type, and determine K_p, K_v, and K_a for the
unity-feedback systems with plant transfer functions given by

    a. $G(s) = \dfrac{10(1 + 2s)}{1 + s + s^2}$

    b. $G(s) = \dfrac{10(s + 2)}{s^2(s^2 + 3s + 1)}$

    c. $G(s) = \dfrac{10(s + 4)^2}{s(s + 1)(s + 2)}$

    d. $G(s) = \dfrac{3s + 1}{s^2} + \dfrac{2}{(s + 1)(s + 2)}$

(§3.3) **3.9** For each of the systems in Problem 3.8, determine the steady-state
error e_{ss} resulting from the following system inputs for $t \geq 0$.

    a.   A unit step input.
    b. $r(t) = t^2$.
    c. $r(t) = \alpha + \beta t$, where α and β are constants.

(§3.3) **3.10** For which of the following functions $F(s)$ does the final-value
theorem apply? Give your reasons. Determine $f(\infty)$ whenever possible.

    a. $F(s) = \dfrac{10}{s^2 + 1}$

    b. $F(s) = \dfrac{20(s - 1)}{s(s + 1)(s + 2)}$

    c. $F(s) = \dfrac{100}{s(s - 1)(s + 2)^2}$

    d. $F(s) = \dfrac{2(s + 3)}{s^3 + 2s^2 + s + 1}$

(§3.3) **3.11** Determine the steady-state error e_{ss} for the system of Figure 3.24 by

    a. Expressing the system in a modified form with a new plant transfer function having negative-unity feedback.

    b. Deriving an expression for $E(s)$ and then using the final-value theorem directly (if applicable).

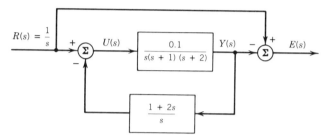

Figure 3.24 The system described in Problem 3.11.

(§3.3) **3.12** Use the state variable approach of Section 3.3 to determine the steady-state error e_{ss} for each of the systems in Figure 3.25.

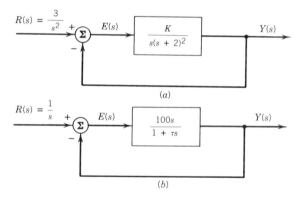

Figure 3.25 Systems for steady-state error analysis in Problem 3.12.

(§3.3) **3.13** Determine the steady-state error e_{ss} for the system in Figure 3.26 by

    a. Finding the appropriate error constant for the cascaded plant combination $G(s)$ and then finding e_{ss}.

    b. Using the state variable approach.

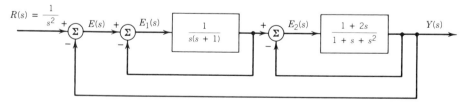

Figure 3.26 A cascaded system with negative unity-feedback described in Problem 3.13.

(§3.3) **3.14**   Determine K_p, K_v, and K_a as functions of **A**, **b**, or **c** in (3.31) for the case of a Type 0 system.

(§3.3) **3.15**   Repeat Problem 3.14 for a Type 1 system using (3.35).

(§3.3) **3.16**   Find e_{ss} for Example 3.4 by using the direct feed-forward form (Table 2.4) for A_r.

(§3.4) **3.17**   Determine the system sensitivity S_α^M for each of the configurations given in Figure 3.27, where M is the system transfer function $Y(s)/R(s)$.

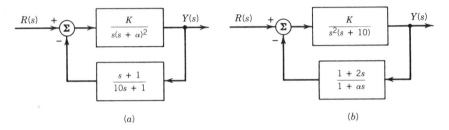

(a)   (b)

Figure 3.27   The closed-loop system configurations described in Problem 3.17.

(§3.4) **3.18**   Determine S_α^M for the feedback system of Figure 3.28 by

    a.   Considering the parameter α as part of the forward path transfer function G, computing S_G^M, and then using $S_\alpha^M = S_G^M S_\alpha^G$.

    b.   Considering α as part of the feedback path transfer function H, computing S_H^M, and using $S_\alpha^M = S_H^M S_\alpha^H$.

    c.   Expressing M as a function of α and computing S_α^M directly.

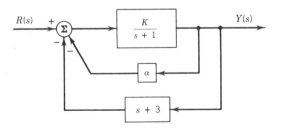

Figure 3.28   The feedback system described in Problem 3.18.

(§3.4) **3.19**   Determine Δy_{ss} due to an incremental variation in β by using the state variable approach for the system in Figure 3.11 to find the sensitivity coefficient $\mathbf{v}_{\beta_{ss}}$ and then using $\Delta y_{ss} = (1 \quad 0)\mathbf{v}_{\beta_{ss}}(\Delta \beta)$. Compare this solution with that given by the transfer function approach in (3.76).

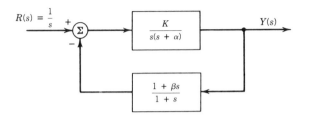

Figure 3.29 The feedback system for Problem 3.20.

(§3.4) **3.20** Use superposition to determine Δy_{ss} due to incremental variations $\Delta \alpha$ and $\Delta \beta$ for the feedback system of Figure 3.29 by

    a. Determining S_{α}^{M} and S_{β}^{M} by the transfer function approach and then solving for Δy_{ss}.

    b. Determining sensitivity coefficient vectors $\mathbf{v}_{\alpha}$ and $\mathbf{v}_{\beta}$ by the state variable approach and then computing Δy_{ss}.

(§3.4) **3.21** Determine the vector differential equation for $\mathbf{v}_{\beta}$ for the system in (3.63) of Example 3.7. Construct a computer simulation diagram for $\mathbf{v}_{\beta}$ similar to the one for $\mathbf{v}_{\alpha}$ in Figure 3.14b. Form a single vector-matrix equation for this fourth-order cascade combination of simulation diagrams for the state $\mathbf{x}$ and the system sensitivity coefficient $\mathbf{v}_{\beta}$.

(§3.4) **3.22** For the positional servomechanism shown in Figure 3.7,

    a. Determine $S_{L_{f}}^{M}$.

    b. Determine S_{V}^{M}.

    c. Find the transfer function between an output torque disturbance $T_{d}(s)$ and the servomechanism output $\theta_{c}(s)$.

    d. Find $\Delta \theta_{c_{ss}}$ due to small variations in L_{f} and V by the state variable approach and compare with the transfer function results in Parts (a) and (b).

(§3.4) **3.23** For the second-order system of Figure 3.30, determine the sensitivity of the closed-loop transfer function to variations in the damping factor α. If $\alpha = \zeta \omega_{n}$, where ζ is the damping coefficient and ω_{n} is the natural frequency, determine the sensitivity of $Y(s)/R(s)$ to variations in ζ and ω_{n}.

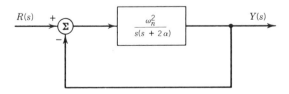

Figure 3.30 The second-order system of Problem 3.23.

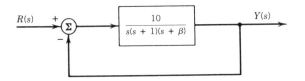

Figure 3.31 The third-order system described in Problem 3.24.

(§3.4) **3.24** Consider the system of Figure 3.31 having a parameter β which is to be varied. If this system is to be approximated by the second-order system of Problem 3.23 with $\omega_n^2 = 10$ and $\alpha = 1$, determine the effect of β on this approximation by calculating the sensitivity of the closed-loop transfer function resulting from variations in β about nominal values $\beta_N = 2$, 10, and 100.

(§§3.3, **3.25** Determine the steady-state error of the system in Figure 3.32 by
3.5)
    a. Using proportional control only ($k_I = 0$).
    b. Using proportional-plus-integral control.

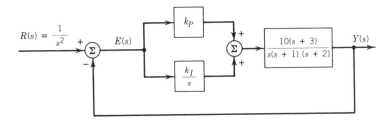

Figure 3.32 The closed-loop system described in Problem 3.25.

(§3.5) **3.26** Determine the closed-loop transfer function for the plant with the negative unity-feedback and PID control in Figure 3.33. Show that derivative control contributes to increased damping for the closed-loop response by comparing the damping coefficient ζ_D with ζ for the original system without PID control.

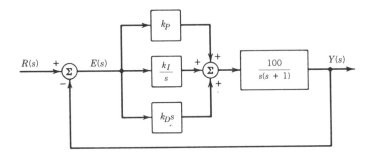

Figure 3.33 The closed-loop system with PID control described in Problem 3.26.

(§§3.5, **3.27**   Consider a positional servomechanism with an amplifier, field-con-
3.6)   trolled dc motor ($L_f = 0$), and potentiometer error detector. Pro-
              vide a detailed block diagram for each of the following control
              schemes, determine $Y(s)/R(s)$ for each case, and describe the
              effects of each closed-loop controller.

a. Positional and tachometer feedback.
b. PID control.
c. SVFB control.

(§3.6) **3.28**   For the given plant transfer functions, construct simulation
              diagrams corresponding to each of the following state variable
              forms and then determine the closed-loop transfer function using
              linear SVFB control for each form from both (3.116) and (3.117).

a. Direct phase variable form.
b. Direct feed-forward form.
c. Uncoupled form.

$$(1) \qquad\qquad G(s) = \frac{s}{s^2 + 7s + 12}$$

$$(2) \qquad\qquad G(s) = \frac{3s + 1}{s(s + 2)(s + 3)}$$

(§3.6) **3.29**   Consider the field-controlled dc motor having a plant transfer
              function given by

$$\frac{\theta_m(s)}{V_f(s)} = \frac{\dfrac{K_T}{J_L L_f}}{s\left(s + \dfrac{R_f}{L_f}\right)\left(s + \dfrac{B_L}{J_L}\right)}$$

Let $L_f = 0.5$, $R_f = 2$, $K_T/J_L = 2$, and $B_L/J_L = 1$. Using the direct
phase variable form of state variables, determine the gain K and the
vector $\mathbf{k}$ for linear state variable feedback such that the closed-loop
system transfer function is

$$\frac{\theta_m(s)}{\theta_r(s)} = \frac{1}{(s + 1)(s + 2)(s + 3)} = \frac{1}{s^3 + 6s^2 + 11s + 6}$$

In mathematical notation, the plant equations are

$$\dot{\mathbf{x}} = A\mathbf{x} + \mathbf{b}v_f(t)$$

$$\theta_m(t) = \mathbf{c}^T\mathbf{x}$$

Let $v_f(t) = K[\theta_r(t) - \mathbf{k}^T\mathbf{x}]$. Find K and $\mathbf{k}$.

(§3.6) **3.30**   Suppose a linear SVFB control law is applied to a linear plant
              described by one state variable form $\mathbf{x}$ and a second linear SVFB
              control law is constructed for another state variable form $\mathbf{x}^*$ for the
              same plant. If the closed-loop pole placement is identical in both
              cases, determine the relationship that exists between the two SVFB
              control vectors $\mathbf{k}$ and $\mathbf{k}^*$.

PART II
CONTROL SYSTEM ANALYSIS

4

Time Response

Assessing the time performance of closed-loop system models is highly important for the control engineer. This analysis provides a quantitative answer to questions related to speed and accuracy of the system time response. Does the robot arm respond quickly enough and without excessive oscillations in performing its assembly task? Does the elevator approach a desired level smoothly, yet not too slowly? Does the scanning electron microscope become accurately positioned as it is adjusted dynamically to a new viewing orientation? Analysis enables the control engineer to determine how well a given system model responds to its input. The need for changes in the model or in the system design can be identified and proper modifications made.

In this part of the book we look at analysis and cover time response in this chapter, stability analysis in the s-plane and in the time domain in Chapter 5, and frequency response in Chapter 6. Initially, in this chapter we examine some performance specifications in the time domain and interpret these specifications for second-order systems. Next, we describe the relationship between the time response of linear time-invariant systems and s-plane closed-loop pole locations. We then present several methods for determining the state transition matrix and the complete time solution. We determine the time solution of the sensitivity equations and the time response of systems having PID and SVFB control. Finally, we conclude the chapter by amplifying earlier results on controllability and observability.

4.1 PERFORMANCE SPECIFICATIONS

The performance of dynamic systems in the time domain can be defined in terms of the time response to standard test inputs. Typical deterministic test inputs include sinusoidal signals, an impulse input, and the unit step, ramp, and parabolic inputs used in Chapter 3 for steady-state error analysis. Generally, performance specifications for the transient part of the time response are identified for a unit step input, but the steady-state performance is measured also with respect to ramp and parabolic inputs. Some selected time response performance specifications [1, 2] are

1. Percent overshoot.
2. Rise time.
3. Delay time.
4. Settling time.
5. Steady-state errors.
6. Constraints on the plant input or state variables.
7. Response sensitivity.

These specifications are described in detail in the following subsections.

4.1.1 Percent Overshoot

Consider the system time response $y(t)$ resulting from a unit step input, as shown in Figure 4.1. The percent overshoot (%OS) is defined as

$$\%OS = 100 \left[\frac{\text{Maximum value of } y(t) - y(\infty)}{y(\infty)} \right] \tag{4.1}$$

The numerator in (4.1) is the maximum overshoot and the denominator is the final value of the response $y(t)$. The 100 is included to convert the ratio to percentage. As a measure of relative stability, a large %OS indicates that the degree of stability is small. Additional stabilizing feedback is generally needed in such cases, unless the loop gain itself can be decreased to reduce the %OS.

4.1.2 Rise Time

The rise time t_r is defined as the time required for the system time response $y(t)$ to move between 10% and 90% of its final value $y(\infty)$. An important specification on the time response, rise time obviously indicates the speed of the time response.

4.1.3 Delay Time

The time required for the system time response $y(t)$ to reach 50% of its final value $y(\infty)$ is referred to as the delay time t_d. As with rise time, delay time is yet another measure of the speed of response.

4.1.4 Settling Time

Settling time is the time required for the system time response $y(t)$ to enter and remain within a small band around the final value. Typically, this band may be

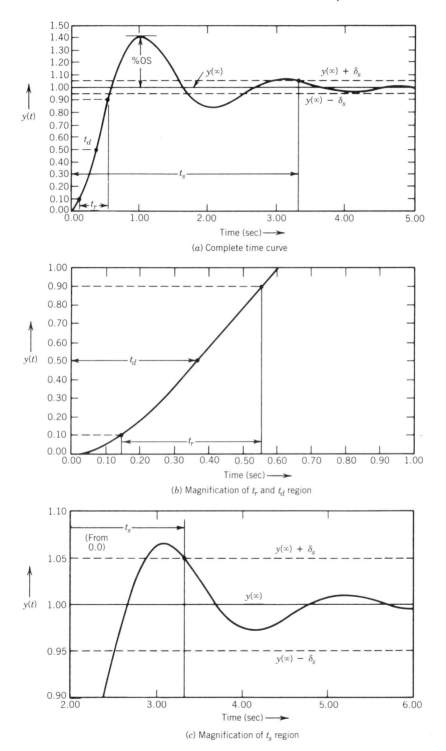

(a) Complete time curve

(b) Magnification of t_r and t_d region

(c) Magnification of t_s region

Figure 4.1 Performance specifications in the time domain.

$\pm 2\%$, $\pm 4\%$, or even $\pm 5\%$ of the final value. The precise percentage figure being used must be identified as part of the settling time specification. Settling time is primarily a measure of the speed of response but, as indicated in [2], is also an indirect measure of the relative stability.

4.1.5 Steady-State Errors

Described already in Section 3.3, steady-state errors are useful as a measure of the system accuracy. A key design requirement often imposed is that the steady-state error for a unit ramp input to a Type 1 system be within some given limit. This case was considered earlier in Figure 3.8. For a unit ramp input, e_{ss} is determined as $1/K_v$, which is $1/K$ from (3.18) and Table 3.2 for a Type 1 system. Consequently, a suitably large loop gain K is required to meet this design specification.

4.1.6 Plant Input and State Constraints

Constraints may be placed on the plant input u or on one or more of the state variables. Constraints on u might result from limitations in available power or torque. An example is the throttle control of an automobile: a full-on throttle is supplying the maximum available input. Constraints on the state variables appear, for example, in limits on velocities or accelerations. An elevator in a high-rise apartment building might be limited in velocity intentionally to prevent structural damage that may result from vibrations at excessive velocities. Further-more, limits on the acceleration of an elevator in a hospital should be an expected part of the design specifications. These are but a few of the many situations where input and state constraints appear as design requirements.

4.1.7 Response Sensitivity

The reduction in system sensitivity due to negative feedback was demonstrated in Section 3.4. It was shown that the sensitivity of the system response can be made much smaller when $|1 + GH|$ is increased. Ordinarily, this increase is realized as a result of making the gain K suitably large. Therefore, the usual cost of reduced sensitivity to parameter and disturbance variations is increased gain.

4.1.8 Performance Measures

The time response performance specifications described above require a large gain K to yield satisfactory values of rise time, delay time, settling time, steady-state errors, and response sensitivity. On the other hand, a suitably small value of gain

K gives a low percent overshoot and, furthermore, usually ensures a better relative stability. Other parameters, such as the elements of series compensation networks described in Chapter 7, can be varied to meet these performance criteria simultaneously.

Traditional design procedures select a high value of K to meet the criteria indicated above for which this choice is needed. Then other parameters are adjusted to satisfy the remaining specifications without violating the criteria already satisfied. The modern approach uses a performance measure to describe the overall quality of the system time response. Typical performance measures are

$$\int_{t_0}^{t_f} e^2 \, dt \qquad \int_{t_0}^{t_f} |e| \, dt$$

$$\int_{t_0}^{t_f} (t - t_0) e^2 \, dt \qquad \int_{t_0}^{t_f} (t - t_0)|e| \, dt$$

$$\int_{t_0}^{t_f} u^2 \, dt \qquad \int_{t_0}^{t_f} |u| \, dt \tag{4.2}$$

The minimization of these performance indices, or cost functionals, yields time responses that generally satisfy the traditional performance specifications quite well. Some of the resulting controllers are nonlinear. However, a class of performance measures that yields linear controllers for linear plants having neither control input nor state variable constraints is given by

$$J = \int_{t_0}^{t_f} (\mathbf{x}^T Q \mathbf{x} + \gamma u^2) \, dt \tag{4.3}$$

where Q is a weighting matrix for the state variables, and γ is a positive weighting constant for the control u.

Although the incorporation of the desirable qualities of a system's time response into a single performance measure is more elegant in a mathematical sense, the same time response can be obtained by either the traditional "performance specification" approach or the modern "performance measure" approach. The relationship between these two approaches and their interpretations in terms of a final design parallels the transfer function versus state variable approaches for modeling and analysis problems.

4.2 SECOND-ORDER SYSTEMS

We now consider a specific class of systems that allows us to see the consequences of the restrictions imposed by time response performance specifications. The selected class is formed by considering second-order linear systems with constant parameters. It is quite appropriate that this particular choice is made. The simpler

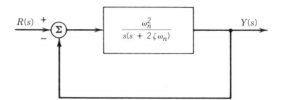

Figure 4.2 A second-order closed-loop system with negative-unity feedback.

class of first-order systems permits only monotonic growth or decay for initial conditions or step inputs. Second-order systems may exhibit either monotonic (overdamped or critically damped) or oscillatory (underdamped) time responses, depending on particular gain settings and parameter values. Furthermore, higher-order systems are often approximated by second-order systems in the initial modeling phase of the control problem. In many of these applications, the second-order model is a sufficiently accurate representation of the physical subsystem [3–5].

We define in this section the different amounts of damping that have traditionally been used to classify second-order system time responses. Several of the performance specifications of the last section are examined for these systems. We especially emphasize the percent overshoot specification and derive a useful formula for %OS for the underdamped case. Proceeding beyond the analysis problem, we apply this formula for tachometer feedback design and then use the same principles to derive a modified version for derivative control.

4.2.1 System Configuration

Consider the second-order unity-feedback system of Figure 4.2. The closed-loop transfer function is given by

$$\frac{Y(s)}{R(s)} = \frac{\omega_n^2}{s^2 + 2\zeta\omega_n s + \omega_n^2} \tag{4.4}$$

where ζ is the damping coefficient and ω_n is the natural frequency. Observe from (4.4) that the low-frequency gain for the closed-loop system is unity. Provided the closed-loop system in Figure 4.2 is asymptotically stable, the steady-state value of $y(t)$ may be determined for a unit step input from the final-value theorem as

$$y_{ss} = \lim_{t \to \infty} y(t) = \lim_{s \to 0} [sY(s)]$$

$$= \lim_{s \to 0} s \left[\frac{\omega_n^2(1/s)}{s^2 + 2\zeta\omega_n s + \omega_n^2} \right] = 1 \tag{4.5}$$

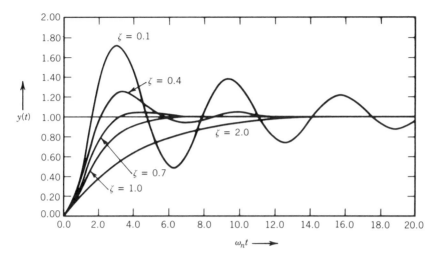

Figure 4.3 Time response curves for the second-order system of Figure 4.2.

Therefore, $y(t)$ approaches the unit step input as t approaches infinity. Thus, the time response curve in steady state resembles the one shown in Figure 4.1. The shape of the transient response depends on the value of the damping coefficient ζ and the natural frequency ω_n.

The poles of the closed-loop transfer function are determined by setting the denominator of (4.4) to zero and solving this characteristic equation for s to yield

$$s_1, s_2 = -\zeta\omega_n \pm \sqrt{[\zeta\omega_n]^2 - \omega_n^2}$$

$$= \omega_n\left[-\zeta \pm \sqrt{\zeta^2 - 1}\right] \tag{4.6}$$

If ζ in (4.6) is greater than 1, there are two real, distinct roots of the characteristic equation. If ζ equals 1, then there are two real, repeated roots, and if ζ is between 0 and 1, there are two complex conjugate roots. These three possibilities have been traditionally classified as overdamped, critically damped, and underdamped cases. These are shown in Figure 4.3, and we will discuss each of them in more detail below. Finally, we should note that ζ equals 0 denotes the undamped case and that values of ζ less than 0 yield two right-half s-plane closed-loop poles, which indicates that the system is unstable.

4.2.2 Overdamped Response

If $\zeta > 1$ in (4.4), then $Y(s)/R(s)$ may be expressed as

$$\frac{Y(s)}{R(s)} = \frac{\gamma_1\gamma_2}{(s + \gamma_1)(s + \gamma_2)} \tag{4.7}$$

where $\gamma_1\gamma_2 = \omega_n^2$ and $\gamma_1 + \gamma_2 = 2\zeta\omega_n$. When we solve for γ_1 and γ_2, we obtain the negatives of the two roots s_1 and s_2 in (4.6). Identifying one of these roots as $-\gamma_1$ and the other as $-\gamma_2$, we arbitrarily select

$$\gamma_1 = \omega_n \left[\zeta - \sqrt{\zeta^2 - 1} \right]$$

$$\gamma_2 = \omega_n \left[\zeta + \sqrt{\zeta^2 - 1} \right] \tag{4.8}$$

For a unit step input, (4.7) becomes

$$Y(s) = \frac{\gamma_1\gamma_2}{s(s + \gamma_1)(s + \gamma_2)}$$

$$= \frac{1}{s} - \frac{\gamma_2/(\gamma_2 - \gamma_1)}{s + \gamma_1} + \frac{\gamma_1/(\gamma_2 - \gamma_1)}{s + \gamma_2} \tag{4.9}$$

Finding inverse Laplace transforms gives

$$y(t) = 1 - \left(\frac{\gamma_2}{\gamma_2 - \gamma_1} \right) e^{-\gamma_1 t} + \left(\frac{\gamma_1}{\gamma_2 - \gamma_1} \right) e^{-\gamma_2 t} \tag{4.10}$$

This overdamped response is indicated in Figure 4.3 for $\zeta > 1$. Different time response curves $y(t)$ shown in Figure 4.3 are functions only of ζ, rather than both ζ and ω_n, since these time-scaled curves are plots of $y(t)$ versus the scaled variable $\omega_n t$.

In solving for $y(t)$ in (4.9) and (4.10), we used a partial fraction expansion for $Y(s)$. Recall that a partial fraction expansion was used earlier in Section 2.4 for expressing the transfer function in a format that isolates individual denominator factors. The resulting uncoupled form of state variable representation has the input entering parallel channels, each describing one state variable, with a weighted sum of these state variables forming the output $y(t)$. Therefore, because each state variable describes one mode of behavior, this uncoupled form is also referred to as the *modal coordinate representation*. These details are described more completely in Section 4.3 in relating state variable and transfer function procedures for the case of real, distinct poles.

4.2.3 Critical Damping

Suppose $\zeta = 1$ in (4.4) and Figure 4.2. From (4.6), the two closed-loop poles are both equal to $-\omega_n$. This condition is referred to as *critical damping*. Therefore, (4.4) becomes

$$\frac{Y(s)}{R(s)} = \frac{\omega_n^2}{(s + \omega_n)^2} \tag{4.11}$$

For a unit step input,

$$Y(s) = \frac{\omega_n^2}{s(s + \omega_n)^2}$$

$$= \frac{1}{s} - \frac{1}{s + \omega_n} - \frac{\omega_n}{(s + \omega_n)^2} \tag{4.12}$$

Again, by taking inverse Laplace transforms, we obtain

$$y(t) = 1 - e^{-\omega_n t} - \omega_n t e^{-\omega_n t} \tag{4.13}$$

which is shown for $\zeta = 1$ in Figure 4.3.

This case of repeated eigenvalues occurs only infrequently, because a precise setting of parameters must be maintained for critical damping to exist. Unlike the overdamped and underdamped cases, which have ranges of ζ, the critically damped case must have ζ equal to a single value—that is, unity. Any variation of ζ from this value, however slight, results in one of the other cases.

4.2.4 Underdamped Response

An underdamped response is obtained when $0 < \zeta < 1$. From (4.6), the closed-loop poles for the system of Figure 4.2 are

$$s_1, s_2 = \omega_n \left[-\zeta \pm j\sqrt{1 - \zeta^2} \right] \tag{4.14}$$

For a unit step input, (4.4) becomes

$$Y(s) = \frac{\omega_n^2}{s(s^2 + 2\zeta\omega_n s + \omega_n^2)}$$

$$= \frac{1}{s} - \frac{s + 2\zeta\omega_n}{s^2 + 2\zeta\omega_n s + \omega_n^2} \tag{4.15}$$

Taking inverse Laplace transforms gives

$$y(t) = 1 - \frac{e^{-\zeta\omega_n t}}{\sqrt{1 - \zeta^2}} \sin(\omega_d t + \theta) \tag{4.16}$$

where θ is a phase angle and ω_d is the damped natural frequency. In terms of ζ

and ω_n, these parameters are

$$\theta = \tan^{-1}\left(\frac{\sqrt{1 - \zeta^2}}{\zeta}\right)$$

$$\omega_d = \omega_n\sqrt{1 - \zeta^2} \qquad (4.17)$$

From the expression for θ in (4.17), we may visualize a right triangle with $\sqrt{1 - \zeta^2}$ as the length of the side opposite the acute angle θ and ζ as the length of the adjacent side. Since the hypotenuse of this right triangle is unity, we may form an alternate expression for θ as

$$\theta = \cos^{-1}(\zeta) \qquad (4.18)$$

Equation (4.16) yields oscillatory responses, as shown in Figure 4.3 for $0 < \zeta < 1$.

Since the two closed-loop poles in (4.14) are complex, the uncoupled and cascade forms of Section 2.4 are not applicable. However, both direct forms—that is, the phase variable and feed-forward forms—represent legitimate choices for state variable representations.

We determine time response performance specifications for the second-order system of Figure 4.2 in the next few subsections.

4.2.5 Percent Overshoot

The percent overshoot performance specification is not applicable to the overdamped or critically damped responses, but it is a very important consideration for the underdamped case. From the definition in (4.1), we need to find the maximum overshoot for $y(t)$ in (4.16). We could do this by taking the derivative of $y(t)$, setting it to zero, and solving for the value of peak time t_p. An alternate procedure is to form $sY(s)$ from (4.15), take its inverse Laplace transform $\dot{y}(t)$, and then, as before, solve for t_p by setting $\dot{y}(t_p) = 0$. This alternate approach yields

$$\dot{y}(t) = \mathscr{L}^{-1}\{sY(s)\} = \mathscr{L}^{-1}\left(\frac{\omega_n^2}{s^2 + 2\zeta\omega_n s + \omega_n^2}\right)$$

$$= \frac{\omega_n}{\sqrt{1 - \zeta^2}} e^{-\zeta\omega_n t} \sin \omega_d t \qquad (4.19)$$

Setting $\dot{y}(t)$ in (4.19) to zero and denoting the corresponding value of t as t^*

yields $\sin \omega_d t^* = 0$, from which $\omega_d t^* = n\pi$, to yield

$$t^* = \frac{n\pi}{\omega_d} \qquad (n = 0, 1, 2, 3, \ldots) \qquad (4.20)$$

Odd values of n correspond to maxima of $y(t)$ and even values to minima. The maximum overshoot (at t_p) corresponds to the first odd value of n $(n = 1)$ in

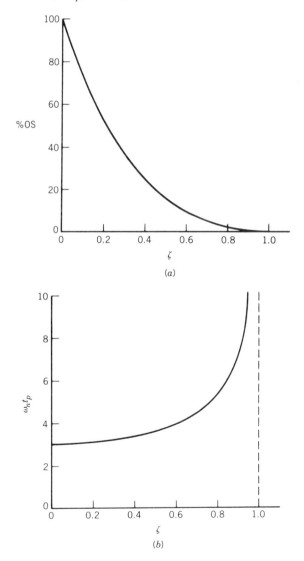

Figure 4.4 Plots of %OS and $\omega_n t_p$ versus ζ for the underdamped case.

(4.20). Thus, from (4.20) and (4.17),

$$t_p = t^*\big|_{n=1} = \frac{\pi}{\omega_d} = \frac{\pi}{\omega_n\sqrt{1-\zeta^2}} \tag{4.21}$$

Substituting (4.21) into (4.16) gives $y(t_p)$ as

$$y(t_p) = 1 - \frac{1}{\sqrt{1-\zeta^2}} e^{-\pi\zeta/\sqrt{1-\zeta^2}} \sin(\pi + \theta) \tag{4.22}$$

Using (4.18) and some trigonometric relationships, we have

$$\sin(\pi + \theta) = -\sin\theta = -\sqrt{1-\cos^2\theta} = -\sqrt{1-\zeta^2} \tag{4.23}$$

Therefore, inserting (4.23) into (4.22) gives

$$y(t_p) = 1 + e^{-\pi\zeta/\sqrt{1-\zeta^2}} \tag{4.24}$$

Finally, since $y(\infty) = 1$ for this Type 1 system, we have from (4.1)

$$\%OS = 100\left[\frac{y(t_p) - y(\infty)}{y(\infty)}\right]$$

$$= 100 e^{-\pi\zeta/\sqrt{1-\zeta^2}} \tag{4.25}$$

The curve of %OS versus ζ is shown in Figure 4.4a. Observe that large values of %OS are obtained for ζ between 0 and approximately 0.2, but the %OS is below 25% for ζ between approximately 0.4 and 1.0. This desired region of ζ, which depends on the specific application, may often be realized by properly adjusting controller or other system parameters. A plot of $\omega_n t_p$ versus ζ is given in Figure 4.4b.

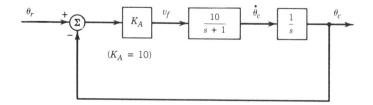

Figure 4.5 The positional servomechanism of Example 4.1.

EXAMPLE 4.1

Consider the block diagram of a positional servomechanism shown in Figure 4.5. All plant and controller parameters are specified. It is assumed that the inductance of the field circuit in the field-controlled dc motor is negligible, which results in a second-order system of the form shown in Figure 4.2. The problem is to determine the %OS resulting from a step input command.

We combine blocks in the plant and substitute the given parameter values to obtain

$$G(s) = \frac{100}{s(s + 1)} \qquad (4.26)$$

Therefore, from Figure 4.2,

$$\omega_n^2 = 100$$

$$2\zeta\omega_n = 1 \qquad (4.27)$$

Thus, $\omega_n = 10$ and $\zeta = 1/(2\omega_n) = 0.05$. Using $\zeta = 0.05$ in (4.25) gives a %OS of 85.4%. Clearly, this value is too large for most applications.

EXAMPLE 4.2

Suppose that in Figure 4.5 all of the closed-loop system gains and parameters except K_A are specified. Determine the range of K_A for which the %OS is 20% or less.

The transfer function $G(s)$ for this condition becomes

$$G(s) = \frac{10K_A}{s(s + 1)} \qquad (4.28)$$

As before, from Figure 4.2, we have $\omega_n^2 = 10K_A$ and $2\zeta\omega_n = 1$, which gives

$$\omega_n = \sqrt{10K_A}$$

$$\zeta = 1/(2\omega_n) = \frac{1}{2\sqrt{10K_A}} \qquad (4.29)$$

We determine the acceptable range of ζ first and then use (4.29) to find the corresponding range of K_A.

Using (4.25), we may write

$$e^{\pi\zeta/\sqrt{1-\zeta^2}} = \left(\frac{100}{\%OS}\right) \tag{4.30}$$

or

$$\frac{\pi\zeta}{\sqrt{1-\zeta^2}} = \ln\left(\frac{100}{\%OS}\right) \tag{4.31}$$

Squaring both sides of (4.31) and solving for ζ gives

$$\zeta = \frac{\ln\left(\dfrac{100}{\%OS}\right)}{\sqrt{\pi^2 + \ln^2\left(\dfrac{100}{\%OS}\right)}} \tag{4.32}$$

Using (4.32) for $\%OS = 20\%$, we find that $\zeta = 0.456$. Since we require the $\%OS$ to be less than or equal to 20%, then ζ must obviously be greater than or equal to 0.456. This conclusion may be obtained by examining (4.32) in greater detail or, more simply, by observing from Figure 4.4a that the $\%OS$ is a monotonically decreasing function of ζ. Interpreting this result in terms of K_A by using (4.29) yields

$$\zeta = \frac{1}{2\sqrt{10K_A}} \geq 0.456 \tag{4.33}$$

which gives the range of K_A as $0 < K_A \leq 0.120$. We show later in this section that other parameters, such as a tachometer feedback gain, may be adjusted to meet the $\%OS$ performance specification. In such a case, the amplifier gain K_A may be set at a much higher value to satisfy other design specifications.

4.2.6 Rise, Delay, and Settling Times

Curves of rise time (t_r), delay time (t_d), and settling time (t_s) versus the damping coefficient ζ are presented for $0 < \zeta \leq 2$ in this subsection. Instead of plotting t_r, t_d, and t_s versus ζ directly, it is more enlightening to form normalized curves of $\omega_n t_r$, $\omega_n t_d$, and $\omega_n t_s$ versus ζ, as shown in Figure 4.6. Points on these curves can be obtained by setting ω_n equal to 1 and then determining t_r, t_d, and t_s according to their definitions. The resulting normalized curves are valid for any value of ω_n.

As an example of how to form these curves, suppose we wish to compute the value of $\omega_n t_r$ for the overdamped case where $\zeta = 1.25$ and $\omega_n = 1$. We find γ_1

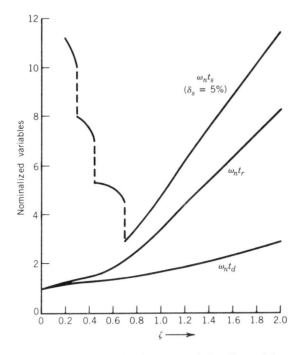

Figure 4.6 Normalized curves of rise time, delay time, and settling time for the second-order system of Figure 4.2.

and γ_2 in (4.7) from (4.8) as 0.5 and 2, respectively. Using these values in (4.10) gives $y(t)$ as

$$y(t) = 1 - \tfrac{4}{3}e^{-0.5t} + \tfrac{1}{3}e^{-2t} \tag{4.34}$$

Let the value of t for which $y(t) = 0.9$ be denoted as $t_{0.9}$. Consequently, from (4.34), $t_{0.9}$ satisfies

$$0.9 = 1 - \tfrac{4}{3}e^{-0.5t_{0.9}} + \tfrac{1}{3}e^{-2t_{0.9}} \tag{4.35}$$

or

$$f(t_{0.9}) = 0.1 - \tfrac{4}{3}e^{-0.5t_{0.9}} + \tfrac{1}{3}e^{-2t_{0.9}} = 0 \tag{4.36}$$

We may solve (4.36) numerically, for example, by using Newton's method,[1] to obtain $t_{0.9} = 5.18$. Similarly, $t_{0.1}$, which satisfies $y(t_{0.1}) = 0.1$, may be determined

---

[1] Newton's method is useful for solving algebraic equations of the form $f(\alpha) = 0$. The method is based on a recursive relationship $\alpha_{N+1} = \alpha_N - f(\alpha_N)/f'(\alpha_N)$, which often converges to an acceptable solution from a suitably chosen starting value α_0 in as few as three or four iterations.

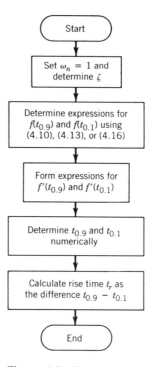

Figure 4.7 Flowchart for calculating the rise time of the time response for the second-order system of Figure 4.2.

as $t_{0.1} = 0.56$ from

$$f(0.1) = 0.9 - \tfrac{4}{3}e^{-0.5t_{0.1}} + \tfrac{1}{3}e^{-2t_{0.1}} = 0 \tag{4.37}$$

The value of rise time t_r is the time difference $t_{0.9} - t_{0.1} = 5.18 - 0.56 = 4.62$. A flowchart for calculating rise time is given in Figure 4.7.

The curves of $\omega_n t_d$ and $\omega_n t_s$ versus ζ in Figure 4.6 are obtained in a similar manner. Each calculation requires the numerical solution of only one algebraic equation for each value of ζ. However, for the underdamped case, we must first narrow the search for t_s by observing maxima and minima of $y(t)$ and noting when one such extremum point lies outside the $1 \pm \delta_s$ band and the next is inside this band. For example, if a minimum point (trough) is outside the band and the next peak is inside, then we search for t_s satisfying

$$y(t_s) = 1 - \delta_s = 1 - \frac{e^{-\zeta\omega_n t_s}}{\sqrt{1 - \zeta^2}} \sin\left(\omega_d t_s + \theta\right) \tag{4.38}$$

On the other hand, if a peak is outside the band and the next trough is inside, then t_s satisfies

$$y(t_s) = 1 + \delta_s = 1 - \frac{e^{-\zeta\omega_n t_s}}{\sqrt{1 - \zeta^2}} \sin(\omega_d t_s + \theta) \tag{4.39}$$

Discontinuities shown in Figure 4.6 for $\omega_n t_s$ with $\delta_s = 5\%$ result from a peak or trough moving inside the 5% band as ζ is increased. Using (4.32), we may easily verify that the maximum peak (and, hence, all peaks and troughs) are inside this band for $\zeta > 0.69$.

4.2.7 Response Sensitivity

We interpret the sensitivity performance specification in this subsection for the second-order system of Figure 4.2. Whereas the discussion of Section 3.4 on system sensitivity was restricted primarily to steady-state values, we examine the transient and steady-state regions of the system response here. Both transfer function and state variable analysis procedures are developed, but the solution of the state variable sensitivity equations is postponed until Section 4.6, which depends on the state transition matrix discussion of Section 4.4.

Let us compute S_ζ^M, defined in (3.46), as

$$S_\zeta^M = \frac{\partial M}{\partial \zeta}\left(\frac{\zeta}{M}\right) = \left[\frac{\partial M}{\partial G}\left(\frac{G}{M}\right)\right]\left[\frac{\partial G}{\partial \zeta}\left(\frac{\zeta}{G}\right)\right]$$

$$= \left[\frac{1}{1 + \dfrac{\omega_n^2}{s(s + 2\zeta\omega_n)}}\right]\left[\frac{-2\omega_n^3 s}{s^2(s + 2\zeta\omega_n)^2}\left(\frac{\zeta}{\dfrac{\omega_n^2}{s(s + 2\zeta\omega_n)}}\right)\right]$$

$$= \frac{-2\zeta\omega_n s}{s^2 + 2\zeta\omega_n s + \omega_n^2} \tag{4.40}$$

Using (3.71) for a unit step input gives corresponding changes $\Delta Y(s)$ due to changes $\Delta\zeta$ as

$$\Delta Y(s) = S_\zeta^M\left(\frac{M}{\zeta}\right)\left(\frac{1}{s}\right)(\Delta\zeta)$$

$$= \frac{-2\omega_n^3(\Delta\zeta)}{\left(s^2 + 2\zeta\omega_n s + \omega_n^2\right)^2} \tag{4.41}$$

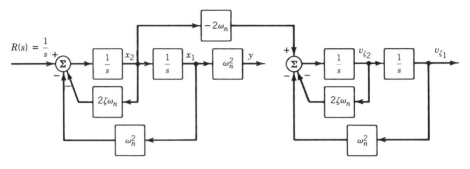

Figure 4.8 Block diagram for response sensitivity in (4.43) and (4.44).

Taking inverse Laplace transforms $(0 < \zeta < 1)$ yields

$$\Delta y(t) = \frac{-(\Delta\zeta)}{\left[\sqrt{1-\zeta^2}\right]^3} e^{-\zeta\omega_n t}\left[\sin\omega_d t - \omega_d t \cos\omega_d t\right] \tag{4.42}$$

From the state variable approach, we may write

$$\dot{x}_1 = x_2$$

$$\dot{x}_2 = -\omega_n^2 x_1 - 2\zeta\omega_n x_2 + r(t)$$

$$y = \omega_n^2 x_1 \tag{4.43}$$

where the direct phase variable form is being used. Forming the state equations for the state sensitivity coefficient yields

$$\dot{v}_{\zeta_1} = v_{\zeta_2}$$

$$\dot{v}_{\zeta_2} = -\omega_n^2 v_{\zeta_1} - 2\zeta\omega_n v_{\zeta_2} - 2\omega_n x_2 \tag{4.44}$$

Figure 4.8 shows the cascaded simulated diagram for (4.43) and (4.44). The time solution of these two sets of equations is described in Section 4.6. However, observe that the Laplace transform relationship for v_{ζ_1} may be obtained by applying block diagram reduction rules to Figure 4.8. Using $\Delta y = \mathbf{c}^T(\Delta\mathbf{x})$, with $\Delta\mathbf{x} = \mathbf{v}_\zeta(\Delta\zeta)$ from (3.61), we again obtain the transform result in (4.41). Similarly, the effects of variations in ω_n can be determined.

4.2.8 Tachometer Feedback and Derivative Control

We consider now two control schemes to aid in reducing the %OS resulting from a step input while maintaining a large velocity error constant. In these cases the proportional feedback control is not altered, but additional control is provided by either tachometer feedback or derivative control.

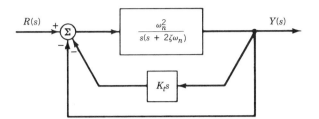

Figure 4.9 Tachometer feedback for the system of
Figure 4.2.

Tachometer feedback is indicated in Figure 4.9. The closed-loop transfer
function is given by

$$\frac{Y(s)}{R(s)} = \frac{\omega_n^2}{s^2 + \left(2\zeta\omega_n + K_t\omega_n^2\right)s + \omega_n^2} \tag{4.45}$$

Observe that the natural frequency of the time response ω_n is unchanged by the
incorporation of tachometer feedback. However, the new damping coefficient ζ_t is
obtained as

$$2\zeta_t\omega_n = 2\zeta\omega_n + K_t\omega_n^2$$

$$\zeta_t = \zeta + \tfrac{1}{2}K_t\omega_n \tag{4.46}$$

Therefore, increasing the tachometer feedback gain K_t increases ζ_t, which results
in a smaller %OS, according to (4.25) and Figure 4.4a. A flowchart for determining K_t is given in Figure 4.10.

The closed-loop transfer function for the derivative control case of Figure 4.11
is given by

$$\frac{Y(s)}{R(s)} = \frac{\omega_n^2(1 + k_Ds)}{s^2 + \left(2\zeta\omega_n + k_D\omega_n^2\right)s + \omega_n^2} \tag{4.47}$$

Notice that (4.47) is not the same form as the transfer function in (4.4) for which
the %OS formula of (4.25) was derived. As shown in Example 4.4 below, the
presence of the closed-loop zero (numerator factor) in (4.47) results in a somewhat higher %OS. We now derive an expression for that value.

For a step input, we have

$$Y(s) = \frac{\omega_n^2 + k_D\omega_n^2s}{s\left(s^2 + 2\zeta_D\omega_ns + \omega_n^2\right)} \tag{4.48}$$

where the new damping coefficient ζ_D for the derivative control case is given by

$$\zeta_D = \zeta + \frac{1}{2}k_D\omega_n \tag{4.49}$$

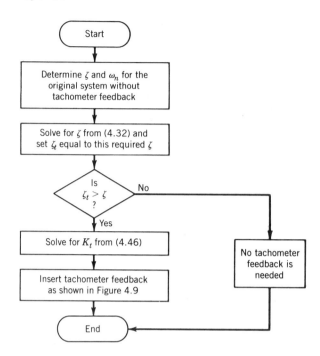

Figure 4.10 Flowchart for determining the tachometer feedback gain K_t in Figure 4.9.

Taking inverse Laplace transforms for $Y(s)$ in (4.48), we obtain

$$y(t) = 1 - \frac{e^{-\zeta_D \omega_n t}}{\sqrt{1 - \zeta_D^2}} \sin(\omega_d t + \theta) + k_D \omega_n \frac{e^{-\zeta_D \omega_n t}}{\sqrt{1 - \zeta_D^2}} \sin \omega_d t \qquad (4.50)$$

where θ and ω_d are defined as

$$\theta = \tan^{-1}\left(\frac{\sqrt{1 - \zeta_D^2}}{\zeta_D}\right)$$

$$\omega_d = \omega_n \sqrt{1 - \zeta_D^2} \qquad (4.51)$$

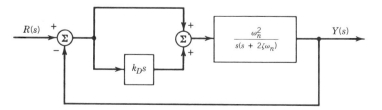

Figure 4.11 Derivative control for the system of Figure 4.2.

Setting $\dot{y}(t_p)$ equal to zero and solving for t_p yields

$$t_p = \frac{1}{\omega_d}(\pi - \alpha) \tag{4.52}$$

where

$$\alpha = \tan^{-1}(\beta) \tag{4.53}$$

and

$$\beta = \frac{k_D \omega_d}{1 - k_D \zeta_D \omega_n} \tag{4.54}$$

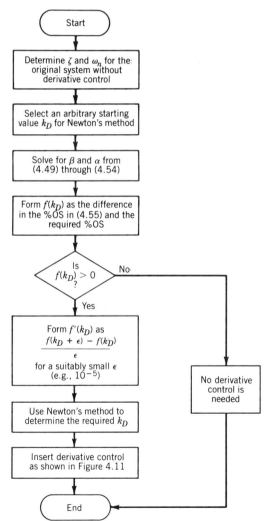

Figure 4.12 Flowchart for determining the derivative control gain k_D in Figure 4.11.

Consequently, by inserting (4.51) through (4.54) into (4.50) and by using the definition of %OS in (4.1) we obtain

$$\%OS = \frac{100 \exp\left[-\zeta_D(\pi - \alpha)/\sqrt{1 - \zeta_D^2}\right]}{\sqrt{1 - \zeta_D^2}\sqrt{1 + \beta^2}}\left(\sqrt{1 - \zeta_D^2} - \beta\zeta_D + \beta k_D\omega_n\right) \quad (4.55)$$

which is considerably more complicated than the %OS formula in (4.25) for the system in Figure 4.2. However, notice that for $k_D = 0$ we have $\beta = 0$ and $\alpha = 0$, which yields, from (4.55), the original %OS formula given in (4.25). Finally, we must recognize that solving (4.55) for k_D in terms of %OS for a given problem is quite a formidable numerical task. A flowchart based on using Newton's method with (4.49) through (4.55) is shown in Figure 4.12. On the other hand, it is straightforward to solve (4.25) for ζ_t, as in (4.32), and subsequently determine K_t in tachometer feedback design (Figure 4.10).

EXAMPLE 4.3

As a first example in this subsection, we wish to verify the time response results of Section 1.5 in Figure 1.11 for tachometer feedback control of a positional servomechanism. Let the plant transfer function be defined as $G(s) = K/s(1 + \tau_m s)$. For $\tau_m = 1$, the closed-loop system transfer function is given by

$$\frac{\theta_c(s)}{\theta_r(s)} = \frac{K}{s^2 + (1 + KK_t)s + K} \quad (4.56)$$

For $K_t = 0$, that is, no tachometer feedback, we have

$$\omega_n = \sqrt{K}$$

$$\zeta = \frac{1}{2\omega_n} = \frac{1}{2\sqrt{K}} \quad (4.57)$$

which yields $\zeta = 0.5$ for $K = 1$ and $\zeta = 0.158$ for $K = 10$. Using (4.25) gives a %OS of 16.3% for $K = 1$ and 60.5% for $K = 10$. If the overshoot of 16.3% is acceptable for a given application, then we may proceed by designing a tachometer feedback control for $\zeta_t = 0.5$, while requiring $K = 10$ to meet other performance specifications. Referring to (4.46) and (4.56), we determine the tachometer gain K_t (for $K = 10$) as

$$2\zeta_t\omega_n = 1 + 10K_t$$

$$K_t = \frac{2\zeta_t\omega_n - 1}{10} = \frac{2(0.5)\sqrt{10} - 1}{10} = 0.216 \quad (4.58)$$

which agrees with the approximate value (0.2) used to obtain the curve in Figure 1.11.

EXAMPLE 4.4

Consider as a second example on tachometer feedback the plant transfer function given by

$$G(s) = \frac{10}{s(s + 2)} \tag{4.59}$$

It is required to find the tachometer feedback gain K_t that will yield a %OS of 10%, to determine the %OS for derivative control with k_D set equal to the K_t just found, and to find k_D for a %OS of 10%.

Initially, we use (4.25) with $\zeta = 2/(2\omega_n) = 1/\sqrt{10} = 0.316$ to obtain a %OS of 35.1% for the original system with only negative-unity feedback. For tachometer feedback, we have

$$\frac{Y(s)}{R(s)} = \frac{10}{s^2 + (2 + 10K_t)s + 10} \tag{4.60}$$

Using (4.32) with the %OS set to 10%, we obtain $\zeta_t = 0.591$. Therefore, from (4.46), we have

$$2\zeta_t\sqrt{10} = 2 + 10K_t$$

$$K_t = \frac{2\zeta_t\sqrt{10} - 2}{10} = 0.174 \tag{4.61}$$

These calculations correspond to those indicated in the flowchart of Figure 4.10.

For derivative control, we use the formula in (4.55) with $k_D = 0.174$. After a somewhat lengthy computation corresponding to an initial evaluation for Newton's method in Figure 4.12, we obtain the %OS as 12.4%. Observe that we approximate $f'(k_D)$ as the incremental slope about the most recent iteration of k_D. Proceeding with Newton's method yields $k_D = 0.206$ in just two iterations.

TABLE 4.1 Numerical Results for Example 4.4

Control Design Strategy	Damping Coefficient	%OS
Proportional control	0.316	35.1
Tachometer feedback ($K_t = 0.174$)	0.591	10.0
Derivative control ($k_D = 0.174$)	0.591	12.4
Derivative control ($k_D = 0.206$)	0.642	10.0

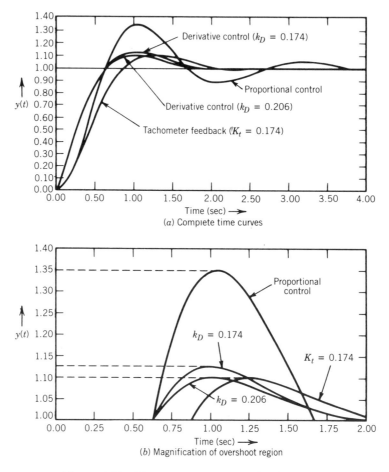

Figure 4.13 Time response curves for Example 4.4.

These calculations are summarized in Table 4.1 with the time response curves shown in Figure 4.13.

Example 4.4 indicates that the additional zero in the closed-loop transfer function having the same denominator parameters results in a somewhat larger %OS. We examine more explicitly the relationships between the time response and the poles and zeros of a transfer function in the following section.

4.3 THE s-PLANE AND TIME RESPONSE

We describe in this section the relationship between the locations of system transfer function poles and zeros in the s-plane and the corresponding time response. This concept is quite important in studies of the state transition matrix in the following section and the complete solution in Section 4.5. A construction of s-plane vectors [6,7] allows us to visualize more clearly the precise contribution of poles and zeros on the time response. We then discuss the modal coordinate representation as a connecting link between state variable and transform analysis viewpoints in solving for the time response.

4.3.1 Effects of s-Plane Pole Locations

Let the Laplace transform of the system output $Y(s)$ be expressed as

$$Y(s) = G_s(s)R(s) = \frac{K(s + z_1) \cdots (s + z_l)}{(s + p_1)(s + p_2)\ldots(s + p_\rho)} \qquad (4.62)$$

where $-z_1, \ldots, -z_l$ are the zeros, and $-p_1, \ldots, -p_\rho$ are the poles of $Y(s)$ with $l \le \rho$. Some of these poles and zeros appear in (4.62) as a result of the system transfer function $G_s(s)$, and the remainder are due to the input $R(s)$. We may

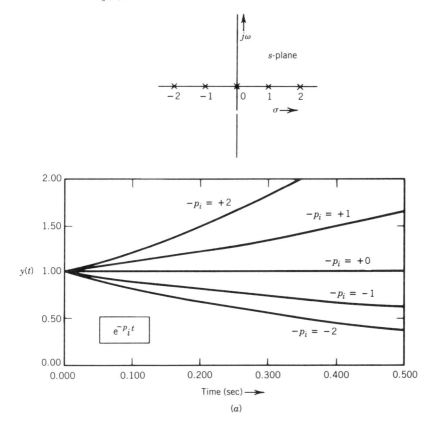

(a)

Figure 4.14 Typical time response curves for various s-plane pole locations.

express (4.62) in the partial fraction format

$$Y(s) = \frac{K_1}{s + p_1} + \frac{K_2}{s + p_2} + \cdots + \frac{K_\rho}{s + p_\rho} + K_{\rho+1} \tag{4.63}$$

where $K_{\rho+1} = K$ if $l = \rho$ and $K_{\rho+1}$ is zero if $l < \rho$ in (4.62). We consider here the time response resulting from poles on the real axis, on the imaginary axis, and in the complex s-plane.

Poles on the Real Axis

Suppose a particular pole $-p_i$ in (4.63) is located somewhere along the real axis, as shown in Figure 4.14a (Page 179). The corresponding sketch of the time response indicates that an exponential curve $e^{-p_i t}$ is obtained. This result arises from the inverse Laplace transformation of the ith term in (4.63). If $-p_i$ is on the negative real axis, the curve decreases exponentially; if $-p_i$ is at the origin in the s-plane, the curve is a constant K_i; and if $-p_i$ is on the positive real axis, the curve increases exponentially, as shown in Figure 4.14a.

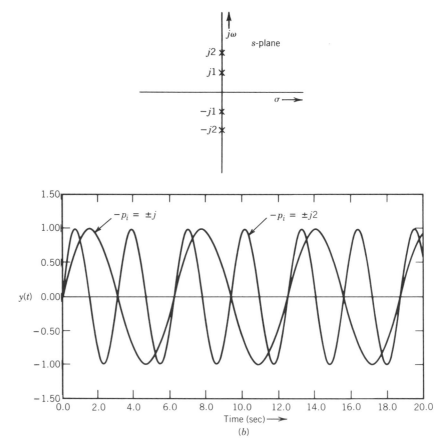

Figure 4.14 (*Continued*) Typical time response curves for various s-plane pole locations.

Poles on the Imaginary Axis

Figure 4.14b shows the time response for poles on the $j\omega$-axis. Provided $Y(s)$ may be formed as the ratio of two polynomials in s having real coefficients, the resulting imaginary axis poles in (4.63) occur in pairs at $+j\omega_i$ and $-j\omega_i$. The frequency of sinusoidal oscillation ω_i is increased when these simple poles are placed farther from the origin in the s-plane, as shown in Figure 4.14b.

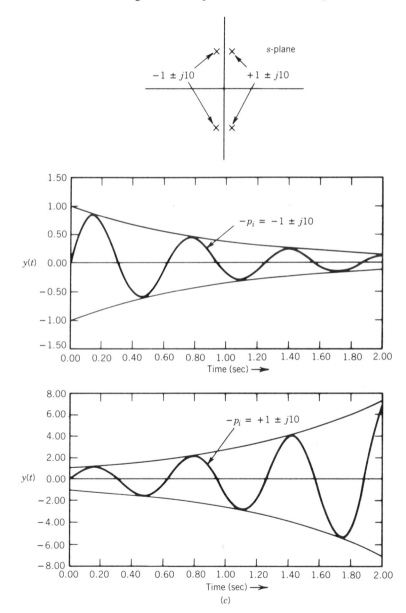

Figure 4.14 (*Continued*) Typical time response curves for various s-plane pole locations.

Poles in the Complex s-Plane

A combination of the two previous cases is obtained when pairs of complex poles appear in (4.63). If the complex pair is in the left-half s-plane, the time response is a damped sinusoid; if in the right-half s-plane, the envelope of the sinusoidal oscillations increases exponentially. Figure 4.14c shows time response curves for two typical complex pole locations.

4.3.2 Residue Evaluation and s-Plane Vectors

Consider a special case of (4.62) in which $Y(s)$ has no zeros and only distinct poles, that is,

$$Y(s) = \frac{K}{(s + p_1)(s + p_2)\dots(s + p_\rho)} \tag{4.64}$$

We wish to express (4.64) by partial fractions as

$$Y(s) = \frac{K_1}{s + p_1} + \frac{K_2}{s + p_2} + \cdots + \frac{K_i}{s + p_i} + \cdots + \frac{K_\rho}{s + p_\rho} \tag{4.65}$$

where the K_i are referred to as the residues at the poles $-p_i$. Multiplying both sides of (4.65) by $(s + p_i)$ gives

$$(s + p_i)Y(s) = \frac{(s + p_i)K_1}{s + p_1} + \cdots + K_i + \cdots + \frac{(s + p_i)K_\rho}{s + p_\rho} \tag{4.66}$$

If we evaluate (4.66) at $s = -p_i$, we obtain

$$[(s + p_i)Y(s)]\big|_{s = -p_i} = K_i \tag{4.67}$$

since all other terms on the right side of (4.66) become zero for $s = -p_i$.

We may use the concept of s-plane vectors to evaluate K_i in (4.67). Any factor $(s + \alpha)$ can be treated as an s-plane vector, as shown in Figure 4.15a. Note that a vector $-\alpha$ from the origin in the s-plane to the root's location at $s = -\alpha$ may be added vectorially to a vector from the root to the point s. The result of this vector addition must yield the vector **s**. Therefore, the vector from the root at $-\alpha$ to the point s must be **s** + **α**, which has a magnitude and an angle associated with it. As for any vector, the magnitude is its length and the angle is measured counterclockwise from the positive real axis.

The application of s-plane vectors to the problem of determining K_i is indicated in Figure 4.15b. The product $(s + p_i) Y(s)$ in (4.67) may be evaluated at $s = -p_i$ by s-plane vectors drawn from the other poles to the point $s = -p_i$.

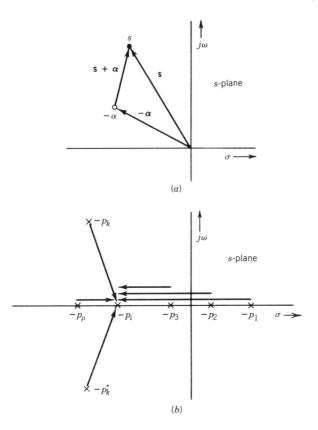

Figure 4.15 Definition of s-plane vectors and their use in determining residues in (4.65).

We obtain

$$K_i = \left[(s + p_i)Y(s)\right]\big|_{s=-p_i}$$

$$= \frac{K}{\displaystyle\prod_{\substack{j=1 \\ (j \neq i)}}^{\rho} (-p_i + p_j)} \tag{4.68}$$

Therefore, the magnitude of K_i is the magnitude of K divided by the product of the lengths of s-plane vectors so constructed, that is,

$$|K_i| = \frac{|K|}{\displaystyle\prod_{\substack{j=1 \\ (j \neq i)}}^{\rho} |-p_i + p_j|} = \frac{|K|}{\begin{array}{c}\text{Product of lengths}\\\text{of } s\text{-plane vectors}\\\text{drawn to } s = -p_i\end{array}} \tag{4.69}$$

Moreover, the angle of K_i in (4.68) is given by

$$\angle K_i = \angle K - \sum_{\substack{j=1 \\ (j \neq i)}}^{\rho} \angle (-p_i + p_j) \tag{4.70}$$

where $\angle K$ is zero if K is positive and $180°$ if K is negative. From (4.69), we note that the magnitude of K_i is increased by the presence of other poles near $-p_i$ and decreased by other poles far from $-p_i$. Although the use of s-plane vectors yields the same results as the algebraic manipulation of the corresponding complex numbers, this characterization does provide a visual aid for recognizing the importance of s-plane pole locations on the output time response.

EXAMPLE 4.5

Determine the time response $y(t)$ by using s-plane vectors to find the residues at the poles of

$$Y(s) = \frac{1}{s(s+1)(s+4)} \tag{4.71}$$

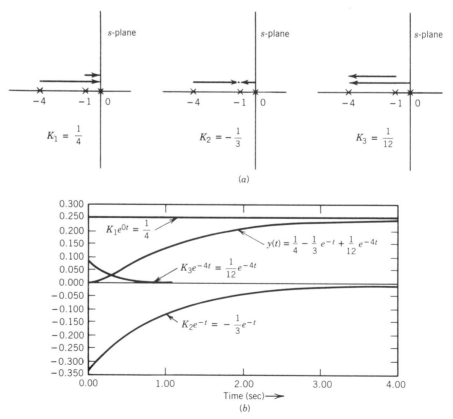

Figure 4.16   Application of s-plane vectors for Example 4.5.

Figure 4.16*a* shows the application of *s*-plane vectors to find K_1, K_2, and K_3 for the poles at 0, -1, and -4, respectively, by using (4.69) and (4.70). These residues are

$$K_1 = \frac{1}{|0 + 1| \cdot |0 + 4|} \angle -(0° + 0°) = \frac{1}{4} \angle 0° = +\frac{1}{4}$$

$$K_2 = \frac{1}{|-1| \cdot |-1 + 4|} \angle -(180° + 0°) = \frac{1}{3} \angle -180° = -\frac{1}{3}$$

$$K_3 = \frac{1}{|-4| \cdot |-4 + 1|} \angle -(180° + 180°) = \frac{1}{12} \angle -360° = +\frac{1}{12} \quad (4.72)$$

Therefore, we may express $y(t)$ as

$$y(t) = K_1 e^{(0)t} + K_2 e^{(-1)t} + K_3 e^{(-4)t}$$

$$= \tfrac{1}{4} - \tfrac{1}{3}e^{-t} + \tfrac{1}{12}e^{-4t} \quad (4.73)$$

The corresponding time response curve for each pole and the composite curve for $y(t)$ is shown in Figure 4.16*b*.

4.3.3 Effects of *s*-Plane Zero Locations

The *s*-plane vector concept may be applied for evaluating more general functions $Y(s)$, as in (4.62), where the poles are distinct. Multiplying both sides of (4.63) by $(s + p_i)$ and evaluating the resulting equation as before at $s = -p_i$ gives

$$K_i = [(s + p_i)Y(s)]\big|_{s = -p_i} = \frac{K \prod\limits_{j=1}^{l}(-p_i + z_j)}{\prod\limits_{\substack{j=1 \\ (j \neq i)}}^{\rho}(-p_i + p_j)} \quad (4.74)$$

The magnitude of K_i is the product of the gain K and the *s*-plane vector lengths drawn from the zeros of $Y(s)$ to the point $s = -p_i$ divided by the product of the vector lengths from the other poles (all except $-p_i$) to $s = -p_i$. Therefore,

$$|K_i| = \frac{|K| \prod\limits_{j=1}^{l}|(-p_i + z_j)|}{\prod\limits_{\substack{j=1 \\ (j \neq i)}}^{\rho}|(-p_i + p_j)|} \quad (4.75)$$

The angle of K_i is given by

$$\underline{/K_i} = \underline{/K} + \sum_{j=1}^{l} \underline{/(-p_i + z_j)} - \sum_{\substack{j=1 \\ (j \ne i)}}^{p} \underline{/(-p_i + p_j)} \qquad (4.76)$$

We see from (4.75) that an effect of s-plane zero locations is to reduce the magnitude of the residue K_i when the zero is near $-p_i$.

EXAMPLE 4.6

As a final example in this section on the use of s-plane vectors, we determine the differences that occur in the time response of a second-order system resulting from derivative control and tachometer feedback. Consider again the plant transfer function of Example 4.4 given by

$$G(s) = \frac{10}{s(s + 2)} \qquad (4.77)$$

which has negative-unity feedback before the addition of either tachometer feedback or derivative control. The description of the closed-loop transfer function for tachometer feedback in (4.60) allows us to express $Y(s)$ for a unit step input as

$$Y(s) = \frac{10}{s[s^2 + (2 + 10K_t)s + 10]} \qquad (4.78)$$

We showed in (4.61) that for a %OS of 10% we require $K_t = 0.174$. Inserting this value of K_t into (4.78) and factoring yields

$$Y(s) = \frac{10}{s(s^2 + 3.74s + 10)}$$

$$= \frac{10}{s(s + 1.87 - j2.55)(s + 1.87 + j2.55)}$$

$$= \frac{K_1}{s} + \frac{K_2}{s + 1.87 - j2.55} + \frac{K_3}{s + 1.87 + j2.55} \qquad (4.79)$$

Figure 4.17a shows the construction of s-plane vectors for determining K_1, K_2, and K_3. We solve for these residues to obtain $K_1 = 1$, $K_2 = 0.62\underline{/143.8°}$, and

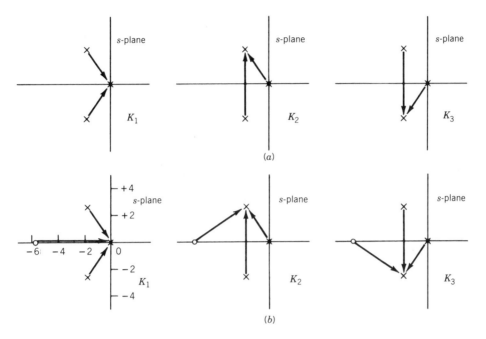

Figure 4.17 Application of s-plane vectors for Example 4.6 involving (a) tachometer feedback and (b) derivative control.

$K_3 = 0.62 \underline{/-143.8°}$. We then invert $Y(s)$ in (4.79) to form $y(t)$ as

$$y(t) = K_1 \exp[(0)t] + K_2 \exp[(-1.87 + j2.55)t] + K_3 \exp[(1.87 - j2.55)t]$$

$$(4.80)$$

Since K_3 is the complex conjugate of K_2, we may combine the second and third terms in (4.80) to obtain

$$K_2 \exp[(-1.87 + j2.55)t] + K_3 \exp[(-1.87 - j2.55)t]$$

$$= |K_2| \exp\left[j\underline{/K_2}\right] \exp[(-1.87 + j2.55)t]$$

$$\quad + |K_2| \exp\left[-j\underline{/K_2}\right] \exp[(-1.87 - j2.55)t]$$

$$= |K_2| \exp[-1.87t]\left\{\exp\left[j\left(2.55t + \underline{/K_2}\right)\right]\right.$$

$$\quad \left. + \exp\left[-j\left(2.55t + \underline{/K_2}\right)\right]\right\}$$

$$= 2|K_2| e^{-1.87t} \cos\left(2.55t + \underline{/K_2}\right)$$

$$= 2(0.62) e^{-1.87t} \cos(2.55t + 143.8°)$$

$$= -1.24 e^{-1.87t} \sin(2.55t + 53.8°) \qquad (4.81)$$

Therefore, $y(t)$ in (4.80) is given by

$$y(t) = 1 - 1.24e^{-1.87t} \sin(2.55t + 53.8°) \qquad (4.82)$$

The corresponding closed-loop transfer function for derivative control is given by

$$\frac{Y(s)}{R(s)} = \frac{10(1 + k_D s)}{s^2 + (2 + 10k_D)s + 10} = \frac{10k_D(s + 1/k_D)}{s^2 + (2 + 10k_D)s + 10} \qquad (4.83)$$

Letting $k_D = 0.174$ and $R(s) = 1/s$, we have

$$Y(s) = \frac{1.74(s + 5.75)}{s(s^2 + 3.74s + 10)}$$

$$= \frac{1.74(s + 5.75)}{s(s + 1.87 - j2.55)(s + 1.87 + j2.55)}$$

$$= \frac{K_1}{s} + \frac{K_2}{s + 1.87 - j2.55} + \frac{K_3}{s + 1.87 + j2.55} \qquad (4.84)$$

Calculating the residues from s-plane vectors in Figure 4.17b yields $K_1 = 1$, $K_2 = 0.50\underline{/177.1°}$, and $K_3 = 0.50\underline{/-177.1°}$. Inverting (4.84) and combining terms as before, we have

$$y(t) = 1 - e^{-1.87t} \sin(2.55t + 87.1°) \qquad (4.85)$$

Time response curves for (4.82) and (4.85) were shown earlier in Figure 4.13. A comparison of these curves reveals that a faster rise time and larger %OS is obtained for derivative control than for tachometer feedback control. Because the only difference between the two cases is the presence of a zero factor for derivative control, we have clearly demonstrated the effect of $(s + 1/k_D)$ on the time response for this second-order system.

4.3.4 Modal Coordinate Representation

We have shown how to determine the output time response $y(t)$ from the s-plane. Characteristics of the time response components are defined by pole locations in the s-plane, and their magnitudes are determined by the residues at those poles. We may relate this transform analysis of the time response directly to the uncoupled form of state variable representation in Section 2.4. If the transfer function poles, or eigenvalues, are real and distinct, then we may express the uncoupled form in vector-matrix notation to yield a diagonal matrix A in

$$\dot{\mathbf{x}} = A\mathbf{x} + \mathbf{b}u$$

$$y = \mathbf{c}^T\mathbf{x} \qquad (4.86)$$

We note that the component state variable equations (4.86) are uncoupled scalar differential equations and, hence, are quite simple to solve for $x_i(t)$, where $i = 1, 2, \ldots, n$. Since each equation represents a single mode of dynamic behavior, we often refer to the uncoupled form of state variables as the modal coordinate representation.

If the transfer function is given to us, we must calculate the residues, or coefficients in the partial fraction expansion, from that function. Each mode—that is, time response due to a particular pole—is present in the output $y(t)$ in an amount specified by the residues. On the other hand, if only the transfer function denominator is specified—that is, only the A matrix in (4.86) is fixed—then we can select those modes of behavior we desire to form the output $y(t)$. We may alternatively suppress those modes of behavior we wish to neglect in the output. This concept of mode selection or suppression is directly related to the observability of the system, as described in Section 2.5. In fact, to suppress any particular mode entirely means that a numerator factor (zero) is used to cancel a corresponding denominator factor (pole). Because of stability considerations in cases of inexact cancellation, this operation is valid only for left-half s-plane poles.

4.4 THE STATE TRANSITION MATRIX

Our goal in this section and the following one is to present methods for determining the time response for linear systems described by state variable equations. We discuss procedures here that relate and interpret state-space and transform approaches to find the state transition matrix. This matrix determines the effect of initial conditions on the time response. In Section 4.5 we use the state transition matrix to form a superposition integral for describing the effect of external inputs on the time response.

The state transition matrix is defined initially and some of its properties are identified. Three methods, representative of the several procedures available in the vast area of linear systems theory [8–11], are then described for finding the matrix for linear time-invariant systems.

4.4.1 Definition and Properties

Consider the unforced ($u = 0$) linear system described by

$$\dot{\mathbf{x}} = A(t)\mathbf{x} \qquad (4.87)$$

The time solution of (4.87) has the form

$$\mathbf{x}(t) = \Phi(t, t_0)\mathbf{x}(t_0) \tag{4.88}$$

where $\Phi(t, t_0)$ is referred to as the state transition matrix. The initial condition $\mathbf{x}(t_0)$ is transferred to the state $\mathbf{x}$ at time t by the matrix $\Phi(t, t_0)$. Obviously, $\Phi(t_0, t_0) = I$, the identity matrix, since the state $\mathbf{x}(t)$ is equal to $\mathbf{x}(t_0)$ at $t = t_0$.

The transition property of the state transition matrix may be expressed as

$$\Phi(t_2, t_0) = \Phi(t_2, t_1)\Phi(t_1, t_0) \tag{4.89}$$

which indicates that if an initial state vector $\mathbf{x}(t_0)$ is transferred to $\mathbf{x}(t_1)$ by $\Phi(t_1, t_0)$ and if $\mathbf{x}(t_1)$ is then transferred to $\mathbf{x}(t_2)$ by $\Phi(t_2, t_1)$, then $\mathbf{x}(t_0)$ may be transferred directly to $\mathbf{x}(t_2)$ by $\Phi(t_2, t_0)$, the product of the two state transition matrices.

An inversion property may be written mathematically as

$$\Phi^{-1}(t_1, t_0) = \Phi(t_0, t_1) \tag{4.90}$$

which may easily be derived from (4.88) by setting $t = t_1$ and solving for $\mathbf{x}(t_0)$ in terms of Φ and $\mathbf{x}(t_1)$.

If the matrix A in (4.87) is constant with time, then the state transition matrix $\Phi(t, t_0)$ is a function only of the difference in t and t_0, that is,

$$\Phi(t, t_0) = \Phi(t - t_0) \tag{4.91}$$

We work almost exclusively with this time-invariant case. Procedures for determining $\Phi(t - t_0)$ are indicated in the following subsection. For convenience, we set t_0 equal to zero and denote the state transition matrix simply as $\Phi(t)$. If t_0 is not zero in some application of interest, we can then reinsert t_0 by replacing t everywhere in $\Phi(t)$ by $t - t_0$.

4.4.2 Methods for Finding $\Phi(t)$

The three methods presented here for solving for the state transition matrix are the

1. Resolvent matrix method.
2. Series method.
3. Cayley-Hamilton method.

The first of these techniques is based on finding $\Phi(s)$ first and then using inverse Laplace transformations to obtain $\Phi(t)$. The other two methods solve for $\Phi(t)$ directly by using either a series expansion of the matrix exponential or matrix function properties.

Resolvent Matrix Method

The essence of this method was described earlier in Section 2.6 as a part of the derivation of the plant transfer function $Y(s)/U(s)$ in terms of the state variable matrix A and the vectors **b** and **c**. To review that development here, we write

$$\dot{\mathbf{x}} = A\mathbf{x} \qquad (4.92)$$

where A is a constant. Taking Laplace transforms throughout in (4.92), we obtain

$$s\mathbf{X}(s) - \mathbf{x}(0) = A\mathbf{X}(s) \qquad (4.93)$$

Rearranging the terms in (4.93) and factoring the result yields

$$s\mathbf{X}(s) - A\mathbf{X}(s) = \mathbf{x}(0)$$

$$sI\mathbf{X}(s) - A\mathbf{x}(s) = \mathbf{x}(0)$$

$$(sI - A)\mathbf{X}(s) = \mathbf{x}(0) \qquad (4.94)$$

Premultiplying the third equation in (4.94) by the inverse of the coefficient matrix $(sI - A)$ yields

$$\mathbf{X}(s) = (sI - A)^{-1}\mathbf{x}(0) \qquad (4.95)$$

Therefore, since $\mathbf{x}(t) = \Phi(t)\mathbf{x}(0)$ and, consequently, $\mathbf{X}(s) = \Phi(s)\mathbf{x}(0)$, we may identify $\Phi(s)$ from (4.95) as

$$\Phi(s) = (sI - A)^{-1} \qquad (4.96)$$

where $\Phi(s)$ is referred to as the *resolvent matrix*. We may then obtain the state transition matrix $\Phi(t)$ from inverse Laplace transformation as

$$\Phi(t) = \mathscr{L}^{-1}\{\Phi(s)\} \qquad (4.97)$$

Alternatively, we can apply an impulsed-integrator procedure to obtain the resolvent matrix without requiring a matrix inverse. We first construct the computer simulation diagram for the system in (4.92) with state variables defined as the outputs of integrators. We may write the unforced state solution $\mathbf{x}(t) =$

$\Phi(t)\mathbf{x}(0)$ in component form as

$$
\begin{pmatrix} x_1(t) \\ x_2(t) \\ \vdots \\ x_i(t) \\ \vdots \\ x_n(t) \end{pmatrix} = \begin{pmatrix} \phi_{11}(t) & \phi_{12}(t) & \cdots & \phi_{1n}(t) \\ \phi_{21}(t) & \phi_{22}(t) & \cdots & \phi_{2n}(t) \\ \vdots & \vdots & & \vdots \\ \phi_{i1}(t) & & & \phi_{in}(t) \\ \vdots & & & \vdots \\ \phi_{n1}(t) & \cdots & & \phi_{nn}(t) \end{pmatrix} \begin{pmatrix} x_1(0) \\ x_2(0) \\ \vdots \\ x_i(0) \\ \vdots \\ x_n(0) \end{pmatrix} \quad (4.98)
$$

Consider the ith component equation

$$
x_i(t) = \phi_{i1}(t)x_1(0) + \cdots + \phi_{ij}(t)x_j(0) + \cdots + \phi_{in}(t)x_n(0) \quad (4.99)
$$

Setting zero initial conditions for every state variable except the jth one, we have

$$
x_i(t) = \phi_{ij}(t)x_j(0) \quad (4.100)
$$

If we insert an initial condition of unity for $x_j(0)$, then the response $x_i(t)$ at the output of the ith integrator is equivalent to $\phi_{ij}(t)$, that is,

$$
x_i(t) = \phi_{ij}(t)x_j(0)
$$

$$
= \phi_{ij}(t)[1]
$$

$$
= \phi_{ij}(t) \quad (4.101)
$$

Initial conditions on integrators in computer simulation diagrams are ordinarily placed on integrator outputs. In particular, if we impose $x_j(0) = 1$, then in Laplace transform notation we are specifying a value of $1/s$ on the integrator output. Using the block diagram transformation rules of Table 2.2, we may instead insert at the integrator input a signal of unit strength that has a Laplace transform of 1. The corresponding time input at the integrator input is a unit impulse $\delta(t)$, as shown in Figure 4.18a. Consequently, from the preceding discussion and (4.101), we may determine $\phi_{ij}(s)$ from the computer simulation diagram by

1. Inserting a unit impulse at the jth integrator input.
2. Setting all other initial conditions to zero.
3. Determining the resulting transfer function measured from that input to the output of the ith integrator.

Figure 4.18b shows this general procedure. Any of the block diagram reduction techniques in Section 2.2 may be used in Step 3. After $\phi_{ij}(s)$ has been found, its Laplace transform inversion yields $\phi_{ij}(t)$, as for the entire matrix in (4.97).

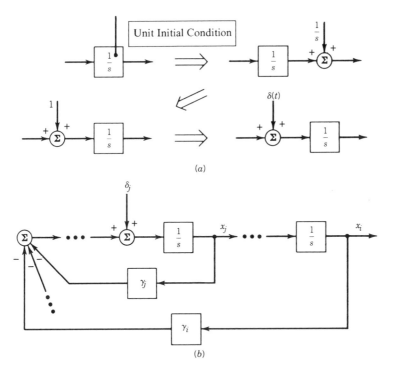

Figure 4.18   Obtaining the resolvent matrix by an impulsed-integrator procedure.

Series Method

Consider the matrix exponential solution to (4.92) given by

$$\mathbf{x}(t) = e^{At}\mathbf{x}(0) \tag{4.102}$$

To show that (4.102) is indeed a solution to (4.92), we use the matrix differentiation rule to form

$$\dot{\mathbf{x}} = \frac{d}{dt}[\mathbf{x}(t)] = \frac{d}{dt}[e^{At}\mathbf{x}(0)]$$

$$= Ae^{At}\mathbf{x}(0) \tag{4.103}$$

Substituting (4.102) into (4.103) yields $\dot{\mathbf{x}} = A\mathbf{x}$. Comparing (4.102) and (4.92) gives

$$\Phi(t) = e^{At} \tag{4.104}$$

We may express e^{At} in a Taylor series about $t = 0$ to give

$$\Phi(t) = e^{At} = \sum_{i=0}^{\infty} \frac{A^i t^i}{i!}$$

$$= I + At + \frac{A^2 t^2}{2!} + \cdots + \frac{A^n t^n}{n!} + \cdots \qquad (4.105)$$

which is the operational equation for the series method of solving for the state transition matrix.

Cayley-Hamilton Method

We state without proof the Cayley-Hamilton theorem as

$$\boxed{\begin{array}{l} \text{Every square matrix } A \text{ satisfies} \\ \text{its own characteristic equation} \\ \qquad \det(sI - A) = 0 \end{array}} \qquad (4.106)$$

The characteristic equation for an nth-order system is a polynomial in s of order n having the general form

$$s^n + \alpha_{n-1}s^{n-1} + \cdots + \alpha_2 s^2 + \alpha_1 s + \alpha_0 = 0 \qquad (4.107)$$

Therefore, the Cayley-Hamilton theorem states that A satisfies

$$A^n + \alpha_{n-1}A^{n-1} + \cdots + \alpha_2 A^2 + \alpha_1 A + \alpha_0 I = 0 \qquad (4.108)$$

To utilize this result in solving for $\Phi(t)$, we may first solve (4.108) for A^n to yield

$$A^n = -\alpha_{n-1}A^{n-1} - \cdots - \alpha_2 A^2 - \alpha_1 A - \alpha_0 I \qquad (4.109)$$

Consequently, $A^{n+1}, A^{n+2}, \ldots$, and all higher powers of A may be expressed in terms of $A^{n-1}, A^{n-2}, \ldots, A^2, A,$ and I. Using this result in (4.105) gives

$$\Phi(t) = e^{At} = I + At + \frac{A^2 t^2}{2!} + \cdots + \frac{A^n t^n}{n!} + \cdots$$

$$= \rho_0(t)I + \rho_1(t)A + \cdots + \rho_{n-1}(t)A^{n-1}$$

$$= \sum_{i=0}^{n-1} \rho_i(t)A^i \qquad (4.110)$$

Thus, we see that $\Phi(t)$ can be expressed in a finite series involving only the first $n - 1$ powers of A. Moreover, the coefficients ρ_i in this series are functions of t.

We may solve for the coefficients $\rho_i(t)$ in (4.110) by examining the corresponding uncoupled form of state variables for the given system. Let this set of state variables be expressed as

$$\dot{\mathbf{z}} = \Lambda \mathbf{z} \tag{4.111}$$

where $\mathbf{z}$ is the state and Λ is the coefficient matrix. If Λ is a diagonal matrix of the form

$$\Lambda = \begin{pmatrix} \lambda_1 & 0 & \cdots & 0 \\ 0 & \lambda_2 & & 0 \\ \vdots & \vdots & & \vdots \\ 0 & & & \lambda_n \end{pmatrix} \tag{4.112}$$

we may write the state transition matrix easily as

$$\Phi_z(t) = e^{\Lambda t} = \begin{pmatrix} e^{\lambda_1 t} & 0 & \cdots & 0 \\ 0 & e^{\lambda_2 t} & \cdots & 0 \\ \vdots & \vdots & & \vdots \\ 0 & \cdots & & e^{\lambda_n t} \end{pmatrix} \tag{4.113}$$

The Cayley-Hamilton theorem yields from (4.110)

$$\Phi_z(t) = e^{\Lambda t} = \sum_{i=0}^{n-1} \rho_i(t) \Lambda^i \tag{4.114}$$

Equating (4.113) and (4.114) gives

$$\begin{pmatrix} e^{\lambda_1 t} & 0 & \cdots & 0 \\ 0 & e^{\lambda_2 t} & & 0 \\ \vdots & & & \vdots \\ 0 & \cdots & & e^{\lambda_n t} \end{pmatrix} = \rho_0 I + \rho_1 \Lambda + \cdots + \rho_{n-1} \Lambda^{n-1} \tag{4.115}$$

Properties of matrix functions and transformations can be used to show that the $\rho_i(t)$ in (4.115) for λ are the same coefficients $\rho_i(t)$ in (4.110) for A. Therefore, the steps in using the Cayley-Hamilton method to find $\Phi(t)$ are

1. Find Λ.
2. Use (4.115) to solve for the $\rho_i(t)$.
3. Form $\Phi(t) = e^{At}$ from (4.110) using the $\rho_i(t)$ just found.

EXAMPLE 4.7

Use the three methods of this section to find the state transition matrix for the transfer function given below when the state variables are expressed in the direct phase variable form. The transfer function is

$$G(s) = \frac{1}{(s+1)(s+2)} = \frac{1}{s^2 + 3s + 2} \qquad (4.116)$$

We may write the state equations as

$$\dot{x}_1 = x_2$$

$$\dot{x}_2 = -2x_1 - 3x_2 + u(t) \qquad (4.117)$$

which yields an A matrix of

$$A = \begin{pmatrix} 0 & 1 \\ -2 & -3 \end{pmatrix} \qquad (4.118)$$

Using (4.96) for the resolvent matrix method gives

$$\Phi(s) = (sI - A)^{-1} = \left[s\begin{pmatrix} 1 & 0 \\ 0 & 1 \end{pmatrix} - \begin{pmatrix} 0 & 1 \\ -2 & -3 \end{pmatrix} \right]^{-1}$$

$$= \begin{pmatrix} s & -1 \\ +2 & s+3 \end{pmatrix}^{-1} = \frac{\begin{pmatrix} s+3 & -2 \\ +1 & s \end{pmatrix}^T}{s(s+3) - (-2)(1)}$$

$$= \begin{vmatrix} \dfrac{s+3}{s^2 + 3s + 2} & \dfrac{1}{s^2 + 3s + 2} \\ \dfrac{-2}{s^2 + 3s + 2} & \dfrac{s}{s^2 + 3s + 2} \end{vmatrix}$$

$$= \begin{vmatrix} \dfrac{2}{s+1} - \dfrac{1}{s+2} & \dfrac{1}{s+1} - \dfrac{1}{s+2} \\ \dfrac{-2}{s+1} + \dfrac{2}{s+2} & \dfrac{-1}{s+1} + \dfrac{2}{s+2} \end{vmatrix} \qquad (4.119)$$

Taking inverse Laplace transforms as indicated in (4.97) yields

$$\Phi(t) = \begin{pmatrix} 2e^{-t} - e^{-2t} & e^{-t} - e^{-2t} \\ -2e^{-t} + 2e^{-2t} & -e^{-t} + 2e^{-2t} \end{pmatrix} \qquad (4.120)$$

Note that $\Phi(0)$ is the identity matrix I.

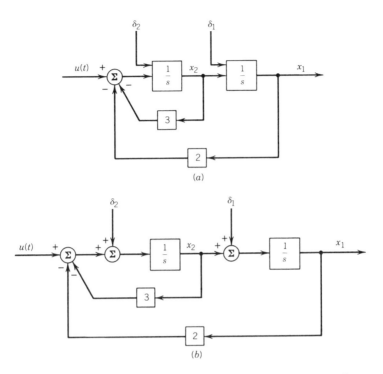

Figure 4.19 Application of the impulsed-integrator procedure for Example 4.7.

Alternatively, the impulsed-integrator procedure can be applied to the computer simulation diagram of Figure 4.19. We use Mason's signal flow technique to determine the appropriate transfer functions. When the unit impulse is applied to the input of the integrator with output x_1, we measure ϕ_{11} at the output x_1 and ϕ_{21} at the output x_2 to obtain

$$\phi_{11}(s) = \frac{\dfrac{1}{s}\left(1 + \dfrac{3}{s}\right)}{1 + \dfrac{3}{s} + \dfrac{2}{s^2}} = \frac{s + 3}{s^2 + 3s + 2} = \frac{2}{s + 1} - \frac{1}{s + 2}$$

$$\phi_{21}(s) = \frac{-2/s^2}{1 + \dfrac{3}{s} + \dfrac{2}{s^2}} = \frac{-2}{s^2 + 3s + 2} = \frac{-2}{s + 1} + \frac{2}{s + 2} \qquad (4.121)$$

When the unit impulse is applied to the input of the integrator with output x_2, we measure ϕ_{12} at the output x_1 and ϕ_{22} at the output x_2. These transfer functions are

$$\phi_{12}(s) = \frac{1/s^2}{1 + \dfrac{3}{s} + \dfrac{2}{s^2}} = \frac{1}{s^2 + 3s + 2} = \frac{1}{s+1} - \frac{1}{s+2}$$

$$\phi_{22}(s) = \frac{1/s}{1 + \dfrac{3}{s} + \dfrac{2}{s^2}} = \frac{s}{s^2 + 3s + 2} = \frac{-1}{s+1} + \frac{2}{s+2} \qquad (4.122)$$

Equation (4.97) is then applied to yield the same $\Phi(t)$ we obtained in (4.120). The series method is a direct application of (4.105) to yield

$$\Phi(t) = e^{At} = \sum_{i=0}^{\infty} \frac{A^i t^i}{i!}$$

$$= I + At + A^2 \frac{t^2}{2} + A^3 \frac{t^3}{6} + \cdots$$

$$= \begin{pmatrix} 1 & 0 \\ 0 & 1 \end{pmatrix} + \begin{pmatrix} 0 & 1 \\ -2 & -3 \end{pmatrix} t$$

$$+ \begin{pmatrix} 0 & 1 \\ -2 & -3 \end{pmatrix}^2 \frac{t^2}{2} + \begin{pmatrix} 0 & 1 \\ -2 & -3 \end{pmatrix}^3 \frac{t^3}{6} + \cdots$$

$$= \begin{pmatrix} 1 - t^2 + t^3 + \cdots & t - \dfrac{3}{2}t^2 + \dfrac{7}{6}t^3 + \cdots \\[2mm] -2t + 3t^2 - \dfrac{14}{6}t^3 + \cdots & 1 - 3t + \dfrac{7}{2}t^2 - \dfrac{15}{6}t^3 + \cdots \end{pmatrix} \qquad (4.123)$$

It is not always easy to sum these series for the components of $\Phi(t)$ to obtain a closed-form expression. However, the inverse problem of expanding each component in (4.120) in a Taylor series is not difficult, and this exercise would verify that (4.120) and (4.123) are in agreement for the finite number of terms being considered in (4.123).

For the Cayley-Hamilton method, we may refer to (4.116) to write the diagonal matrix Λ as

$$\Lambda = \begin{pmatrix} -1 & 0 \\ 0 & -2 \end{pmatrix} \tag{4.124}$$

Equation (4.115) becomes for this example

$$\begin{pmatrix} e^{-t} & 0 \\ 0 & e^{-2t} \end{pmatrix} = \rho_0(t) \begin{pmatrix} 1 & 0 \\ 0 & 1 \end{pmatrix} + \rho_1(t) \begin{pmatrix} -1 & 0 \\ 0 & -2 \end{pmatrix} \tag{4.125}$$

which yields the two equations for $\rho_0(t)$ and $\rho_1(t)$ as

$$e^{-t} = \rho_0(t) - \rho_1(t)$$

$$e^{-2t} = \rho_0(t) - 2\rho_1(t) \tag{4.126}$$

The solution of (4.126) is

$$\rho_0(t) = 2e^{-t} - e^{-2t}$$

$$\rho_1(t) = e^{-t} - e^{-2t} \tag{4.127}$$

Therefore, (4.110) yields the state transition matrix as

$$\Phi(t) = e^{At} = \rho_0(t)I + \rho_1(t)A$$

$$= \rho_0(t) \begin{pmatrix} 1 & 0 \\ 0 & 1 \end{pmatrix} + \rho_1(t) \begin{pmatrix} 0 & 1 \\ -2 & -3 \end{pmatrix}$$

$$\Phi(t) = \begin{pmatrix} \rho_0(t) & \rho_1(t) \\ -2\rho_1(t) & \rho_0(t) - 3\rho_1(t) \end{pmatrix} \tag{4.128}$$

Substituting (4.127) into (4.128) yields the same matrix in (4.120).

We recommend the use of digital computer routines to help avoid mistakes in the calculation of the state transition matrix, especially for large-scale systems. Melsa and Jones [13], for example, have developed a general set of computer programs for computational assistance in understanding linear control theory. Specialized programs are also useful for selected applications, such as those systems described by sparse matrices. Matrix functional properties, which form the basis of other methods not described here for obtaining $\Phi(t)$, can be particularly important in the development of these programs.

4.5 THE COMPLETE SOLUTION

The state transition matrix, determined by several methods in the last section, is used here to yield the complete solution of the state variable equations. We show that a superposition integral is obtained for that part of the solution due to an external input. Adding this integral to that part already found in Section 4.4, which involves only the state transition matrix acting on the state vector of initial conditions, gives the complete solution. Relationships are identified between this approach and results on Laplace transformations and convolution of the impulse response. The superposition integral is applied in the following sections to determine the solution of the state sensitivity equations and to develop controllability and observability matrices.

4.5.1 Laplace Transformations

Consider the forced linear time-invariant system given by

$$\dot{\mathbf{x}} = A\mathbf{x} + \mathbf{b}u(t)$$

$$y = \mathbf{c}^T\mathbf{x} \tag{4.129}$$

Taking Laplace transforms and solving for $\mathbf{X}(s)$, we have

$$\mathbf{X}(s) = (sI - A)^{-1}[\mathbf{x}(0) + \mathbf{b}u(s)]$$

$$= \Phi(s)\mathbf{x}(0) + \Phi(s)\mathbf{b}u(s) \tag{4.130}$$

Therefore, we may write the complete solution $\mathbf{x}(t)$ as

$$\mathbf{x}(t) = \mathcal{L}^{-1}\{\Phi(s)\}\mathbf{x}(0) + \mathcal{L}^{-1}\{\Phi(s)\mathbf{b}u(s)\}$$

$$\mathbf{x}(t) = \Phi(t)\mathbf{x}(0) + \mathcal{L}^{-1}\{\Phi(s)\mathbf{b}u(s)\} \tag{4.131}$$

Invoking the property that the product of two Laplace-transformed functions has an inverse Laplace transform equal to the convolution of the two corresponding time functions gives

$$\mathbf{x}(t) = \Phi(t)\mathbf{x}(0) + \int_0^t \Phi(t - \tau)\mathbf{b}u(\tau)\, d\tau \tag{4.132}$$

which is a specialized form of the complete solution for the case $t_0 = 0$. The corresponding result for arbitrary t_0 is given by

$$\mathbf{x}(t) = \Phi(t - t_0)\mathbf{x}(t_0) + \int_{t_0}^t \Phi(t - \tau)\mathbf{b}u(\tau)\, d\tau \tag{4.133}$$

Moreover, if A or $\mathbf{b}$ are functions of t, the general, time-varying solution is

$$\mathbf{x}(t) = \Phi(t, t_0)\mathbf{x}(t_0) + \int_{t_0}^{t} \Phi(t, \tau)\mathbf{b}u(\tau) \, d\tau \qquad (4.134)$$

The integrals in (4.132), (4.133), and (4.134) are referred to as *superposition integrals*, because they represent the contribution of the input $u(t)$, which is superimposed onto the homogeneous or fundamental solution resulting only from initial conditions.

4.5.2 Relationships with Impulse Response Convolution

Once $\mathbf{x}(t)$ has been found, we may form the output $y(t)$ as

$$y(t) = \mathbf{c}^T\mathbf{x}(t) = \mathbf{c}^T\Phi(t - t_0)\mathbf{x}(t_0)$$

$$+ \int_{t_0}^{t} \mathbf{c}^T\Phi(t - \tau)\mathbf{b}u(\tau) \, d\tau \qquad (4.135)$$

This value of $y(t)$ must be identical to that obtained by convolving the system impulse response function $g(t)$ with the input $u(t)$ over all time, that is,

$$y(t) = \int_{-\infty}^{t} g(t - \tau)u(\tau) \, d\tau \qquad (4.136)$$

Moreover, we may express $y(t)$ in (4.136) as the sum of two integrals, that is,

$$y(t) = \int_{-\infty}^{t_0} g(t - \tau)u(\tau) \, d\tau + \int_{t_0}^{t} g(t - \tau)u(\tau) \, d\tau \qquad (4.137)$$

Comparing the second integral in (4.137) with the integral term in (4.135) gives

$$g(t - \tau) = \mathbf{c}^T\Phi(t - \tau)\mathbf{b} \qquad (4.138)$$

Using (4.138) in the first integral in (4.137), we may write

$$\int_{-\infty}^{t_0} g(t - \tau)u(\tau) \, d\tau = \mathbf{c}^T\int_{-\infty}^{t_0} \Phi(t - \tau)\mathbf{b}u(\tau) \, d\tau$$

$$= \mathbf{c}^T\int_{-\infty}^{t_0} \Phi(t - t_0)\Phi(t_0 - \tau)\mathbf{b}u(\tau) \, d\tau$$

$$= \mathbf{c}^T\Phi(t - t_0)\int_{-\infty}^{t_0} \Phi(t_0 - \tau)\mathbf{b}u(\tau) \, d\tau \qquad (4.139)$$

where the transition property of the state transition matrix has been used.

Therefore, we set

$$\mathbf{x}(t_0) = \int_{-\infty}^{t_0} \Phi(t_0 - \tau)\mathbf{b}u(\tau) \, d\tau \qquad (4.140)$$

We see that the first integral in (4.137) not only provides the contribution of the input $u(t)$ over the range $(-\infty, t_0)$, but also transfers the initial state so determined in (4.140) to the present state $\mathbf{x}$ at time t. The second integral in (4.137) simply gives the contribution of the input over the range (t_0, t).

This development has related and interpreted the complete solution of the state equations in terms of Laplace transformations, the convolution integral involving the impulse response function, and the superposition integral (itself a convolution) in terms of the state transition matrix.

EXAMPLE 4.8

Consider again the linear plant of Example 4.7, that is,

$$\dot{x}_1 = x_2$$

$$\dot{x}_2 = -2x_1 - 3x_2 + u(t)$$

$$y = x_1 \qquad (4.141)$$

where $u(t)$ is given as e^{3t} for all $t > -\infty$. The problem is to determine $y(t)$ by the three methods presented above.

(a) *Impulse Response Convolution Integral.* We can easily obtain the impulse response function $g(t - \tau)$ from the plant transfer function

$$G(s) = \frac{1}{s^2 + 3s + 2} = \frac{1}{s + 1} - \frac{1}{s + 2}$$

$$g(t - \tau) = e^{-(t-\tau)} - e^{-2(t-\tau)} \qquad (4.142)$$

Using (4.142) in (4.136) gives

$$y(t) = \int_{-\infty}^{t} \left[e^{-(t-\tau)} - e^{-2(t-\tau)} \right] e^{3\tau} \, d\tau$$

$$= \frac{1}{20} e^{3t} \qquad (4.143)$$

which is valid for all t.

(b) Superposition Integral. The complete solution of the state equation in (4.141) may be obtained by using (4.133) to yield

$$\mathbf{x}(t) = \begin{pmatrix} \phi_{11} & \phi_{12} \\ \phi_{21} & \phi_{22} \end{pmatrix} \mathbf{x}(t_0) + \int_{t_0}^{t} \begin{pmatrix} \phi_{11}(t-\tau) & \phi_{12}(t-\tau) \\ \phi_{21}(t-\tau) & \phi_{22}(t-\tau) \end{pmatrix} \begin{pmatrix} 0 \\ 1 \end{pmatrix} e^{3\tau} d\tau \quad (4.144)$$

where $\Phi(t - t_0)$ was found in Example 4.7 as

$$\Phi(t - t_0) = \begin{pmatrix} 2e^{-(t-t_0)} - e^{-2(t-t_0)} & e^{-(t-t_0)} - e^{-2(t-t_0)} \\ -2e^{-(t-t_0)} - 2e^{-2(t-t_0)} & -e^{-(t-t_0)} + 2e^{-2(t-t_0)} \end{pmatrix} \quad (4.145)$$

Simplifying (4.144) gives

$$\mathbf{x}(t) = \begin{pmatrix} \phi_{11}x_1(t_0) + \phi_{12}x_2(t_0) + \int_{t_0}^{t} \phi_{12}(t-\tau)e^{3\tau} d\tau \\ \phi_{21}x_1(t_0) + \phi_{22}x_2(t_0) + \int_{t_0}^{t} \phi_{22}(t-\tau)e^{3\tau} d\tau \end{pmatrix} \quad (4.146)$$

Now we may determine $\mathbf{x}(t_0)$ from (4.140) as

$$\mathbf{x}(t_0) = \int_{-\infty}^{t_0} \begin{pmatrix} \phi_{11}(t_0-\tau) & \phi_{12}(t_0-\tau) \\ \phi_{21}(t_0-\tau) & \phi_{22}(t_0-\tau) \end{pmatrix} \begin{pmatrix} 0 \\ 1 \end{pmatrix} e^{3\tau} d\tau$$

$$= \begin{pmatrix} \int_{-\infty}^{t_0} \left[e^{-(t_0-\tau)} - e^{-2(t_0-\tau)}e^{3\tau} \right] d\tau \\ \int_{-\infty}^{t_0} \left[-e^{-(t_0-\tau)} + 2e^{-2(t_0-\tau)}e^{3\tau} \right] d\tau \end{pmatrix}$$

$$= \begin{pmatrix} \dfrac{1}{20}e^{3t_0} \\ \dfrac{3}{20}e^{3t_0} \end{pmatrix} \quad (4.147)$$

Equivalently, $\mathbf{x}(t_0)$ may be found simply by letting an original t_0 in (4.146) be

$-\infty$, for which $x_1(t_0) = x_2(t_0) = 0$, and letting the time t in (4.146) be some t_0, which serves as an initial time for a subsequent time period of interest (t_0, t). Both procedures yield the $\mathbf{x}(t_0)$ in (4.147). Observe that (4.143) and (4.147) agree, because $\mathbf{x}$ is composed of y and $\dot{y}$.

Using (4.147) in (4.146) to find $\mathbf{x}(t)$ for all $t \geq t_0$ gives

$$x_1(t) = [2e^{-(t-t_0)} - e^{-2(t-t_0)}]\left(\frac{1}{20}e^{3t_0}\right)$$

$$+ [e^{-(t-t_0)} - e^{-2(t-t_0)}]\left(\frac{3}{20}e^{3t_0}\right)$$

$$+ \int_{t_0}^{t} [e^{-(t-\tau)} - e^{-2(t-\tau)}]e^{3\tau}\,d\tau$$

$$= \frac{1}{20}e^{3t}$$

$$x_2(t) = [-2e^{-(t-t_0)} + 2e^{-2(t-t_0)}]\left(\frac{1}{20}e^{3t}\right)$$

$$+ [-e^{-(t-t_0)} + 2e^{-2(t-t_0)}]\left(\frac{3}{20}e^{3t_0}\right)$$

$$+ \int_{t_0}^{t} [-e^{-(t-\tau)} + 2e^{-2(t-\tau)}]e^{3\tau}\,d\tau$$

$$= \frac{3}{20}e^{3t} \tag{4.148}$$

Again, since $x_1 = y$ and $x_2 = \dot{y}$, we have complete agreement between (4.148) and (4.143).

(c) *Laplace Transformations.* We use either one of the two previous methods to determine $\mathbf{x}(t_0)$, where $t_0 = 0$, since the Laplace transform is defined for positive t only. The result in (4.147) for $t_0 = 0$ gives

$$\mathbf{x}(0) = \begin{pmatrix} 1/20 \\ 3/20 \end{pmatrix} \tag{4.149}$$

From (4.130) and the results of Example 4.7, we have

$$
\mathbf{X}(s) = \begin{pmatrix} \dfrac{s+3}{s^2+3s+2} & \dfrac{1}{s^2+3s+2} \\[4mm] \dfrac{-2}{s^2+3s+2} & \dfrac{s}{s^2+3s+2} \end{pmatrix} \begin{pmatrix} \dfrac{1}{20} \\[4mm] \dfrac{3}{20} \end{pmatrix}
$$

$$
+ \begin{pmatrix} \dfrac{s+3}{s^2+3s+2} & \dfrac{1}{s^2+3s+2} \\[4mm] \dfrac{-2}{s^2+3s+2} & \dfrac{s}{s^2+3s+2} \end{pmatrix} \begin{pmatrix} 0 \\ 1 \end{pmatrix} \begin{pmatrix} \dfrac{1}{s-3} \end{pmatrix}
$$

$$
= \begin{pmatrix} \dfrac{\dfrac{1}{20}(s+3)+\dfrac{3}{20}(1)+\dfrac{1}{s-3}}{s^2+3s+2} \\[6mm] \dfrac{\dfrac{1}{20}(-2)+\dfrac{3}{20}(s)+\dfrac{s}{s-3}}{s^2+3s+2} \end{pmatrix}
$$

$$
= \begin{pmatrix} \dfrac{1/20}{s-3} \\[4mm] \dfrac{3/20}{s-3} \end{pmatrix}
$$

$$
\mathbf{x}(t) = \begin{pmatrix} \dfrac{1}{20}\,e^{3t} \\[4mm] \dfrac{3}{20}\,e^{3t} \end{pmatrix} \qquad (t \geq 0) \tag{4.150}
$$

which agrees with the other two methods over the positive range of t for which (4.150) is applicable.

In summary, we have determined the complete solution to the state equations as the sum of a term involving the external input in a superposition integral and a term involving the state vector of initial conditions. Both of these terms depend on the state transition matrix. We compared the superposition integral approach with results on Laplace transformations and the convolution of the impulse response. A review of Section 2.3 on the concept of state variables is recommended at this point. The evolution of the time solution of $x(t)$ from $-\infty$ to some t_0 and then from t_0 to t can be understood more easily in view of the relationships developed here between the superposition integral and the convolution of the impulse response.

4.6 SENSITIVITY EQUATIONS AND ERROR PROPAGATION

We utilize the state transition matrix and associated superposition integral in this section to find the time response of the sensitivity equations. We then use these solutions to obtain linear approximations of the propagated errors due to inaccurate plant parameters or system inputs. An important consequence of this error propagation analysis is that we are able to specify the number of significant digits to retain in subsequent calculations.

4.6.1 Time Response of Sensitivity Equations

Consider the state equations and corresponding sensitivity equations describing a linear plant and variations in the matrix A.

$$\dot{x} = Ax + bu(t)$$

$$\dot{v}_\alpha = Av_\alpha + \left(\frac{\partial A}{\partial \alpha} \right) x \tag{4.151}$$

A convenient way to solve for $v_\alpha(t)$ is first to determine $x(t)$ as

$$x(t) = \Phi(t - t_0)x(t_0) + \int_{t_0}^{t} \Phi(t - \tau)bu(\tau)\, d\tau \tag{4.152}$$

and then form $v_\alpha(t)$ as

$$v_\alpha(t) = \Phi(t - t_0)v_\alpha(t_0) + \int_{t_0}^{t} \Phi(t - \tau)\left(\frac{\partial A}{\partial \alpha} \right) x(\tau)\, d\tau \tag{4.153}$$

Substituting from (4.152) for $\mathbf{x}(t)$ in (4.153) gives

$$\mathbf{v}_\alpha(t) = \Phi(t - t_0)\mathbf{v}_\alpha(t_0) + \int_{t_0}^{t} \Phi(t - \tau_1)\left(\frac{\partial A}{\partial \alpha}\right)$$

$$\times \left\{ \Phi(\tau_1 - t_0)\mathbf{x}(t_0) + \int_{t_0}^{\tau_1} \Phi(\tau_1 - \tau_2)\mathbf{b}u(\tau_2) \, d\tau_2 \right\} d\tau_1 \quad (4.154)$$

We observe that the eigenvalues of the matrix A occur in both state and sensitivity equations. Since $\mathbf{x}(t)$ serves as an input to the sensitivity equations, the eigenvalues of A occur with a multiplicity of two in $\mathbf{v}_\alpha(t)$. If we trace the forward path from the input of the state equations to the output of the sensitivity equations, we encounter a cascaded combination of two blocks, each having the eigenvalues of A.

A second procedure for solving (4.151) for $\mathbf{v}_\alpha(t)$ is to form the $2n$-vector

$$\mathbf{z}(t) = \begin{pmatrix} \mathbf{x}(t) \\ \mathbf{v}_\alpha(t) \end{pmatrix}$$

In terms of $\mathbf{z}(t)$, (4.151) becomes

$$\dot{\mathbf{z}}(t) = \begin{pmatrix} A & 0 \\ \dfrac{\partial A}{\partial \alpha} & A \end{pmatrix} \mathbf{z} + \begin{pmatrix} \mathbf{b} \\ 0 \end{pmatrix} u(t) \qquad (4.155)$$

The solution for $\mathbf{z}(t)$ is

$$\mathbf{z}(t) = \Phi_z(t - t_0)\mathbf{z}(t_0) + \int_{t_0}^{t} \Phi_z(t - \tau)\begin{pmatrix} \mathbf{b} \\ 0 \end{pmatrix} u(\tau) \, d\tau \qquad (4.156)$$

where $\Phi_z(t - t_0)$ is the state transition matrix associated with the $2n$ by $2n$ coefficient matrix in (4.155). Evidently, the previous method in (4.152) to (4.154) involving sequential operations is simpler to calculate than $\mathbf{z}(t)$ in (4.155) and (4.156).

EXAMPLE 4.9

Determine the time response of the sensitivity equations for variations in the parameter α for the first-order plant equation given by

$$\dot{x} = -\alpha x + u(t) \qquad (4.157)$$

where $x(0) = 0$ and $u(t)$ is a unit step input applied at $t = 0$. The corresponding sensitivity equation is

$$\dot{v}_\alpha = -\alpha v_\alpha - x(t) \qquad (4.158)$$

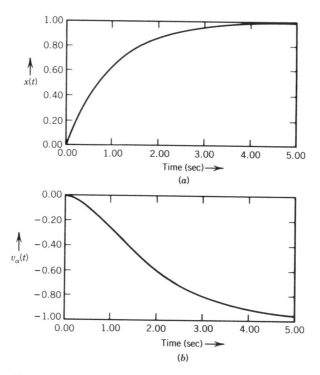

Figure 4.20 Time curves of $x(t)$ and $v_\alpha(t)$ for Example 4.9.

Solving (4.157) for $x(t)$ gives

$$x(t) = e^{-\alpha t}x(0) + \int_0^t e^{-\alpha(t-\tau)}(1) \, d\tau$$

$$= \frac{1}{\alpha}(1 - e^{-\alpha t}) \qquad (4.159)$$

The solution of (4.158) with $v_\alpha(0) = 0$ is

$$v_\alpha(t) = e^{-\alpha t}v_\alpha(0) + \int_0^t e^{-\alpha(t-\tau)}(-1)\left[\left(\frac{1}{\alpha}\right)(1 - e^{-\alpha\tau})\right] d\tau$$

$$= \frac{1}{\alpha}te^{-\alpha t} - \frac{1}{\alpha^2}(1 - e^{-\alpha t}) \qquad (4.160)$$

We see that a term involving $te^{-\alpha t}$, corresponding to a double eigenvalue at $-\alpha$, appears in $v_\alpha(t)$ as predicted. Time plots of both $x(t)$ and $v_\alpha(t)$ are shown in Figure 4.20 for $\alpha = 1$.

4.6.2 Error Propagation

Sensitivity analysis leads to a linear approximation of the errors in the system output due to incremental variations in system inputs and parameters. We may be interested in an entire output time function or only in a particular output value—for example, the percent overshoot. When such output values are related to the varying parameter α by a given function $g(\alpha)$, we may express the error e_g due to some variation $\Delta\alpha = \alpha - \alpha_N$ as

$$e_g = g(\alpha) - g(\alpha_N) \tag{4.161}$$

where $g(\alpha_N)$ is a nonlinear function of the nominal design parameter value α_N. Obviously, we can compute both $g(\alpha)$ and $g(\alpha_N)$ for some fixed $\Delta\alpha$, but we prefer to examine the error between these two functions for an unspecified $\Delta\alpha$. Expanding $g(\alpha)$ in a Taylor series about α_N and substituting into (4.161) gives

$$e_g = \left[g(\alpha_N) + \frac{\partial g(\alpha)}{\partial \alpha}\bigg|_{\alpha=\alpha_N} (\Delta\alpha) + \frac{1}{2}\frac{\partial^2 g(\alpha)}{\partial \alpha^2}\bigg|_{\alpha=\alpha_N} (\Delta\alpha)^2 + \cdots \right] - g(\alpha_N)$$

$$= \frac{\partial g(\alpha)}{\partial \alpha}\bigg|_{\alpha=\alpha_N} (\Delta\alpha) + \frac{1}{2}\frac{\partial^2 g(\alpha)}{\partial \alpha^2}\bigg|_{\alpha=\alpha_N} (\Delta\alpha)^2 + \cdots \tag{4.162}$$

Neglecting second- and higher-order terms for sufficiently small $\Delta\alpha$ in (4.161) yields a linear approximation for e_g as

$$e_g = \frac{\partial g(\alpha)}{\partial \alpha}\bigg|_{\alpha=\alpha_N} (\Delta\alpha) \tag{4.163}$$

Therefore, the error resulting from variations in the parameter α from a nominal design value α_N can be approximated by (4.163).

We may also use (4.163) to determine how many digits to retain in numerical calculations that involve the parameter α. Suppose α is correct only to within d_α significant digits, which gives an error within $\pm 0.5 \times 10^{-d_\alpha}$. The corresponding value of $g(\alpha)$ is correct to within $\pm 0.5 \times 10^{-d_e}$ (or d_e significant digits). Using (4.163), we have

$$\pm 0.5 \times 10^{-d_e} = \frac{\partial g(\alpha)}{\partial \alpha}\bigg|_{\alpha=\alpha_N} (\pm 0.5 \times 10^{-d_\alpha}) \tag{4.164}$$

Taking logarithms to the base 10 of both sides and requiring that d_e must be an integer gives

$$d_e = \left[d_\alpha - \log_{10} \frac{\partial g(\alpha)}{\partial \alpha} \Big|_{\alpha=\alpha_N} \right]_* \tag{4.165}$$

where $[\cdot]_*$ denotes the largest integer contained within its brackets. Depending on the magnitude of

$$\frac{\partial g(\alpha)}{\partial \alpha} \Big|_{\alpha=\alpha_N}$$

we see that the number of significant digits may be greater than, less than, or the same as the corresponding number specified in the parameter data.

If the entire output time function is of interest, then the effect of variations in a system parameter α may be identified by expressing $y(t, \alpha)$ in a Taylor series form as

$$y(t, \alpha) = y(t, \alpha_N) + \frac{\partial y(t, \alpha)}{\partial \alpha} \Big|_{\alpha=\alpha_N} (\Delta\alpha)$$

$$+ \frac{1}{2} \frac{\partial^2 y(t, \alpha)}{\partial \alpha^2} \Big|_{\alpha=\alpha_N} (\Delta\alpha)^2 + \cdots \tag{4.166}$$

and defining the error $e(t)$ due to changes in α from α_N as

$$e(t) = y(t, \alpha) - y(t, \alpha_N)$$

$$= \frac{\partial y(t, \alpha)}{\partial \alpha} \Big|_{\alpha=\alpha_N} (\Delta\alpha) + \frac{1}{2} \frac{\partial^2 y(t, \alpha)}{\partial \alpha^2} \Big|_{\alpha=\alpha_N} (\Delta\alpha)^2 + \cdots \tag{4.167}$$

Neglecting terms of second and higher order in (4.167) gives a linear approximation for $e(t)$ as

$$e(t) = \frac{\partial y(t, \alpha)}{\partial \alpha} \Big|_{\alpha=\alpha_N} (\Delta\alpha) \tag{4.168}$$

Observe that if $y = \mathbf{c}^T \mathbf{x}$, then (4.168) becomes

$$e(t) = \mathbf{c}^T \frac{\partial \mathbf{x}(t, \alpha)}{\partial \alpha} \Big|_{\alpha=\alpha_N} (\Delta\alpha)$$

$$= \mathbf{c}^T \mathbf{v}_\alpha(t)(\Delta\alpha) \tag{4.169}$$

which relates the solution of the sensitivity equations to error propagation. Analogous to (4.165), the number of significant digits to be retained in the

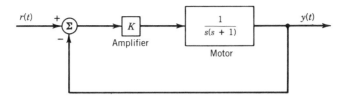

Figure 4.21 A transfer-function block diagram of the simplified positional servomechanism in Example 4.10.

calculation of $e(t)$ is given by

$$d_e(t) = \left[d_\alpha - \log_{10} \mathbf{c}^T \mathbf{v}_\alpha(t) \right]_* \qquad (4.170)$$

which may vary with time.

EXAMPLE 4.10

For the system in Figure 4.21, let us consider the sensitivity and resulting errors in the percent overshoot for a step input due to variations in the amplifier gain K.
From Figure 4.21, we identify $\omega_n = \sqrt{K}$ and $2\zeta\omega_n = 1$, from which

$$\zeta = \frac{1}{2\omega_n} = \frac{1}{2\sqrt{K}} \qquad (4.171)$$

The percent overshoot is given by

$$g(K) = \%OS = 100 \exp\left(-\pi\zeta/\sqrt{1 - \zeta^2} \right)$$

$$= 100 \exp\left[-(\pi/2\sqrt{K})/\sqrt{1 - (1/4K)} \right] \qquad (4.172)$$

Table 4.2 shows values of $g(K)$ and $\partial g(K)/\partial K$ for several values of K. The values of $\partial g(K)/\partial K$ were obtained by using the approximation

$$\frac{\partial g(K)}{\partial K} = \frac{g(K + \varepsilon) - g(K - \varepsilon)}{2\varepsilon} \qquad (4.173)$$

for $\varepsilon = 10^{-3}$. The value of ε to be used in (4.173), generally obtained by trial and

TABLE 4.2   Least Significant Digits for Percent Overshoot

K	$g(K)$	$\dfrac{\partial g(K)}{\partial K}$	d_g
0.300	0.09	6.2	$d_K - 1$
1.000	16.3	19.7	$d_K - 2$
10.00	60.5	1.6	$d_K - 1$
50.0	80.0	0.2	d_K
100.	85.4	0.065	$d_K + 1$
1000.	95.15	0.004	$d_K + 2$

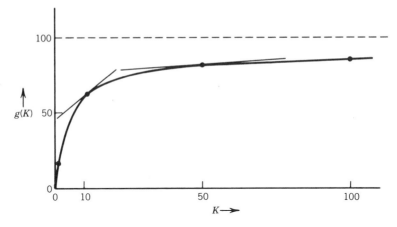

Figure 4.22 A plot of percent overshoot $g(K)$ versus K for Example 4.10.

error, should be suitably small to yield a good approximation. Also shown in Table 4.2 are the values of d_g, the number of significant digits to be retained in the calculation of $g(K)$, as a function of the accuracy specified (d_K as the position of the least significant digit) for the amplifier gain K. Figure 4.22 shows a sketch of $g(K)$ versus K, from which we obtain a visual indication of the errors involved in approximating the percent overshoot in the neighborhood of some nominal design value by the tangent to the curve at that nominal value.

EXAMPLE 4.11

Consider again the system of Example 4.9. Using the results in Figure 4.20, we compare in Figure 4.23 approximate time curves of $x_N(t) + v_\alpha(t)\,\Delta\alpha$ for two

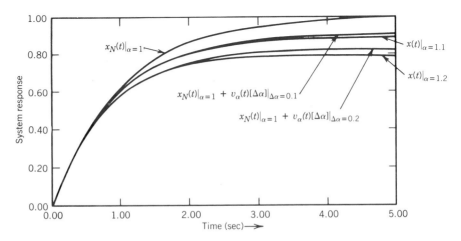

Figure 4.23 Comparisons between variations from a nominal time solution and the exact solution.

A0600422 **JOHN WILEY & SONS, INC.**

ONE WILEY DRIVE, SOMERSET, NEW JERSEY 08873 (201) 469-4400

05/09/86 SENT WITH COMPLIMENTS OF: M. FERNANDEZ 070

QUAN	AUTHOR AND SHORT TITLE	ISBN	LOCATION
1	ROWLAND CONTROL	0471-03276-X	8178A 2907
	POSTAGE	2.15	

PACKING LIST

FORM NO. 1068 (REV. 6-84)

values of $\Delta\alpha$ (0.1 and 0.2), obtained by varying from the nominal design curve $x_N(t)$, with exact curves of $x(t)$. As expected, we see that the accuracy of this approximation decreases as $\Delta\alpha$ increases.

In summary, we have reexamined sensitivity analysis in this section by using the state transition matrix to solve the sensitivity equations developed earlier. Error propagation studies have revealed how to determine the number of significant digits to be retained in numerical calculations using finite-precision parameter data. We have shown cases to illustrate linear approximations in these studies.

4.7 PID AND LINEAR SVFB CONTROL TIME RESPONSES

Time response for linear systems having proportional-integral-derivative (PID) control are compared in this section with those having state variable feedback (SVFB) control. Expanding on the development of Section 3.5 involving system modeling with PID control, we now determine time solutions using either the state transition matrix and superposition integral or computer simulations. With these concepts we can demonstrate more clearly that PID control and linear SVFB control may be used to meet design requirements such as percent over-shoot, rise time, settling time, and steady-state error.

4.7.1 Time Response for PID Control

We developed the dynamic model equations for PID control in Section 3.5 as

$$\dot{\mathbf{x}}_a = A_a \mathbf{x}_a + \mathbf{b}_a e$$

$$\begin{pmatrix} \dot{\mathbf{x}} \\ \dot{x}_{n+1} \end{pmatrix} = \begin{pmatrix} A & k_I \mathbf{b} \\ \mathbf{0}^T & 0 \end{pmatrix} \begin{pmatrix} \mathbf{x} \\ x_{n+1} \end{pmatrix} + \begin{pmatrix} \mathbf{b}' \\ 1 \end{pmatrix} e$$

$$y = \mathbf{c}_a^T \mathbf{x}_a = x_1 \qquad\qquad (3.111)$$

where A, $\mathbf{b}$, and $\mathbf{b}'$ in (3.111) are defined as

$$A = \begin{pmatrix} -\alpha_{n-1} & 1 & \cdots & 0 \\ -\alpha_{n-2} & 0 & \cdots & 0 \\ \vdots & \vdots & & \vdots \\ -\alpha_1 & 0 & \cdots & 1 \\ -\alpha_0 & 0 & \cdots & 0 \end{pmatrix} \qquad \mathbf{b} = \begin{pmatrix} k_P\beta_{n-1} \\ k_P\beta_{n-2} \\ \vdots \\ k_P\beta_1 \\ k_P\beta_0 \end{pmatrix}$$

$$\mathbf{b}' = \begin{vmatrix} k_D\beta_{n-2} + k_P\beta_{n-1} - k_D\beta_{n-1}\alpha_{n-1} \\ k_D\beta_{n-3} + k_P\beta_{n-2} - k_D\beta_{n-1}\alpha_{n-2} \\ \vdots \\ k_D\beta_0 + k_P\beta_1 - k_D\beta_{n-1}\alpha_1 \\ k_P\beta_0 - k_D\beta_{n-1}\alpha_0 \end{vmatrix} \tag{4.174}$$

The matrix A and vectors $\mathbf{b}$ and $\mathbf{b}'$ are based on using the direct feed-forward form of plant state variables. We may express the error e as

$$e = r(t) - y = r(t) - \mathbf{c}_a^T\mathbf{x}_a$$

$$= r(t) - x_1 \tag{4.175}$$

Therefore, we may substitute (4.175) into (3.111) to obtain

$$\dot{\mathbf{x}}_a = A_a\mathbf{x}_a + \mathbf{b}_a\left[r(t) - \mathbf{c}_a^T\mathbf{x}_a\right]$$

$$= \left(A_a - \mathbf{b}_a\mathbf{c}_a^T\right)\mathbf{x}_a + \mathbf{b}_a r(t)$$

$$= A_a'\mathbf{x}_a + \mathbf{b}_a r(t) \tag{4.176}$$

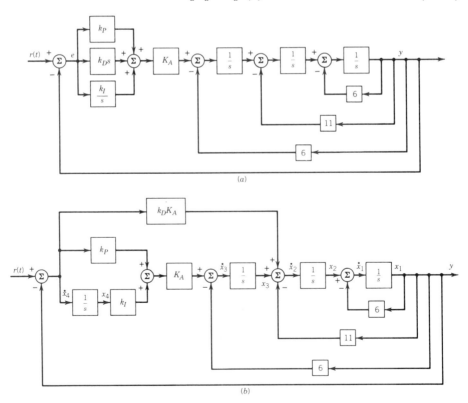

Figure 4.24 The closed-loop system having PID control in Example 4.12.

We may determine $\mathbf{x}_a(t)$ in terms of the state transition matrix and superposition integral as

$$\mathbf{x}_a(t) = \Phi_{\mathbf{x}_a}(t - t_0)\mathbf{x}_a(t_0)$$

$$+ \int_{t_0}^{t} \Phi_{\mathbf{x}_a}(t - \tau)\mathbf{b}_a r(\tau)\, d\tau \tag{4.177}$$

However, determining whether given time response performance specifications can be satisfied by adjusting PID controller parameters is often not an easy task. We encourage the control engineering student to perform a computer simulation of (4.176) for this purpose. Generally, we expect to be able to meet percent overshoot specifications by adjusting k_P and k_D and to satisfy steady-state error requirements by properly selecting k_I.

EXAMPLE 4.12

For the closed-loop system having PID control in Figure 4.24a, determine the effects on the time response for $y(t)$ when k_P, k_D, and k_I are varied. State variables identified in Figure 4.24b obey the state equations given by

$$\dot{x}_1 = -6x_1 + x_2$$

$$\dot{x}_2 = -(11 + K_A k_D)x_1 + x_3 + K_A k_D r(t)$$

$$\dot{x}_3 = -(6 + K_A k_P)x_1 + K_A k_I x_4 + K_A k_P r(t)$$

$$\dot{x}_4 = -x_1 + r(t) \tag{4.178}$$

Assume a unit step input, and use a value of 50 for K_A.

Using only proportional control ($k_D = k_P = 0$), we show in Figure 4.25a that we can obtain a suitable percent overshoot by decreasing k_P from 1.0 to 0.2, but the rise time and steady-state error are excessive. We next show computer simulation results in Figure 4.25b for $k_P = 0.4$ and derivative control gains of $k_D = 0$, 0.1, 0.2, 0.5, and 1.0. The rise time decreased as k_D increased, but the percent overshoot was smallest for $k_D = 0.2$. However, the steady-state error remained excessive. We show the result of adding integral control in Figure 4.25c. The steady-state error is reduced to zero for $k_I = 0.1$, 0.3, and 0.5. A settling time (for $\delta_s = 0.05$) of approximately 2.3 seconds is obtained for $k_I = 0.5$. Therefore, a PID controller with gains $k_P = 0.4$, $k_D = 0.2$, and $k_I = 0.5$ yields a moderately acceptable system time response. Further gain adjustments can be made as desired for fine tuning.

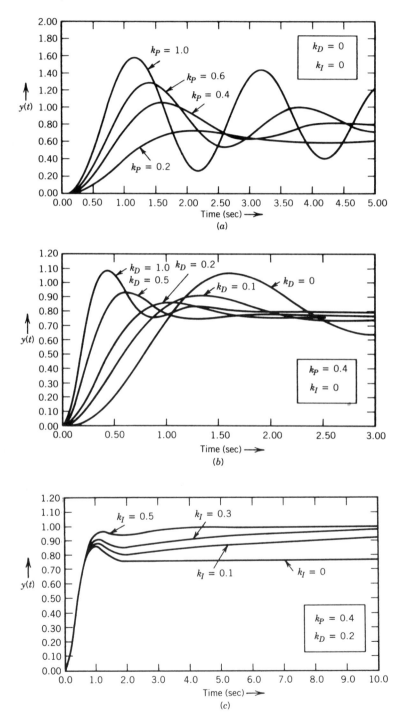

Figure 4.25 Computer simulation results for Example 4.12 with $K_A = 50$.

4.7.2 Time Response for Linear SVFB Control

We determined the state equations for linear SVFB control in Section 3.6 as

$$\dot{\mathbf{x}} = A\mathbf{x} + \mathbf{b}K\left[r(t) - \mathbf{k}^T\mathbf{x}\right]$$

$$= [A - K\mathbf{b}\mathbf{k}^T]\mathbf{x} + K\mathbf{b}r(t)$$

$$= A_k\mathbf{x} + K\mathbf{b}r(t) \tag{3.114}$$

where $\mathbf{k}$ is the n-vector of controller gains to be specified in the design procedure. We showed that these n values $k_1, k_2, \ldots, k_n$ can be selected to achieve any desired closed-loop pole placement.

The time solution for $\mathbf{x}(t)$ in (3.114) is given by

$$\mathbf{x}(t) = \Phi_\mathbf{x}(t - t_0)\mathbf{x}(t_0) + \int_{t_0}^{t}\Phi_\mathbf{x}(t - \tau)K\mathbf{b}r(\tau)\,d\tau \tag{4.179}$$

If the transfer function corresponding to a desired output time response for a given input can be found, we can equate this transfer function to one containing the linear SVFB controller gains, that is, either to

$$\frac{Y(s)}{R(s)} = K\mathbf{c}^T(sI - A_k)^{-1}\mathbf{b} \tag{3.116}$$

or to

$$\frac{Y(s)}{R(s)} = \frac{K\mathbf{c}^T(sI - A)^{-1}\mathbf{b}}{1 + K\mathbf{k}^T(sI - A)^{-1}\mathbf{b}} \tag{3.117}$$

which were determined earlier in Section 3.6.

When the desired output time response is provided as a curve of $y(t)$ versus t, we may decide to use a computer simulation to determine an approximation for the desired transfer function of order n. Transfer function poles can be identified by carefully adjusting parameters until a suitably close approximation is achieved. Such a procedure allows us to meet percent overshoot, rise time, and settling time performance specifications, but we are able to satisfy steady-state error specifications, at least in some cases, only by incorporating integral control into the feedback system design.

EXAMPLE 4.13

For the closed-loop system having SVFB control in Figure 4.26, determine the effects on the time response for $y(t)$ when k_1, k_2, and k_3 are varied. The state

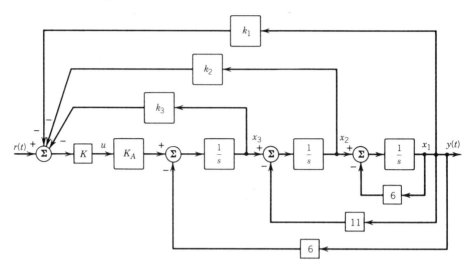

Figure 4.26 SVFB control for Example 4.13.

variable equations are given by

$$\dot{x}_1 = -6x_1 + x_2$$

$$\dot{x}_2 = -11x_1 + x_3$$

$$\dot{x}_3 = -6x_1 + K_A u$$

$$u = K\left[r(t) - k_1 x_1 - k_2 x_2 - k_3 x_3\right] \qquad (4.180)$$

As in Example 4.12, assume a unit step input and zero initial conditions. Let $K = 1$ and $K_A = 50$.

Figure 4.27 shows the required time response curves as k_1, k_2, and k_3 are varied. In Figure 4.27a, we show the results obtained by varying k_1 with $k_2 = k_3 = 0$. We see that the system time response becomes more oscillatory as k_1 is varied from 0 to 1. Arbitrarily selecting $k_1 = 0.2$, we next show the results of varying k_2 (with $k_3 = 0$) in Figure 4.27b. Finally, we show the effects of varying k_3 in the time curves of Figure 4.27c with $k_1 = 0.2$ and $k_2 = 0.05$. Forming the closed-loop transfer function $Y(s)/R(s)$ and using $R(s) = 1/s$, we determine that a zero steady-state error is obtained for $k_3 = 0.0345$. The corresponding time response curve in Figure 4.27c indicates a percent overshoot of approximately 18%, a rise time of less than 1 second, and a settling time of less than 2 seconds. The wide variation in time response curves obtained in Figure 4.27 is possible because we are able to place the closed-loop system poles at arbitrary locations in the s-plane by using linear SVFB control.

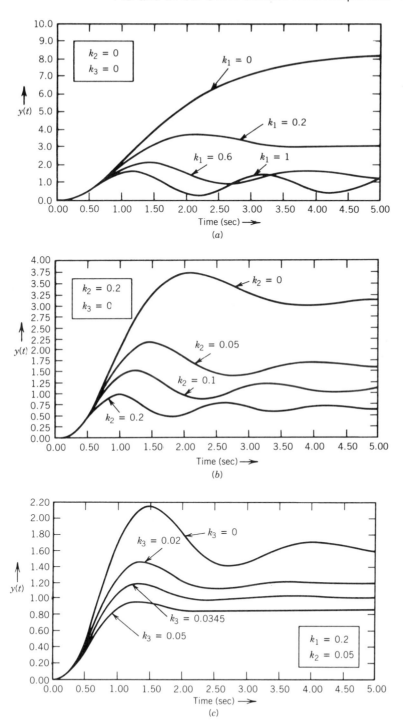

Figure 4.27 Time response curves for Example 4.13 with $K = 1$ and $K_A = 50$.

4.8 CONTROLLABILITY AND OBSERVABILITY REVISITED

In this section we apply the state transition matrix and the superposition integral to develop the controllability and observability matrices of Section 2.5. We first formalize the definition of controllability, derive the controllability matrix, and present procedures for constructing a finite control to achieve the transfer indicated in the definition. We next define observability formally, derive the observability matrix, and show that initial states cannot be determined uniquely from the output time curve when the plant is unobservable.

4.8.1 Controllability Matrix

Although we discussed the concept of state controllability earlier in Section 2.5, we now provide the following formal definition.

Definition 4.1. A plant is state controllable if the initial state $x(t_0)$ can be transferred to any arbitrary state $x(t_1)$ in a finite time $(t_1 - t_0)$ by using some control u.

To apply this definition to linear time-invariant systems, we again consider the single-input, single-output plant given by

$$\dot{x} = Ax + bu(t)$$

$$y = c^T x \tag{4.129}$$

The complete solution for $x(t)$ in (4.129) at time t_1 was determined in Section 4.5 as

$$x(t_1) = \Phi(t_1 - t_0)x(t_0) + \int_{t_0}^{t_1} \Phi(t_1 - \tau)bu(\tau)\, d\tau. \tag{4.133}$$

We define a vector $x^*(t_1)$ as the difference between $x(t_1)$ and $\Phi(t_1 - t_0)x(t_0)$, that is,

$$x^*(t_1) = x(t_1) - \Phi(t_1 - t_0)x(t_0) \tag{4.181}$$

Using (4.181) in (4.133) gives

$$x^*(t_1) = \int_{t_0}^{t_1} \Phi(t_1 - \tau)bu(\tau)\, d\tau \tag{4.182}$$

We see that the control $u(t)$ that transfers the plant from a zero initial state to the state $x^*(t_1)$ in time $t_1 - t_0$ is the same as the control that transfers the plant from $x(t_0)$ to $x(t_1)$ in the same time period. Therefore, we may use (4.182) instead of (4.133) to derive the controllability matrix.

From the Cayley-Hamilton method discussed in Section 4.4, the state transition matrix for (4.129) may be expressed as

$$\Phi(t_1 - \tau) = e^{A(t_1 - \tau)} = \sum_{i=0}^{n-1} \rho_i(t_1 - \tau) A^i \tag{4.183}$$

Using (4.183) in (4.182) gives

$$\mathbf{x}^*(t_1) = \int_{t_0}^{t_1} \sum_{i=0}^{n-1} \rho_i(t_1 - \tau) A^i \mathbf{b} u(\tau) \, d\tau \tag{4.184}$$

which may also be expressed as

$$\mathbf{x}^*(t_1) = (\mathbf{b}, A\mathbf{b}, \ldots, A^{(n-1)}\mathbf{b}) \begin{pmatrix} \int_{t_0}^{t_1} \rho_0(t_1 - \tau) u(\tau) \, d\tau \\ \int_{t_0}^{t_1} \rho_1(t_1 - \tau) u(\tau) \, d\tau \\ \vdots \\ \int_{t_0}^{t_1} \rho_{n-1}(t_1 - \tau) u(\tau) \, d\tau \end{pmatrix} \tag{4.185}$$

To transfer from the zero initial state to an arbitrary state $\mathbf{x}^*(t_1)$, the determinant of the coefficient matrix in (4.185) must be nonzero. Moreover, if this coefficient matrix has a nonzero determinant, it will always be possible to find a control $u(t)$, which results in the indicated transfer. We refer to the coefficient matrix in (4.185) as the controllability matrix; we have shown that the plant in (4.129) is state controllable if and only if the controllability matrix given by

$$M \triangleq (\mathbf{b}, A\mathbf{b}, \ldots, A^{(n-1)}\mathbf{b}) \tag{4.186}$$

has a nonzero determinant. This completes the proof of Theorem 2.1.

4.8.2 Construction of a Finite Control

For cases where the time increment $t_1 - t_0$ is not infinitesimal, we may form a finite control $u(t)$ as

$$u(t) = \mathbf{b}^T e^{-A^T t} R^{-1} e^{-A t_1} \mathbf{x}^*(t_1) \tag{4.187}$$

where R is a nonsingular matrix defined by

$$R = \int_{t_0}^{t_1} e^{-A\tau} \mathbf{b} \mathbf{b}^T e^{-A^T \tau} \, d\tau \tag{4.188}$$

To show that $u(t)$ transfers the plant state from $\mathbf{0}$ to $\mathbf{x}^*(t_1)$, we substitute (4.187) into (4.182) to yield

$$\mathbf{x}^*(t_1) = \int_{t_0}^{t_1} e^{A(t_1-\tau)} \mathbf{b} \left[\mathbf{b}^T e^{-A^T\tau} R^{-1} e^{-At_1} \mathbf{x}^*(t_1) \right] d\tau$$

$$= e^{At_1} \left[\int_{t_0}^{t_1} e^{-A\tau} \mathbf{b}\mathbf{b}^T e^{-A^T\tau} d\tau \right] R^{-1} e^{-At_1} \mathbf{x}^*(t_1) \qquad (4.189)$$

Inserting R into (4.189) and noting that $RR^{-1} = I$ gives

$$\mathbf{x}^*(t_1) = e^{At_1} R R^{-1} e^{-At_1} \mathbf{x}^*(t_1)$$

$$= e^{At_1} I e^{-At_1} \mathbf{x}^*(t_1) = \mathbf{x}^*(t_1) \qquad (4.190)$$

Therefore, we have verified that $u(t)$ in (4.187) results in the desired state transfer.[2]

EXAMPLE 4.14

Determine the control function $u(t)$ that transfers the state of the plant described by

$$\dot{x} = ax + bu \qquad (4.191)$$

from $x(t_0) = 0$ to $x(t_1) = 1$, where a and b are constant scalars.
We first determine R in (4.188) as

$$R = \int_{t_0}^{t_1} b^2 e^{-2a\tau} d\tau$$

$$= -\frac{b^2}{2a} \left(e^{-2at_1} - e^{-2at_0} \right) \qquad (4.192)$$

Therefore, using (4.187) to solve for $u(t)$, we obtain

$$u(t) = \left(\frac{-2a}{b} \right) \left(\frac{e^{-a(t+t_1)}}{e^{-2at_1} - e^{-2at_0}} \right) x(t_1) \qquad (4.193)$$

for the control function on the range $t_0 \le t \le t_1$. If we further specify $a = -1$, $b = 1$, and $t_0 = 0$, we then have

$$u(t) = \frac{2e^{-(t_1-t)}}{1 - e^{-2t_1}} \qquad (4.194)$$

---

[2] The control $u(t)$ for making the desired transfer is not unique. For example, Wiberg [12] describes the construction of another control that contains impulses.

4.8.3 Observability Matrix

We provide the following formal definition of observability.

Definition 4.2. A plant is observable if its initial state $x(t_0)$ can be determined uniquely from the output $y(t)$ on the range $t_0 \le t \le t_1$.

We apply this definition to linear, time-invariant systems by considering (4.129) with $b = 0$. Under this condition, we substitute (4.183) and (4.133) into (4.129) to give

$$y(t) = c^T \Phi(t - t_0) x(t_0)$$

$$= c^T \left[\sum_{i=0}^{n-1} \rho_i(t - t_0) A^i \right] x(t_0)$$

$$= \sum_{i=0}^{n-1} \rho_i(t - t_0) \left[(A^{iT}c)^T x(t_0) \right] \qquad (4.195)$$

We define the n by n observability matrix Q as

$$Q \triangleq \left(c, A^T c, (A^2)^T c, \dots, (A^{n-1})^T c \right) \qquad (4.196)$$

Using (4.196) in (4.195) gives

$$y(t) = \rho^T(t - t_0) Q^T x(t_0) \qquad (4.197)$$

where ρ is an n-vector with components $\rho_0, \rho_1, \dots, \rho_{n-1}$. Premultiplying both sides of (4.197) by $\rho(t - t_0)$ and integrating from t_0 to t_1 gives

$$\int_{t_0}^{t_1} y(t) \rho(t - t_0) \, dt = \left[\int_{t_0}^{t_1} \rho(t - t_0) \rho^T(t - t_0) \, dt \right] Q^T x(t_0) \qquad (4.198)$$

Solving for $Q^T x(t_0)$, we have

$$Q^T x(t_0) = \left[\int_{t_0}^{t_1} \rho(t - t_0) \rho^T(t - t_0) \, dt \right]^{-1} \left[\int_{t_0}^{t_1} y(t) \rho(t - t_0) \, dt \right] \qquad (4.199)$$

which is permissible since the indicated inverse exists. Therefore, we see that the determinant of the coefficient matrix Q^T in (4.199), and hence of Q, must be nonzero to enable us to determine $x(t_0)$ uniquely. This completes the proof of Theorem 2.2.

EXAMPLE 4.15

We examine the observability of the linear plant of Example 2.11 according to the above equations. The observability matrix Q for the direct phase variable form in (2.102) becomes

$$Q = (\mathbf{c}, A^T\mathbf{c}) = \begin{pmatrix} z_1 & -2 \\ 1 & z_1 - 3 \end{pmatrix} \tag{4.200}$$

which has a nonzero determinant if and only if z_1 does not equal either $+1$ or $+2$. With no input, the output $y(t)$ due only to initial conditions $\mathbf{x}(t_0)$ may be expressed as

$$y(t) = \mathbf{c}^T\mathbf{x}(t) = \mathbf{c}^T\Phi(t - t_0)\mathbf{x}(t_0)$$

$$= \begin{pmatrix} z_1 \\ 1 \end{pmatrix}^T \begin{pmatrix} 2e^{-(t-t_0)} - e^{-2(t-t_0)} & e^{-(t-t_0)} - e^{-2(t-t_0)} \\ -2e^{-(t-t_1)} + 2e^{-2(t-t_0)} & -e^{-(t-t_0)} + 2e^{-2(t-t_0)} \end{pmatrix} \begin{pmatrix} x_1(t_0) \\ x_2(t_0) \end{pmatrix} \tag{4.201}$$

Using (4.197), we may alternatively write (4.201) as

$$y(t) = \boldsymbol{\rho}^T(t - t_0)Q^T\mathbf{x}(t_0)$$

$$= \begin{pmatrix} 2e^{-(t-t_0)} - e^{-2(t-t_0)} \\ e^{-(t-t_0)} - e^{-2(t-t_0)} \end{pmatrix}^T \begin{pmatrix} z_1 & -2 \\ 1 & z_1 - 3 \end{pmatrix}^T \begin{pmatrix} x_1(t_0) \\ x_2(t_0) \end{pmatrix} \tag{4.202}$$

where the components of $\boldsymbol{\rho}(t - t_0)$ are given in (4.127). Suppose we arbitrarily let $x_1(t_0) = x_2(t_0) = +1$ and consider the unobservable case where $z_1 = +1$. Therefore, from either (4.201) or (4.202), $y(t)$ becomes

$$y(t) = 2e^{-2(t-t_0)} \tag{4.203}$$

Premultiplying (4.202) throughout by $\boldsymbol{\rho}(t - t_0)$, integrating from t_0 to t_1, and solving for $Q^T\mathbf{x}(t_0)$ yields

$$Q^T\mathbf{x}(t_0) = \begin{pmatrix} z_1 & 1 \\ -2 & z_1 - 3 \end{pmatrix} \begin{pmatrix} x_1(t_0) \\ x_2(t_0) \end{pmatrix}$$

$$= \left[\int_{t_0}^{t_1} \begin{pmatrix} \rho_0^2(t - t_0) & \rho_0(t - t_0)\rho_1(t - t_0) \\ \rho_0(t - t_0)\rho_1(t - t_0) & \rho_1^2(t - t_0) \end{pmatrix} dt \right]^{-1}$$

$$\times \left[\int_{t_0}^{t_1} y(t) \begin{pmatrix} \rho_0(t - t_0) \\ \rho_1(t - t_0) \end{pmatrix} dt \right] \tag{4.204}$$

Substituting $y(t)$ from (4.203) into (4.204) and evaluating for $z = +1$ gives

$$\begin{pmatrix} 1 & 1 \\ -2 & -2 \end{pmatrix} \begin{pmatrix} x_1(t_0) \\ x_2(t_0) \end{pmatrix} = \begin{pmatrix} 2 \\ -4 \end{pmatrix} \tag{4.205}$$

which reduces to

$$x_1(t_0) + x_2(t_0) = 2 \tag{4.206}$$

For this unobservable case, we cannot determine $x_1(t_0)$ and $x_2(t_0)$ uniquely.

In general, the rank of the observability matrix Q specifies the number of independent equations for the n initial conditions $x_1(t_0), x_2(t_0), \ldots, x_n(t_0)$. For a second-order plant with rank $\{Q\} = 1$, as indicated for $z_1 = +1$ in (4.204) through (4.206), the initial conditions lie somewhere on the straight line in (4.206). For an nth-order plant, a line of initial conditions will be obtained for rank $\{Q\} = n - 1$, a plane for rank $\{Q\} = n - 2$, and so forth, for yet lower values of rank $\{Q\}$. These concepts are directly extendable to multivariable plants.

SUMMARY

We have determined the time response of linear feedback systems in this chapter in terms of the state transition matrix and have evaluated this time response by using selected time response performance specifications, such as percent overshoot, rise time, settling time, and steady-state errors. We have investigated the time solution of the sensitivity equations as well as the time response of systems having PID and SVFB control. Finally, we have used state transition matrix properties to prove two theorems from Chapter 2 on controllability and observability.

REFERENCES

1. *IEEE Standard Dictionary of Electrical and Electronics Terms*, 2nd ed. Frank Jay, editor in chief. New York: The Institute of Electrical and Electronics Engineers, 1977.
2. J. L. Melsa and D. G. Schultz. *Linear Control Systems*. New York: McGraw-Hill, 1969.
3. C. M. Close and D. K. Frederick. *Modeling and Analysis of Dynamic Systems*. Boston, Mass.: Houghton Mifflin, 1978.

4. D. P. Maki and M. Thompson. *Mathematical Models and Applications.* Englewood Cliffs, N.J.: Prentice-Hall, 1973.

5. J. M. Smith. *Mathematical Modeling and Digital Simulation for Engineers and Scientists.* New York: Wiley, 1977.

6. B. C. Kuo. *Automatic Control Systems,* 4th ed. Englewood Cliffs, N.J.: Prentice-Hall, 1982.

7. J. J. D'Azzo and C. H. Houpis. *Linear Control System Analysis and Design: Conventional and Modern,* 2nd ed. New York: McGraw-Hill, 1981.

8. K. Ogata. *State Space Analysis of Control Systems.* Englewood Cliffs, N.J.: Prentice-Hall, 1967.

9. T. Kailath. *Linear Systems.* Englewood Cliffs, N.J.: Prentice-Hall, 1980.

10. L. Padulo and M. A. Arbib. *System Theory.* Philadelphia, Pa.: W. B. Saunders Co., 1974.

11. G. M. Swisher. *Introduction to Linear Systems Analysis.* Champaign, Ill.: Matrix Publishers, 1978.

12. D. M. Wiberg. *Theory and Problems of State Space and Linear Systems,* Schaum's Outline Series. New York: McGraw-Hill, 1971.

13. J. L. Melsa and S. K. Jones. *Computer Programs for Computational Assistance in the Study of Linear Control Theory,* 2nd ed. New York: McGraw-Hill, 1973.

PROBLEMS

(§4.1) **4.1** Determine the percent overshoot, rise time, delay time, and settling time (for $\delta_s = 5\%$) for each of the two time response curves given in Figure 4.28.

(§4.1) **4.2** Evaluate by graphical means each of the performance measures (J) defined below for the two time response curves of Problem 4.1.

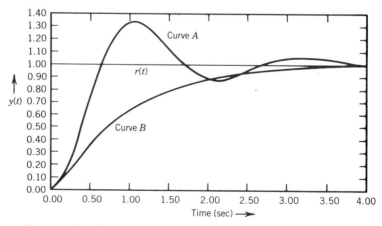

Figure 4.28 Two time response curves for Problem 4.1.

Indicate which response curve yields the smaller value for each J. Sketch $e(t)$ and $\dot{e}(t)$ as a preliminary step, where the error $e(t)$ is defined as $r(t) - y(t)$.

a. $J_1 = \int_0^\infty |e(t)| \, dt$

b. $J_2 = \int_0^\infty e^2(t) \, dt$

c. $J_3 = \int_0^\infty [e^2(t) + \dot{e}^2(t)] \, dt$

(§4.1) **4.3** Consider the performance measure given by

$$J = \int_0^\infty \left[e^2(t) + \alpha \dot{e}^2(t) \right] dt$$

where α is a positive scalar constant. Determine a value of α for which Curve A in Problem 4.1 yields the smaller value for J and another value of α for which J for Curve B is smaller.

(§4.2) **4.4** Classify the time responses for the closed-loop systems of Figure 4.29 as either overdamped, critically damped, or underdamped. Identify ζ and ω_n for each underdamped system.

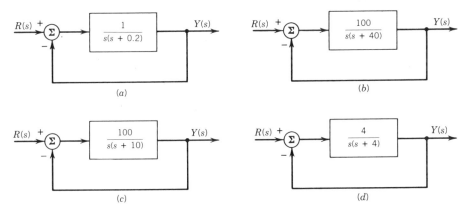

Figure 4.29 Closed-loop second-order systems for Problem 4.4.

(§4.2) **4.5** Determine the maximum overshoot (Max OS), time of the maximum overshoot (t_p), and percent overshoot (%OS) for each of the underdamped systems of Problem 4.4. Observe that not all of these systems are underdamped.

(§4.2) **4.6** Determine expressions for $y(t)$ for each of the systems in Problem 4.4. Using Figure 4.3 as a guide, sketch curves of $y(t)$ versus t for each of these systems.

(§4.2) **4.7** Consider the block diagram of a dc positional servomechanism given in Figure 4.30. Determine the range of K_A for which the %OS is 15% or less.

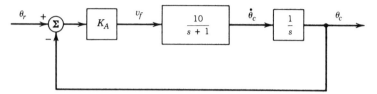

Figure 4.30 The block diagram for a position servomechanism for Problem 4.7.

(§4.2) **4.8** For the closed-loop system of Figure 4.31,

    a. Determine the %OS due to a step input for $K = 10$.
    b. Find the range of K for which the %OS is less than or equal to 10%.
    c. Repeat (b) for 15%, 20%, and 30%, and sketch a curve of K versus %OS.

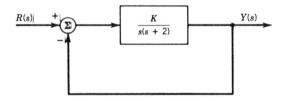

Figure 4.31 The closed-loop system for Problem 4.8.

(§4.2) **4.9** For the dc positional servomechanism of Problem 4.7, determine the %OS for $K_A = 20$, with no tachometer feedback, and then determine the tachometer gain K_t to yield a %OS of 15% with $K_A = 20$.

(§4.2) **4.10** Consider the system that has both the position and velocity fed back in Figure 4.32. Find that value of β that yields a %OS that is only one-fifth as large as the %OS obtained for $\beta = 0$.

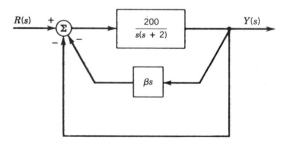

Figure 4.32 A system with position and velocity feedback for Problem 4.10.

(§§3.4, **4.11** For the system given in Figure 4.2, determine the state sensitivity
4.2)   coefficient equations for variations in ω_n, and construct a block
                  diagram for x_1, x_2, $v_{\omega_{n1}}$, and $v_{\omega_{n2}}$ similar to the one in Figure 4.8 for
                  variations in ζ. (Do *not* solve these resulting differential equations.)

(§4.2) **4.12** Determine the %OS for the system of Figure 4.33 for the following
                  control schemes:

   a.   With only proportional control as shown.
   b.   With proportional control and tachometer feedback, where K_t
        $= 0.5$.
   c.   With proportional plus derivative control, where $k_D = 0.5$.

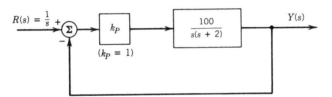

Figure 4.33   The system for Problem 4.12.

(§4.3) **4.13** Write the time expression for $y(t)$ corresponding to

   a.   s-plane poles at -1, -2, and -3 with residues of 10, 20, and
        50, respectively.
   b.   s-plane poles at 0 and -10 with residues of 3 and 2, respec-
        tively.
   c.   s-plane poles at -2, $-1 + j2$, and $-1 - j2$ with residues of 5,
        $2 + j1$, and $2 - j1$, respectively.

(§4.3) **4.14** For each of the following time responses $y(t)$, determine the
                  s-plane poles and the residues at those poles.

   a. $y(t) = 1 - 0.5e^{-t} + 0.8e^{-2t}$
   b. $y(t) = 2e^{-3t} + 5e^{-2t}$
   c. $y(t) = 1 - 0.2e^{-2t} \sin(3t + 60°)$

(§§4.2, **4.15** Consider the closed-loop system having negative-unity feedback in
4.3)   Figure 4.34. If we specify that the natural frequency ω_n is 10
                  radians/second and the %OS for a unit step input is 12%, determine

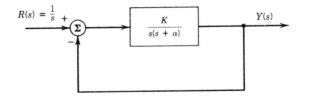

Figure 4.34   The system for Problem 4.15.

a. The closed-loop poles of $Y(s)$ in the s-plane and the residues at those poles.

b. The output time response $y(t)$ and provide a sketch of $y(t)$ versus t.

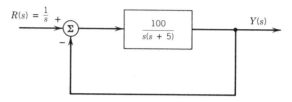

Figure 4.35 The system for Problem 4.16.

(§§4.2, 4.3) **4.16** Determine and sketch $y(t)$ versus t for the system of Figure 4.35

a. When tachometer feedback ($K_t = 0.1$) is added.
b. When derivative control ($k_D = 0.1$) is added.

Use s-plane vectors to determine the residues at the poles.

(§4.3) **4.17** Determine values of β_1 or β_2 for each of the following cases such that the output y has all modes of behavior suppressed except those having eigenvalues more negative than -3.

a.
$$\dot{x}_1 = x_2$$
$$\dot{x}_2 = -8x_1 - 6x_2 + u$$
$$y = \beta_1 x_1 + x_2$$

b.
$$\dot{x}_1 = x_2$$
$$\dot{x}_2 = x_3$$
$$\dot{x}_3 = -10x_1 - 17x_2 - 8x_3 + 12u$$
$$y = \beta_1 x_1 + \beta_2 x_2 + 2x_3$$

c.
$$\dot{x}_1 = -2x_1 + u$$
$$\dot{x}_2 = x_3$$
$$\dot{x}_3 = -12x_2 - 8x_3 + 3u$$
$$y = \beta_1 x_1 + \beta_2 x_2 + x_3$$

(§4.3) **4.18** For the cases given in Problem 4.17, determine values of β_1 or β_2 to select only a single mode of behavior with an eigenvalue of -2 in the output y.

(§4.4) **4.19** Determine the state transition matrix $\Phi(t)$ for each of the following unforced systems by using

  a. Resolvent matrix method.
  b. Series method.
  c. Cayley-Hamilton method.

  (1)
  $$\dot{x}_1 = x_2$$
  $$\dot{x}_2 = -6x_1 - 5x_2$$

  (2)
  $$\dot{x}_1 = -3x_1 + x_2$$
  $$\dot{x}_2 = -2x_1$$

(§4.4) **4.20** Repeat Problem 4.19 for the unforced systems described by

  a.
  $$\dot{x}_1 = -6x_1 + x_2$$
  $$\dot{x}_2 = -11x_1 + x_3$$
  $$\dot{x}_3 = -6x_1$$

  b.
  $$\dot{x}_1 = x_2$$
  $$\dot{x}_2 = -x_1 - 2x_2$$

(§4.5) **4.21** Determine $y(t)$ on the range $t > 0$ for each of the systems given below by

  a. Impulse response convolution integral.
  b. Superposition integral.
  c. Laplace transformations.

  (1)
  $$\dot{x}_1 = -x_1 + e^{2t}$$
  $$\dot{x}_2 = -5x_2 + e^{2t}$$
  $$y = x_1 - x_2$$

  (2)
  $$\dot{x}_1 = x_2$$
  $$\dot{x}_2 = -6x_1 - 5x_2 + e^{2t}$$
  $$y = 4x_1$$

Observe that the input e^{2t} is applied for all time $t \geq -\infty$. Solve for $x(t_0)$ using (4.140) for an arbitrary t_0 in Part (b) and for $t_0 = 0$ in Part (c).

(§4.5) **4.22** Repeat Problem 4.21 for the system described by

$$\dot{x}_1 = x_2$$

$$\dot{x}_2 = x_3$$

$$\dot{x}_3 = -6x_1 - 11x_2 - 6x_3 + u(t)$$

$$y = x_1$$

where $u(t)$ is a unit step input applied at $t = 0$ and $x_1(0) = x_2(0) = x_3(0) = 1$.

(§4.6) **4.23** Consider the second-order system described by

$$\dot{x} = \begin{pmatrix} 0 & 1 \\ -\alpha & -3 \end{pmatrix} x - \begin{pmatrix} 0 \\ 1 \end{pmatrix} u(t)$$

where $u(t)$ is a unit step input for $t \geq 0$ and $x(0) = \begin{pmatrix} 0 \\ 0 \end{pmatrix}$. Use (4.154) to solve for $v_\alpha(t)$.

(§4.6) **4.24** Rework Example 4.10 to show variations in the %OS, that is, $g(K)$, if the plant transfer function is given by

$$G(s) = \frac{K}{s(s + 2)}$$

Provide tabular results for different K, as in Table 4.2, and plot $g(K)$ versus K, as in Figure 4.22.

(§4.6) **4.25** Repeat Problem 4.24 to show variations in the damping coefficient $\zeta(K)$ as the gain K varies over several values that result in an underdamped system. Provide tabular results and a plot of $\zeta(K)$ versus K.

(§4.7) **4.26** Determine a closed-form analytical expression for $x_1(t)$ for the system described by (4.178) with $k_P = 0.4$, $k_D = 0.2$, and $k_I = 0.5$. Assume a unit step input and zero initial conditions. Let $K_A = 50$. Plot your result versus time and compare with the corresponding time curve in Figure 4.25c.

(§4.7) **4.27** Similar to (4.178), express the unity-feedback system having plant transfer function $G(s)$ and PID control in the direct feed-forward form of state variables. Let $G(s)$ be given by

$$G(s) = \frac{K_A}{s(s + 1)(s + 2)}$$

(§4.7) **4.28**   Repeat Problem 4.27 if SVFB control is to be used.

(§4.7) **4.29**   Solve for $x_1(t)$ in (4.180) if k_1, k_2, and k_3 are selected to yield a closed-loop system response with eigenvalues at -1, -2, and -4. Assume a unit step input and zero initial conditions. Let $K = 1$ and $K_A = 50$.

(§4.8) **4.30**   Determine a control function $u(t)$ that transfers the state of the plant described by

$$\dot{x} = -2x + 3u(t)$$

from $x(0) = 0$ to $x(1) = 1$. Sketch curves of $u(t)$ and $x(t)$ versus t.

(§§2.5,   **4.31**   Consider the second-order system described by
4.8)

$$\dot{x}_1 = x_2$$

$$\dot{x}_2 = x_1 + 2x_2 + u$$

$$y = x_1 + 3x_2$$

Is this system state controllable? Is it observable?

5

Stability Analysis in the s-Plane

A dynamic system is said to be asymptotically stable if it tends to resume its given operating condition after a small perturbation in values of the system state variables. For example, an unpowered roller coaster tends to move to a low point along its track in the absence of prohibitive frictional forces. An automobile "cruise" control permits the vehicle to resume its preset speed following a momentary period of braking or acceleration. An automatic pilot brings the aircraft back to its intended course after temporary disturbances due to unexpected turbulence. Stability analysis enables the control engineer to ensure that these systems behave properly when such perturbations occur.

The stability analysis problem is inherently related to the design problem for linear time-invariant systems. The transform solution to both problems for such systems is to place the closed-loop poles in desired locations in the s-plane. The corresponding state variable approach produces a suitable time response for the system state variables. This chapter begins a two-chapter study of stability analysis. Results in the s-plane are presented here and compared in Chapter 6 with frequency-domain stability analysis procedures.

Following a brief description of the definition and properties of stability, we present a criterion to determine stability by using only the characteristic polynomial coefficients in a convenient table. Construction rules for the root locus method are first described for negative feedback systems and then extended to the positive feedback case. Root contours, obtained by varying a parameter other than the open-loop gain, form a generalization of the root locus method.

5.1 STABILITY DEFINITION AND PROPERTIES

For the linear time-invariant system described by

$$\dot{\mathbf{x}} = A\mathbf{x} \tag{5.1}$$

the characteristic equation may be expressed as

$$\det(sI - A) = 0 \tag{5.2}$$

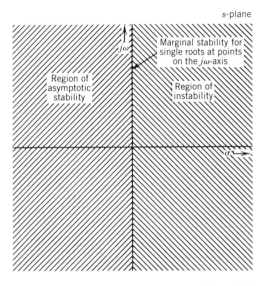

Figure 5.1   Regions of stability and instability in the *s*-plane for linear time-invariant systems.

The roots of the characteristic equation are not only the eigenvalues of the matrix *A* but also the poles of the transfer function between input and output. Thus, the roots of the characteristic equation for a closed-loop system are the closed-loop poles.

Definition 5.1. The system in (5.1) is

- Asymptotically stable if and only if all the roots of the characteristic equation for *A* lie in the left-half *s*-plane.
- Marginally stable if some of the roots occur as a single root at the origin in the *s*-plane or as single pairs of roots at points on the *jω*-axis and the remaining roots are in the left-half *s*-plane.
- Unstable if one or more roots occur in the right-half *s*-plane or if multiple roots or pairs of roots occur at a point on the *jω*-axis.

Figure 5.1 indicates these regions of stability in the *s*-plane.

In terms of the impulse response function $g(t)$ for linear time-invariant systems, asymptotic stability is guaranteed if and only if

$$\int_0^\infty |g(\tau)|\, d\tau < \infty \tag{5.3}$$

In words, a linear time-invariant system is asymptotically stable if and only if the impulse response is absolutely integrable over the infinite range $(0, \infty)$.

It is instructive to consider the associated stability problem for the nonlinear system [1–8] described by

$$\dot{\mathbf{x}} = \mathbf{f}(\mathbf{x}) \tag{5.4}$$

We may determine the equilibrium state x_e from

$$f(x_e) = 0 \tag{5.5}$$

where x_e is simply the static operating point determined by linearization in Section 1.4. As before, we define δx as

$$\delta x = x(t) - x_e \tag{5.6}$$

Linearized variational equations for δx can be expressed in the form of (5.1) for each equilibrium state x_e as

$$\delta \dot{x} = \left. \frac{\partial f}{\partial x} \right|_{x = x_e} \delta x \tag{5.7}$$

As in (5.2), roots of the resulting characteristic equations enable us to determine the stability of each x_e. For the linear time-invariant system in (5.1), there is only one equilibrium state ($x_e = 0$). The asymptotic stability of this single equilibrium state is guaranteed if and only if all roots of the characteristic equation lie in the left-half *s*-plane or, equivalently, if the system response remains bounded for all bounded inputs. Therefore, we often omit reference to the "stability of the equilibrium state x_e" and simply examine the "stability of the system." Observe once again that the stability of a linear system depends only on the dynamics of the system and not its input. The *s*-plane properties defined in this section lead to operational procedures for determining the stability of a given linear time-invariant system.

5.2 THE ROUTH-HURWITZ CRITERION

We present here a procedure for determining the stability of a linear time-invariant system by examining its characteristic polynomial directly. If we were somehow able to factor this polynomial, we could easily plot the root locations in the *s*-plane. According to the results of the last section, the system is asymptotically stable if all roots are in the left-half *s*-plane. However, we now describe a criterion that does not require the factorization of the characteristic polynomial but rather utilizes the polynomial coefficients in an easily applied algorithm to predict stability [9,10].

The Routh-Hurwitz criterion yields both necessary and sufficient conditions for asymptotic stability. In other words, the system is asymptotically stable if and only if the coefficients of the characteristic equation satisfy this criterion. Initially, we present a preliminary Hurwitz test as a necessary condition for asymptotic stability. Then we discuss the Routh-Hurwitz criterion, including two special cases which sometimes occur in completing the Routh table.

5.2.1 Hurwitz Test

Let the characteristic polynomial of the transfer function of the system be written as

$$f(s) = s^n + \alpha_{n-1}s^{n-1} + \cdots + \alpha_2 s^2 + \alpha_1 s + \alpha_0 \qquad (5.8)$$

A necessary condition for asymptotic stability is that all coefficients in (5.8) must have the same sign and none may be zero. This test by inspection is referred to as the Hurwitz test.

We note that a necessary condition for marginal stability is that the Hurwitz test as stated above is satisfied except that all odd or all even coefficients may be zero. If only even coefficients are present, then we might have single pairs of roots of points on the $j\omega$-axis. If only odd coefficients are present, we also have a root at the origin ($s = 0$). Only a necessary condition, the Hurwitz test for marginal stability may be satisfied and yet the system is unstable. All we have guaranteed is that the failure of the Hurwitz test implies instability.

5.2.2 The Routh Table

Hurwitz [9] derived a necessary and sufficient condition for asymptotic stability based on the evaluation of certain determinants involving the characteristic polynomial coefficients. These numerical computations are quite cumbersome, especially for higher-order systems. Earlier, Routh [10] had developed a tabular form that involves only simple sequential calculations. The interpretation of the Hurwitz determinants by means of calculations in a Routh array or table is referred to as the Routh-Hurwitz criterion.

The Routh table for a general polynomial is provided in Table 5.1. The coefficients of the characteristic polynomial in (5.8) are arranged in the first two rows of the table as shown. Assuming for the moment that α_{n-1} is nonzero, the

TABLE 5.1   General Form of the Routh Table

s^n	1	α_{n-2}	α_{n-4}	$\cdots$
s^{n-1}	α_{n-1}	α_{n-3}	α_{n-5}	$\cdots$
s^{n-2}	$P_{n-2,1}$	$P_{n-2,2}$	$P_{n-2,3}$	$\cdots$
s^{n-3}	$P_{n-3,1}$	$P_{n-3,2}$	$P_{n-3,3}$	$\cdots$
$\vdots$	$\vdots$	$\vdots$		
s^2	$P_{2,1}$	$P_{2,2}$		
s^1	$P_{1,1}$			
s^0	$P_{0,1}$			

next row of the table may be determined from

$$
\rho_{n-2,1} = (-1) \frac{\begin{vmatrix} 1 & \alpha_{n-2} \\ \alpha_{n-1} & \alpha_{n-3} \end{vmatrix}}{\alpha_{n-1}} = \frac{\alpha_{n-1}\alpha_{n-2} - (1)\alpha_{n-3}}{\alpha_{n-1}}
$$

$$
\rho_{n-2,2} = (-1) \frac{\begin{vmatrix} 1 & \alpha_{n-4} \\ \alpha_{n-1} & \alpha_{n-5} \end{vmatrix}}{\alpha_{n-1}} = \frac{\alpha_{n-1}\alpha_{n-4} - (1)\alpha_{n-5}}{\alpha_{n-1}} \tag{5.9}
$$

$$
\vdots
$$

Once any row has been completed, the elements of the following row may be determined from the previous two rows in the same manner, provided no division by zero is implied. This possibility will be discussed in special cases later in the section.

The interpretation of the Routh table is easy to remember:

> The number of roots in the right-half *s*-plane is equal to the number of sign changes in the first column of the Routh table.

This condition is known as the Routh-Hurwitz criterion. We do not provide a derivation of this criterion, but we will be using the Routh table extensively as an operational tool for stability analysis.

EXAMPLE 5.1

For the given characteristic polynomials, use the Routh-Hurwitz criterion to determine the number of roots in the right-half *s*-plane.

a. $f(s) = s^4 + 6s^3 + 13s^2 + 12s + 4$
b. $f(s) = s^5 + 5s^4 + 2s^3 + 18s^2 + 45s + 25$
c. $f(s) = s^5 - 4s^4 + 4s^3 + 9s^2 - 12s + 4$

As a preliminary step, we observe that only the polynomials in (a) and (b) pass the Hurwitz test. Sign changes among the coefficients for the polynomial in (c) indicate instability for the corresponding closed-loop system. Additional information regarding root locations for each of the three characteristic polynomials can be obtained from the Routh table.

The coefficients of the polynomial in (a) are placed in the first two rows of the Routh array in Table 5.2 The entry in the first column of the next row—that is,

TABLE 5.2 The Routh Table for Example 5.1(a)

s^4	1	13	4
s^3	6	12	
s^2	$\rho_{2,1} = 11$	$\rho_{2,2} = 4$	
s^1	$\rho_{1,1} = 9.8$		
s^0	$\rho_{0,1} = 4$		

the s^2 row—is formed from the entries in the first two columns of the previous two rows as

$$\rho_{2,1} = (-1)\frac{\begin{vmatrix} 1 & 13 \\ 6 & 12 \end{vmatrix}}{6} = (-1)\left(\frac{1(12) - 6(13)}{6}\right) = 11 \qquad (5.10)$$

The entry in the second column of the s^2 row is

$$\rho_{2,2} = (-1)\frac{\begin{vmatrix} 1 & 4 \\ 6 & 0 \end{vmatrix}}{6} = (-1)\left(\frac{1(0) - 6(4)}{6}\right) = 4 \qquad (5.11)$$

Observe that a zero is assumed whenever no entry appears in a particular location, such as the absence of an entry in the third column of the s^3 row. Moreover, when the entry in the last column of a row is nonzero but the entry immediately below is missing (or zero), then that nonzero entry automatically appears one column earlier in the second row following, as indicated by the arrows in Table 5.2. For example, the $+4$ entry in the s^4 row, column 3, has a missing entry below it in the s^3 row. As calculated in (5.11), this $+4$ automatically appears as $\rho_{2,2}$ in column 2 of the s^2 row.

Continuing with Table 5.2, we have

$$\rho_{1,1} = (-1)\frac{\begin{vmatrix} 6 & 12 \\ 11 & 4 \end{vmatrix}}{11} = \frac{108}{11} = 9.8 \qquad (5.12)$$

Finally, $\rho_{0,1}$ is $+4$, formed from $\rho_{2,2}$ as an automatic entry because $\rho_{1,2}$ is zero. Observe that all entries in the first column of the Routh array are positive. Since there are no sign changes in this first column, the Routh-Hurwitz criterion guarantees that there are no roots in the right-half of the s-plane. This result from the Routh table is consistent with our a priori knowledge for this fourth-order contrived polynomial, which has two roots at -1 and two roots at -2. The student should verify that the product $(s + 1)^2(s + 2)^2$ does yield the given polynomial.

TABLE 5.3   The Routh Table for Example 5.1(b)

s^5	1	2	45
s^4 >	5	18	25
s^3 >	-1.6	40	
s^2	143	25 ←	
s^1	40.3		
s^0	25 ←		

The Routh table for the characteristic polynomial in (b) is shown in Table 5.3. The successive formation of entries beyond the first two rows yields

$$\rho_{3,1} = (-1)\frac{\begin{vmatrix} 1 & 2 \\ 5 & 18 \end{vmatrix}}{5} = -1.6 \qquad \rho_{3,2} = (-1)\frac{\begin{vmatrix} 1 & 45 \\ 5 & 25 \end{vmatrix}}{5} = 40$$

$$\rho_{2,1} = (-1)\frac{\begin{vmatrix} 5 & 18 \\ -1.6 & 40 \end{vmatrix}}{-1.6} = 143 \qquad \rho_{2,2} = (-1)\frac{\begin{vmatrix} 5 & 25 \\ -1.6 & 0 \end{vmatrix}}{-1.6} = 25$$

$$\rho_{1,1} = (-1)\frac{\begin{vmatrix} -1.6 & 40 \\ 143 & 25 \end{vmatrix}}{143} = 40.3 \qquad \rho_{0,1} = (-1)\frac{\begin{vmatrix} 143 & 25 \\ 40.3 & 0 \end{vmatrix}}{40.3} = 25 \quad (5.13)$$

Observe that $\rho_{2,2}$ and $\rho_{0,1}$ may be obtained automatically as indicated by the arrows in Table 5.3. The two sign changes in the first column of this table mean that there are two roots in the right-half of the s-plane. The factors of this contrived polynomial in (b) are $(s - 1 - j2)(s - 1 + j2)(s + 1)^2(s + 5)$, which may easily be verified.

Table 5.4 shows the Routh table for the polynomial in (c). Four sign changes in the first column indicate that there are four roots in the right-half of the s-plane. Moreover, we can further identify that the fifth root is in the left-half of the s-plane. If the fifth root were on the $j\omega$-axis, it would have to be at the origin, because roots off the real axis must appear in complex conjugate pairs. A single root at the origin occurs if and only if α_0 in (5.8) is zero. Thus, the single fifth root must lie somewhere on the negative real axis. The four right-half s-plane roots may be either four real roots, two real and two complex roots, or four complex roots.

TABLE 5.4   The Routh Table for Example 5.1(c)

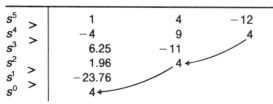

s^5 >	1	4	-12
s^4 >	-4	9	4
s^3	6.25	-11	
s^2	1.96	4 ←	
s^1 >	-23.76		
s^0 >	4 ←		

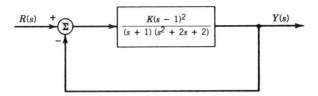

Figure 5.2 The system for stability investigation in Example 5.2.

We note that the polynomials in (a) and (b) have the remaining roots in the left-half s-plane. Specifically, we showed above that (a) and (b) have none and two right-half s-plane roots, respectively. If some of the remaining roots for these polynomials were on the $j\omega$-axis, then the second special case described in the following subsection would be indicated. Because this condition that all elements of a row are zero is not indicated, we conclude that the fourth-order polynomial in (a) has four left-half s-plane roots and the fifth-order polynomial in (b) has the remaining three roots in the left-half s-plane. Therefore, since we have specified that these three polynomials are characteristic polynomials—that is, denominators of closed-loop transfer functions—the corresponding system in (a) is asymptotically stable (all four poles in the left-half s-plane) and the systems in (b) and (c) are both unstable (at least one right-half s-plane pole in each case).

EXAMPLE 5.2

Use the Routh-Hurwitz criterion to determine the range of K for stability for the closed-loop system in Figure 5.2.

The characteristic equation is given by

$$1 + G(s)H(s) = 1 + \frac{K(s-1)^2}{(s+1)(s^2+2s+2)} = 0 \qquad (5.14)$$

Multiplying by the denominator of $G(s)H(s)$ in (5.14) and grouping terms yields the characteristic polynomial

$$f(s) = s^3 + (K+3)s^2 + (4-2K)s + K + 2 = 0 \qquad (5.15)$$

The Routh array is shown in Table 5.5. We calculate $\rho_{1,1}$ as

$$\rho_{1,1} = \frac{-\begin{vmatrix} 1 & 4-2K \\ K+3 & K+2 \end{vmatrix}}{K+3} = \frac{(-1)[(K+2)(1) - (K+3)(4-2K)]}{K+3}$$

$$= \frac{-2K^2 - 3K + 10}{K+3} \qquad (5.16)$$

The value of $\rho_{0,1}$ is easily determined as $K + 2$. The entries in the first column of the Routh array must be positive for asymptotic stability—that is, $\rho_{2,1} > 0$,

TABLE 5.5   The Routh Array for Example 5.2

s^3	1	$4 - 2K$
s^2	$K + 3$	$K + 2$
s^1	$\rho_{1,1}$	
s^0	$\rho_{0,1}$	

$\rho_{1,1} > 0$, and $\rho_{0,1} > 0$. These conditions are satisfied by setting

$$\rho_{2,1} = K + 3 > 0$$

$$\rho_{1,1} = \frac{-2K^2 - 3K + 10}{K + 3} > 0$$

$$\rho_{0,1} = K + 2 > 0 \tag{5.17}$$

The first and third inequalities hold for $K > -2$. We sketch the numerator of $\rho_{1,1}$ versus K in Figure 5.3 and note that this function is positive for $(-3 - \sqrt{89})/4 < K < (-3 + \sqrt{89})/4$. This range is approximately $-3.11 < K < 1.61$. Since we also require $K > -2$ because of the other inequalities in (5.17), we form the intersection of the allowed regions for K. Therefore, the range of K for the asymptotic stability of the closed-loop system in Figure 5.2 is approximately

$$-2 < K < 1.61 \tag{5.18}$$

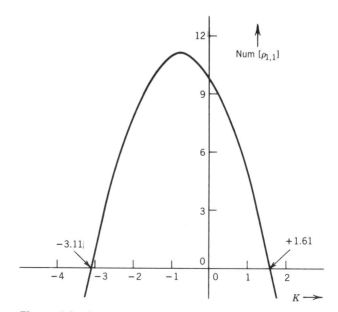

Figure 5.3   Sketch of the numerator of $\rho_{1,1}$ versus K in (5.16) and Table 5.5.

We can show by the methods of the following subsection that closed-loop poles appear on the $j\omega$-axis for $K = -2$ and $K \cong 1.61$.

5.2.3 Special Cases for the Routh Table

The problem occurring when a zero appears in the first column of the Routh table is discussed here. This situation requires a special consideration because division by zero is indicated when blindly applying the former rules. Two possibilities may be encountered:

1. A zero appears in the first column of a row and at least one nonzero element appears elsewhere in the same row.
2. All elements of a particular row are zero.

In the first case, we simply replace the zero in the first column by an arbitrary small positive number ε and proceed as before to complete the table. In the second case, we form an auxiliary polynomial $f_a(s)$ corresponding to the last row and the particular row having all zeros. We then replace the row that would have contained all zeros by the coefficients of the derivative of the auxiliary polynomial with respect to s. This second possibility occurs if there are roots of the original characteristic polynomial either on the $j\omega$-axis or symmetrically located on the real axis about the origin in the s-plane. These $j\omega$-axis or real-axis roots are also roots of the auxiliary polynomial. The second case also occurs when a quad of four complex roots appear symmetrically about the origin in the s-plane. Details of procedures for these special cases are illustrated in the following examples.

EXAMPLE 5.3

Use the Routh-Hurwitz criterion to determine the number of right-half s-plane roots for the given characteristic polynomials.

a. $f(s) = s^3 + s + 1$
b. $f(s) = s^3 + 2s^2 + s + 2$
c. $f(s) = s^5 + s^4 + 4s + 4$

We see at the outset that the polynomials in (a) and (c) fail the Hurwitz test and instability is indicated. The Routh table can provide additional information on root locations.

The Routh array in Table 5.6 for the polynomial in (a) shows the need for the first special case immediately in the second row. We set $\rho_{2,1} = \varepsilon$, where ε is an

TABLE 5.6   The Routh Array for Example 5.3(*a*)

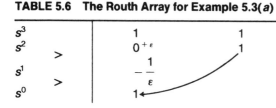

arbitrary small positive value. Calculating $\rho_{1,1}$ and $\rho_{0,1}$ gives

$$\rho_{1,1} = (-1)\frac{\begin{vmatrix} 1 & 1 \\ \varepsilon & 1 \end{vmatrix}}{\varepsilon} = 1 - \frac{1}{\varepsilon} \cong -\frac{1}{\varepsilon}$$

$$\rho_{0,1} = (-1)\frac{\begin{vmatrix} \varepsilon & 1 \\ -1/\varepsilon & 0 \end{vmatrix}}{-1/\varepsilon} = 1 \tag{5.19}$$

We note two sign changes in the first column of the Routh array because only $\rho_{1,1} < 0$. Thus, there are two roots in the right-half *s*-plane for this third-order polynomial. We observe that the second special case must occur if any roots appear on the *jω*-axis. Since the second case is not indicated in Table 5.6, we conclude that the remaining root is in the left-half *s*-plane.

Table 5.7 shows the Routh array for the polynomial in (b). If we try to form $\rho_{1,1}$ from the two previous rows, we obtain

$$\rho_{1,1} = (-1)\frac{\begin{vmatrix} 1 & 1 \\ 2 & 2 \end{vmatrix}}{2} = 0 \tag{5.20}$$

Because there are no nonzero entries in the s^1 row, the second special case is indicated. We form an auxiliary polynomial $f_a(s)$ from the s^2 row as

$$f_a(s) = 2s^2 + 2 \tag{5.21}$$

Taking the derivative of $f_a(s)$ yields

$$f_a'(s) = \frac{df_a(s)}{ds} = 4s \tag{5.22}$$

TABLE 5.7   The Routh Array for Example 5.3(*b*)

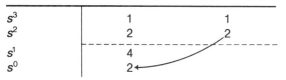

TABLE 5.8 The Routh Array for Example 5.3(c)

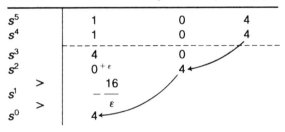

s^5	1	0	4
s^4	1	0	4
s^3	4	0	
s^2	$0 + \varepsilon$	4	
s^1	$\dfrac{16}{\varepsilon}$		
s^0	4		

We indicate the occurrence of the second special case by a dashed line following the s^2 row. Inserting for $\rho_{1,1}$ the coefficient of s^1 in (5.22), that is, $+4$, we may then complete the Routh array in Table 5.7 by setting $\rho_{0,1} = 2$. There are no sign changes in the first column and, consequently, there are no right-half s-plane roots of the polynomial in (b).

The dashed line indicating the occurrence of the second special case at the s^2 row means that there are two roots either on the $j\omega$-axis or symmetrically located about the origin on the real axis. These two roots may be determined by factoring $f_a(s)$ in (5.21) to yield $(s + j)(s - j)$. Furthermore, we note that $f_a(s)$ is itself a factor of the original polynomial in (b). The roots of that original polynomial are $+j$, $-j$, and -2. The corresponding closed-loop system may be classified as marginally stable.

The Routh array for the polynomial in (c) is shown in Table 5.8. Since the s^5 and s^4 rows are identical, the s^3 row would contain all zero entries if the general rules of the previous subsection were applied blindly. We place dashes after the s^4 row, indicating the second special case, and form an auxiliary polynomial as

$$f_a(s) = \rho_{4,1}s^4 + \rho_{4,2}s^2 + \rho_{4,3} \tag{5.23}$$

where $\rho_{4,1} = 1$, $\rho_{4,2} = 0$, and $\rho_{4,3} = 4$. Its derivative $f_a'(s)$ is $4\rho_{4,1}s^3 + 2\rho_{4,2}s$, which yields entries of $4\rho_{4,1} = 4$ and $2\rho_{4,2} = 0$ in the s^3 row, as shown in Table 5.8. Completing the table yields two sign changes in the first column. We note that the factors of $f_a(s)$ are $(s + 1 + j)$, $(s + 1 - j)$, $(s - 1 + j)$, and $(s - 1 - j)$. Therefore, a quad of the corresponding four roots are formed in the s-plane with a fifth root somewhere at $(s = -1$ for this particular case) in the left-half s-plane.

EXAMPLE 5.4

For the closed-loop system of Figure 5.4, determine the range of K for asymptotic stability, the value of K to place closed-loop poles on the $j\omega$-axis, and the location of those $j\omega$-axis poles.

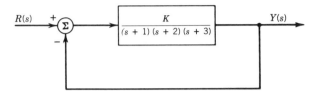

Figure 5.4 System for Example 5.4.

We form the characteristic equation from

$$1 + G(s)H(s) = 1 + \frac{K}{(s+1)(s+2)(s+3)} = 0 \qquad (5.24)$$

Simplifying yields

$$s^3 + 6s^2 + 11s + 6 + K = 0 \qquad (5.25)$$

The Routh array is given in Table 5.9. To determine the range of K for asymptotic stability, we require that all first column entries are positive, that is,

$$\rho_{1,1} = \frac{60 - K}{6} > 0 \Rightarrow K < +60$$

$$\rho_{0,1} = K + 6 > 0 \Rightarrow K > -6 \qquad (5.26)$$

The intersection of the ranges of K specified in (5.26) yields the range for asymptotic stability as

$$-6 < K < +60 \qquad (5.27)$$

Consider the result of setting K equal to $+60$. For $K > 60$, $\rho_{1,1}$ is negative and, consequently, there are two sign changes in the first column. We note that $\rho_{1,1} = 0$ when $K = 60$, and the second special case occurs. We may write the auxiliary equation as

$$f_a(s) = 6s^2 + K + 6|_{K=60} = 6s^2 + 66 \qquad (5.28)$$

The roots of $f_a(s) = 0$ are also roots of the original polynomial in (5.25). Since these roots occur on the $j\omega$-axis, we denote the corresponding values of s as

TABLE 5.9 The Routh Array for Example 5.4

s^3	1	11
s^2	6	$K + 6$
s^1	$\rho_{1,1} = \dfrac{60 - K}{6}$	
s^0	$\rho_{0,1} = K + 6$	

$\pm j\omega^*$, and we have

$$6(j\omega^*)^2 + 66 = 0 \qquad (5.29)$$

from which $\omega^* = \sqrt{11} \cong 3.32$.

For $K < -6$, $\rho_{0,1}$ is negative and there is a single sign change in the first column of the Routh array. We again recognize the second special case, form an auxiliary equation of $f_a(s) = \rho_{1,1}s^1$, and set $s = j\omega^*$ to obtain $\omega^* = 0$. Thus, we conclude that for $K = -6$ a closed-loop pole occurs at the origin of the s-plane and for $K = +60$ closed-loop poles are on the $j\omega$-axis at $s = \pm j\sqrt{11}$. The system is marginally stable for these values of K.

5.3 THE ROOT LOCUS METHOD

The root locus method enables us to sketch the locations of the closed-loop poles as some system parameter is varied [11–13]. We restrict ourselves in this section and the next one to positive variations in the open-loop gain K in the negative feedback configuration, as shown in Figure 5.5.

5.3.1 The Root Locus Concept

Consider as a standard form for root locus construction the open-loop transfer function given by

$$GH(s) = \frac{K(s + z_1)(s + z_2) \cdots (s + z_z)}{(s + p_1)(s + p_2)(s + p_3) \cdots (s + p_p)} \qquad (5.30)$$

where there are z finite zeros and p finite poles of $GH(s)$. We write the characteristic equation for the system of Figure 5.5 as

$$1 + GH(s) = 0 \qquad (5.31)$$

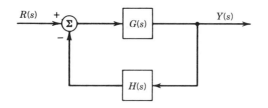

Figure 5.5 Feedback diagram for basic root locus construction.

and search for all points s in the s-plane satisfying (5.31). Therefore, we require

$$GH(s) = -1 \qquad (5.32)$$

There are two conditions implied by (5.32). These conditions relate to the magnitude and phase (or angle) of $GH(s)$, that is,

$$|GH(s)| = 1 \qquad (5.33)$$

$$\underline{/GH(s)} = (2k + 1)\pi \qquad (k = 0, \pm 1, \pm 2, \ldots) \qquad (5.34)$$

In words, (5.33) states that the magnitude of the open-loop transfer function must be unity.[1] The gain K may be adjusted to satisfy this requirement, provided (5.34) holds. Thus, the root locus method may be characterized from (5.34) as determining all points in the s-plane such that the phase of the open-loop transfer function $GH(s)$ is π radians or any angle obtained by $2k\pi$ (where $k = 0, \pm 1, \pm 2, \ldots$) radians of rotation beyond π (in either direction) to yield this same terminal position. Once the points for which $GH(s) = \pi$ radians in the s-plane are found, K may be determined by using (5.30) in (5.33) to give

$$\frac{K|(s + z_1)(s + z_2) \cdots|}{|(s + p_1)(s + p_2)(s + p_3) \cdots|} = 1 \qquad (5.35)$$

from which

$$K = \frac{|s + p_1| \cdot |s + p_2| \cdot |s + p_3| \cdots}{|s + z_1| \cdot |s + z_2| \cdots} \qquad (5.36)$$

In terms of the factors in (5.30) and the corresponding s-plane vectors, we require from (5.34) that points on the root locus for $0 < K < +\infty$ satisfy

$$\underline{/GH(s)} = \sum_{i=1}^{z} \underline{/s + z_i} - \sum_{i=1}^{p} \underline{/s + p_i} = (2k + 1)\pi \qquad (5.37)$$

Moreover, (5.33) and, more precisely, (5.36) yield

$$K = \frac{\text{Product of vector lengths from poles}}{\text{Product of vector lengths from zeros}} \qquad (5.38)$$

Alternatively, we could satisfy (5.33) first and then (5.34); both conditions must hold regardless of the order in which they are considered. In either case, we could

---

[1] For the root locus method, we combine the transfer functions of the (open-loop) plant $G(s)$ and the feedback controller $H(s)$ in Figure 5.5 to form the open-loop transfer function $GH(s)$. Henceforth, the terms "open-loop zeros" and "open-loop poles" will be used to refer to the zeros and poles of $GH(s)$.

utilize a digital computer, possibly with associated computer graphics, to search for those points in the s-plane satisfying (5.33) and (5.34), or (5.37) and (5.38). Instead, we emphasize here the development of some basic root locus construction rules derived from (5.33) and (5.34) and their application to obtain root loci sketches.

5.3.2 Basic Construction Rules

Six root locus construction rules are provided in Table 5.10. We discuss in this subsection the first three of these rules and then work some carefully selected examples that require only the application of these rules. Rule 4 is described more fully in the next subsection, and Rules 5 and 6 are presented in Section 5.4.

RULE 1: Starting and Ending Points

Root loci start ($K = 0$) on the open-loop poles and end ($K \rightarrow \infty$) on the open-loop zeros.

Rule 1 may be proved directly from (5.36). When $s = -p_1, -p_2, -p_3, \ldots$, the value of K is 0. Therefore, the root locus plot starts ($K = 0$) at the open-loop poles. Furthermore, since $K = \infty$ for $s = -z_1, -z_2, \ldots$, we see that the root locus plots end on the open-loop zeros. The number of separate loci is the maximum of the number of finite poles (p) and the number of finite zeros (z). For most systems of interest, z is less than or equal to p.

RULE 2: Root Locus Segments on the Real Axis

Root loci occur on a particular segment of the real axis if and only if there are an odd number of total poles and zeros of the open-loop transfer function lying to the right of that segment.

Equation (5.37) may be used to prove Rule 2, which determines the sections of the real axis where roots occur. If a point s on the real axis is to the right of a zero z_i or a pole p_i, the s-plane vector drawn to that point from this zero or pole has an angle of 0 radians. If a point s is to the left of a zero, the associated s-plane vector from the zero to the point s contributes π radians to the net angle of $GH(s)$. If the point s is to the left of a pole on the real axis, the associated s-plane vector from the pole to the point s contributes $-\pi$ radians to the net angle of $GH(s)$. Consequently, we must have an odd number of the sum of finite poles and finite zeros to the right of the point s to yield a net angle of π radians

---

TABLE 5.10   Basic Rules for Root Locus Construction

---

1. **STARTING AND ENDING POINTS**

   Root locus plots start ($K = 0$) on the open-loop poles and end ($K \to \infty$) on the open-loop zeros.

2. **ROOT LOCUS SEGMENTS ON THE REAL AXIS**

   Root loci occur on a particular segment of the real axis if and only if there are an odd number of total poles and zeros of the open-loop transfer function lying to the right of that segment.

3. **IMAGINARY AXIS INTERSECTIONS**

   Use the Routh-Hurwitz criterion to determine $j\omega$-axis crossings of the root locus plots. Both the gain K^* and the value of ω^* may be found from the Routh table.

4. **ASYMPTOTES (FOR $p \neq z$)**

   Root locus plots are asymptotic to straight lines with angles given by

   $$\theta_A = \frac{(2k + 1)\pi}{p - z}$$

   as s approaches infinity. These straight lines intersect at a point σ_I on the real axis specified by

   $$\sigma_I = \frac{\sum \text{Poles of } GH(s) - \sum \text{Zeros of } GH(s)}{p - z}$$

   where p is the number of finite poles and z is the number of finite zeros of $GH(s)$.

5. **ANGLES OF DEPARTURE AND ARRIVAL**

   Assume a point s_θ arbitrarily near the pole (for departure) or the zero (for arrival) and then apply the fundamental angle relationship

   $$\sum \text{Angles of zeros of } GH(s) - \sum \text{Angles of poles of } GH(s) = (2k + 1)\pi$$

6. **BREAKAWAY POINTS**

   Breakaway points may be determined by expressing the characteristic equation for the gain K as a function of s and then solving for the breakaway points s_B from

   $$\left. \frac{dK(s)}{ds} \right|_{s = s_B} = 0$$

for $GH(s)$. This point s may be anywhere along the section where Rule 2 is satisfied.

RULE 3: Imaginary Axis Intersections

Use the Routh-Hurwitz criterion to determine $j\omega$-axis crossings of the root locus plots. Both the gain K^* and the value of ω^* may be found from the Routh table.

The s^2 row of the Routh table may be used to determine the value of ω^* for $j\omega$-axis crossings. Rule 3 is based on the second special case of the Routh table where $j\omega$-axis poles are encountered. Recall that all elements of a particular row are zero in this case. We wish to consider this situation when it happens for the s^1 row. Thus, the s^2 row may be used to write the auxiliary polynomial as indicated in Section 5.2. The procedure is to set the entry in the s^1 row to zero and solve for the gain K^* which yields roots on the $j\omega$-axis. This value of K^* is then substituted into the auxiliary polynomial $f_a(s)$, which is equal to zero. Using $s = j\omega^*$, we solve for ω^*. These operations were demonstrated earlier in Example 5.4. These first three root locus construction rules are illustrated in the examples that follow.

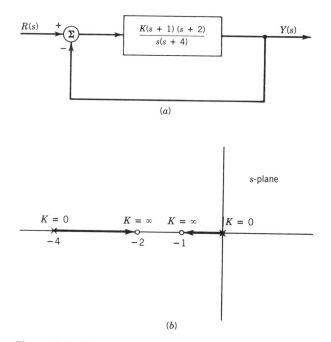

Figure 5.6 Closed-loop system diagram and s-plane root locus construction for Example 5.5.

EXAMPLE 5.5

We plot the location of the poles and zeros of $GH(s)$ from Figure 5.6a in the *s*-plane of Figure 5.6b. The root loci for the closed-loop system start on the poles and end on the zeros by Rule 1. We determine by Rule 2 that root loci occur on the real axis between the pole at -4 and the zero at -2 and between the zero at -1 and the pole at the origin. The complete root locus plot is the simple construction shown in Figure 5.6b.

EXAMPLE 5.6

The two poles and two zeros of $GH(s)$ from Figure 5.7a are plotted in the *s*-plane of Figure 5.7b. We find from Rule 2 that root loci occur on the real axis between

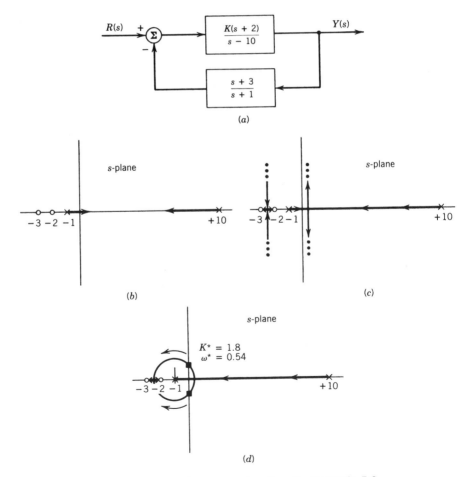

Figure 5.7   Root locus construction for Example 5.6.

TABLE 5.11   The Routh Array for Example 5.6

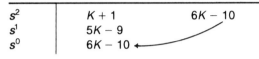

s^2	$K + 1$	$6K - 10$
s^1	$5K - 9$	
s^0	$6K - 10$	

the poles and between the zeros. From Rule 1, the root loci begin ($K = 0$) on the poles at -1 and $+10$. As K increases, these loci move toward each other along the real axis and at some point become equal. At this point the roots break away from the real axis to form two complex closed-loop poles, as shown in Figure 5.7c. At yet some larger value of K, the roots re-enter onto the real axis somewhere between the two zeros at -2 and -3. From this re-entry point the two roots separate and move along the real axis toward the zeros as $K \to \infty$. We reserve the details of how to compute these breakaway and re-entry points (Rule 6) until the next section.

We examine the Routh array to determine whether any $j\omega$-axis crossings occur. If the breakaway point on the real axis is to the right of the origin, then the two loci move across the $j\omega$-axis toward the re-entry point between -2 and -3. On the other hand, if the breakaway point occurs between the origin and -1, then there are no $j\omega$-axis crossings. The Routh array is given in Table 5.11. The element $\rho_{1,1}$ is positive for $K > 9/5 = 1.8$ and $\rho_{0,1}$ is positive for $K > 10/6 = 1.67$. Therefore, the root locus segment starting at -1 crosses into the right-half s-plane at the origin. The root is at the origin for $K = 1.67$. A breakaway point occurs for some slightly larger K, and the two loci cross the $j\omega$-axis moving into the left-half s-plane for $K^* = 1.8$. We form the auxiliary equation from the s^2 row as

$$f_a(s) = (K + 1)s^2 + (6K - 10)|_{K^* = 1.8} \tag{5.39}$$

Thus, ω^* satisfies

$$(K + 1)(j\omega^*)^2 + (6K - 10)|_{K^* = 1.8} = 0 \tag{5.40}$$

from which $\omega^* = 0.54$ radians. The complete root locus construction is shown in Figure 5.7d.

As a final remark on this example, we observe that root loci for second-order systems always occur as straight lines, circles, or segments thereof. Thus, the oval curve in Figure 5.7d can be shown to be a circle.

5.3.3 Asymptotes of Root Loci as $K \rightarrow \infty$

RULE 4: Asymptotes (for $p \neq z$)

Root locus plots are asymptotic to straight lines with angles given by

$$\theta_A = \frac{(2k + 1)\pi}{p - z}$$

as s approaches infinity. These straight lines intersect at a point σ_I on the real axis specified by

$$\sigma_I = \frac{\Sigma \text{ Poles of } GH(s) - \Sigma \text{ Zeros of } GH(s)}{p - z}$$

where p is the number of finite poles and z is the number of finite zeros of $GH(s)$.

Rule 4 provides an expression for the angle of the asymptotes as $K \rightarrow \infty$. When the number of finite poles (p) exceeds the number of finite zeros (z), one or more of the loci tend toward zeros at ∞ as $K \rightarrow \infty$. For example, if $p - z = 3$, the denominator of $GH(s)$ is of order three greater than the numerator, and we have asymptotes of $\pi/3$, π, and $-\pi/3$. The chart in Table 5.12 should be helpful for simple cases. Rule 4 also gives an expression for the intersection of the asymptotes with the real axis. For brevity, we omit the proof of Rule 4 here; it is based on approximating (5.30) for large K by retaining only s^z and s^{z-1} terms in the numerator and s^p and s^{p-1} terms in the denominator of $GH(s)$. The two examples which follow use the first four root locus construction rules.

TABLE 5.12 Angles of Asymptotes for Certain Values of $(p - z)$

$p - z$	Angles of Asymptotes (θ_A)
0	No Asymptote
1	π
2	$\pi/2, -\pi/2$
3	$\pi/3, \pi, -\pi/3$
4	$\pi/4, 3\pi/4, -\pi/4, -3\pi/4$
5	$\pi/5, 3\pi/5, \pi, -\pi/5, -3\pi/5$
$\vdots$	$\vdots$

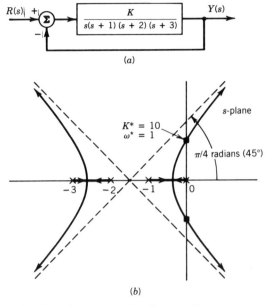

Figure 5.8   Root locus construction for Example 5.7.

EXAMPLE 5.7

The open-loop poles from $GH(s)$ in Figure 5.8a are plotted in the s-plane of Figure 5.8b. By Rules 1 and 2, the loci begin on the four poles, with the segment on the left formed by the loci at -2 and -3 approaching each other along the real axis, forming a breakaway point, and moving toward the asymptotes as $K \to \infty$. The segment on the right between -1 and the origin on the real axis is formed similarly, as shown in Figure 5.8b.

The asymptotes by Rule 4 (and Table 5.12) are

$$\theta_A = \frac{(2k+1)\pi}{p-z} = \frac{(2k+1)\pi}{4} \tag{5.41}$$

which yields $\pm\pi/4$ and $\pm3\pi/4$ radians. The intersection (σ_I) of these asymptotes with the real axis is given by

$$\sigma_I = \frac{-3-2-1-0}{p-z} = -\frac{6}{4} = -1.5 \tag{5.42}$$

Table 5.13 shows the Routh array. The entry $\rho_{1,1}$ is zero for $K^* = 10$, and ω^* is

TABLE 5.13   The Routh Array for Example 5.7

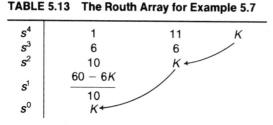

$$f_a(s) = 10s^2 + K$$

$$10(j\omega^*)^2 + K|_{K^*=10} = 0 \Rightarrow \omega^* = 1 \tag{5.43}$$

Therefore, the sketch in Figure 5.8*b* is complete except for determining the exact location of the breakaway points (Rule 6 to be discussed in Section 5.4).

EXAMPLE 5.8

The root locus plot of closed-loop poles in the s-plane as K varies from 0 to $+\infty$ is shown in Figure 5.9*b* for the closed-loop system in Figure 5.9*a*. There are two breakaway points and one re-entry point for root loci on the real axis. Two of the loci approach asymptotes of $\pm\pi/2$ radians with $\sigma_I = -3.5$, and the others approach the finite zeros at -2 and -3. There are no $j\omega$-axis crossings. The oval curve on the right is approximately circular in shape, though not precisely a circle since the system is not second order.

Figure 5.10 provides a number of root locus construction examples, indicating only shapes and omitting details such as $j\omega$-axis crossings and exact breakaway locations. Observe that zeros tend to attract root loci toward them and poles tend to repel them. These plots are included here to encourage the student to ponder the many possibilities that can occur in root locus construction. A broader class of problems having complex open-loop poles or zeros is considered in the following section.

5.4 ANGLES OF DEPARTURE AND BREAKAWAY POINTS

This section continues with the development and illustration of the root locus construction rules listed in Table 5.10. We emphasize Rules 5 and 6 in the five examples presented here.

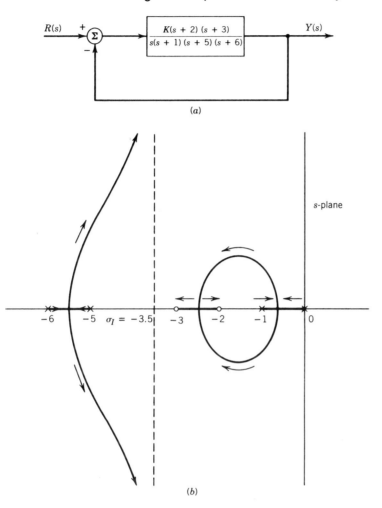

Figure 5.9 Root locus construction for Example 5.8.

RULE 5: Angles of Departure and Arrival

Assume a point s_θ arbitrarily near the pole (for departure) or the zero (for arrival) and then apply the fundamental angle relationship

$$\Sigma \text{ Angles of zeros of } GH(s) - \Sigma \text{ Angles of poles of } GH(s) = (2k + 1)\pi$$

Rule 5 is a direct application of the basic angle requirement for s-plane roots of the characteristic equation. The angle of $GH(s)$ must be π radians or any

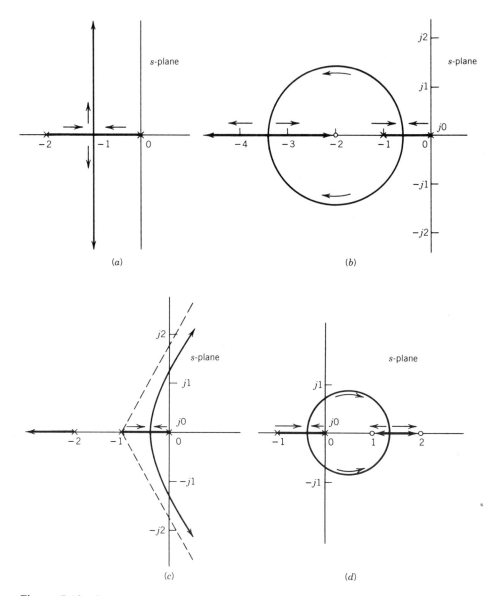

Figure 5.10 Root locus examples involving systems having real open-loop poles and zeros.

equivalent angle (by rotation) that has the same terminal side. A point on the root locus that is an infinitesimal distance away from a particular pole (or zero) is selected, and s-plane vectors are drawn from all poles and zeros to that point. We then use Rule 5 to compute the difference between the sum of the angles for the zeros and the sum of the angles for the poles and equate this difference to $(2k + 1)\pi$, where $k = 0, \pm1, \pm2, \ldots$. The angle of departure from the particular pole (or angle of arrival at the zero) is the only unknown in this equation.

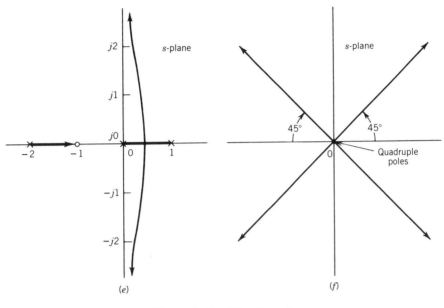

Figure 5.10 (*Continued*)

Examples 5.9 and 5.10 illustrate angle of departure and angle of arrival calculations.

RULE 6: Breakaway Points

Breakaway points may be determined by expressing the characteristic equation for the gain K as a function of s and then solving for the breakaway point(s) s_B from

$$\left.\frac{dK(s)}{ds}\right|_{s=s_B} = 0$$

Rule 6 presents one of many acceptable ways to calculate breakaway and re-entry points. We note that, when two root loci come together along the real axis, there is a relative maximum value of K at the breakaway point. Conversely, when two loci have "broken onto" the real axis, that is, formed a re-entry point, the value of K at that point is a relative minimum. Thus, by forming an algebraic expression for $K(s)$ from the characteristic equation and by setting its derivative (with respect to s) to zero, we obtain this relative extremum in either case. Examples 5.11 and 5.12 demonstrate the calculation of breakaway and re-entry points.

The fifth example of this section illustrates the interdependence of all six root locus construction rules. For each of the first four examples, Rules 5 and 6 provide supporting details for root locus construction, but a rough sketch of the

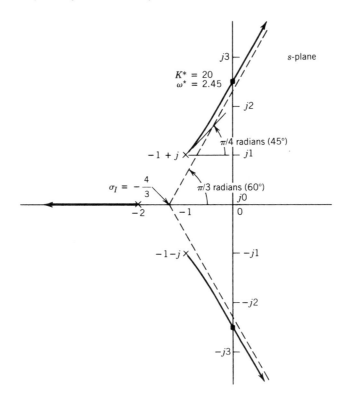

Figure 5.11 Root locus plot for Example 5.9.

root locus could be made by using only Rules 1 through 4. However, Example 5.13 requires the calculation of breakaway points even before the shape of the rough sketch of the root locus can be determined.

EXAMPLE 5.9

Consider the plant transfer function given by

$$GH(s) = \frac{K}{(s + 2)(s^2 + 2s + 2)} \tag{5.44}$$

which appears in a negative feedback configuration, as shown in Figure 5.5. We plot the open-loop pole locations at -2, $-1 + j$, and $-1 - j$ in the *s*-plane of Figure 5.11 and proceed to sketch the root locus as K varies from 0 to $+\infty$.

We first apply Rules 1 and 2 to note that the three roots start at the given poles and one root (the one at -2) moves along that section of the real axis that is to the left of -2. Next, we use Rule 4 to determine that the angles of the asymptotes for $p - z = 3 - 0 = 3$ are $+\pi/3$, $-\pi/3$, and π radians with an intersection of

TABLE 5.14   The Routh Array for Example 5.9

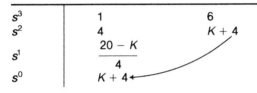

s^3	1	6
s^2	4	$K + 4$
s^1	$\dfrac{20 - K}{4}$	
s^0	$K + 4$	

these asymptotes with the real axis given by

$$\sigma_I = \frac{\Sigma \text{ Poles} - \Sigma \text{ Zeros}}{p - z} = \frac{(-2) + (-1 + j) + (-1 - j)}{3 - 0} = -\frac{4}{3} \quad (5.45)$$

The rough sketch shown in Figure 5.11 requires only Rules 1, 2, and 4; the root locus that starts on the pole at -2 moves to the left along the real axis toward infinity at an angle of π radians, and the root loci that start on the two complex poles move into the right-half s-plane toward asymptotes of $\pi/3$ and $-\pi/3$ radians, that is, 60° and $-60°$.

Rules 3 and 5 provide supporting details for the sketch. In applying Rule 3, we form the Routh array in Table 5.14 for the characteristic equation given by

$$1 + \frac{K}{(s + 2)(s^2 + 2s + 2)} = 0$$

$$s^3 + 4s^2 + 6s + K + 4 = 0 \quad (5.46)$$

We see that the entry in the s^1 row is zero when $K = K^* = 20$. Using the auxiliary equation from the s^2 row with $s = j\omega^*$ gives

$$4(j\omega^*)^2 + (K^* + 4) = 0 \quad (5.47)$$

which yields $\omega^* = \sqrt{6} \cong 2.45$ radians per second.

Finally, we compute the angle of departure (θ_x) from the pole at $s = -1 + j$. Referring to Rule 5, we need to evaluate

$$\Sigma \Big/ \text{Zeros of } GH(s) - \Sigma \Big/ \text{Poles of } GH(s) = (2k + 1)\pi \quad (5.48)$$

We consider a point on the locus leaving the pole at $-1 + j$ which is only an infinitesimal distance from this pole. The angle θ_x is formed by the s-plane vector drawn from $s = -1 + j$ to this point. The angles of s-plane vectors from the other two poles are $\pi/2$ radians for the pole at $-1 - j$ and $\pi/4$ radians for the pole at -2. Since there are no open-loop zeros, (5.48) becomes

$$-\left(\theta_x + \frac{\pi}{2} + \frac{\pi}{4}\right) = (2k + 1)\pi \quad (5.49)$$

which yields $\theta_x = \pi/4 - 2k\pi$. We may set k equal to any positive or negative integer in (5.49) to obtain θ_x. Using $k = 0$ gives $\theta_x = \pi/4$ radians. Since the root loci appear symmetrically about the real axis in the *s*-plane, we observe that the angle of departure from the pole at $-1 - j$ is $-\pi/4$ radians.

Inevitably, some students are concerned with whether the root locus crosses the asymptote of $\pi/3$ radians or always remains on the same side of this asymptote. Not only does this root locus remain above the asymptote, as shown in Figure 5.11, but we can also show the same result for the larger class of linear plants of the form given in (5.44), but with poles at $-a$, $-1 + j$, and $-1 - j$ for $0 < a < 1 + \sqrt{3}$. Moreover, when $a > 1 + \sqrt{3}$ the root locus that starts at $-1 + j$ remains below the asymptote of $\pi/3$ radians. The student should verify these results, including the special case for $a = 1 + \sqrt{3}$ where the root locus from the pole at $-1 + j$ coincides with the asymptote of $\pi/3$ radians. Although this exercise for a restricted form might be interesting, we can show other cases of plants with more general pole/zero configurations for which the root locus does cross the asymptote one or more times.

EXAMPLE 5.10

To illustrate calculations for both angles of departure from poles and angles of arrival at zeros, we consider the open-loop plant with transfer function given by

$$GH(s) = \frac{K(s^2 + 2s + 2)}{(s + 2)(s^2 - 2s + 2)} \qquad (5.50)$$

A sketch of the root locus is shown in Figure 5.12. Root loci start on the right-half

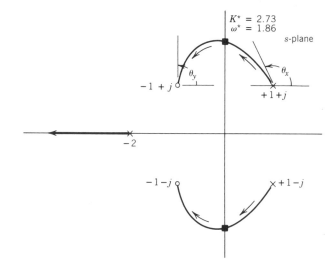

Figure 5.12   Root locus plot for Example 5.10.

TABLE 5.15 The Routh Array for Example 5.10

s^3	1	$2K - 2$
s^2	K	$2K + 4$
s^1	$\dfrac{2K^2 - 4K - 4}{K}$	
s^0	$2K + 4$	

s-plane poles at $1 + j$ and $1 - j$ and move across the $j\omega$-axis to end $(K \to \infty)$ on the zeros at $-1 + j$ and $-1 - j$. The locus of points for the third root starts at $s = -2$ and moves to the left along the real axis to infinity. Rules 2 and 4 apply to designate this section of the real axis for the root locus and an angle of π radians for the asymptote, since $p - z = 3 - 2 = 1$. With regard to $j\omega$-axis crossings (Rule 3), Table 5.15 shows the Routh array from which we compute $K^* \cong 2.73$ and $\omega^* \cong 1.86$ radians per second.

We let θ_x denote the angle of departure from the pole at $1 + j$, as shown in Figure 5.12. Using Rule 5, we obtain

$$\Sigma\underline{/\text{Zeros of } GH(s)} \;-\; \Sigma\underline{/\text{Poles of } GH(s)} \;=\; (2k + 1)\pi$$

$$\left(0 + \frac{\pi}{4}\right) - \left(\theta_x + \frac{\pi}{2} + \tan^{-1}\left(\frac{1}{3}\right)\right) = (2k + 1)\pi$$

$$\therefore \theta_x = 2.03 \text{ radians } (116.6°) \quad (5.51)$$

by using $k = -1$. We designate θ_y as the angle of arrival at the zero at $-1 + j$. From Rule 5, we have

$$\Sigma\underline{/\text{Zeros of } GH(s)} \;-\; \Sigma\underline{/\text{Poles of } GH(s)} \;=\; (2k + 1)\pi$$

$$\left(\theta_y + \frac{\pi}{2}\right) - \left(\pi + \frac{3\pi}{4} + \frac{\pi}{4}\right) = (2k + 1)\pi$$

$$\therefore \theta_y = \frac{\pi}{2} \text{ radians } (90°) \quad (5.52)$$

which was obtained by using $k = 0$. From symmetry about the real axis, the angle of departure from the pole at $1 - j$ is -2.03 radians and the angle of arrival at the zero at $-1 - j$ is $-\pi/2$ radians.

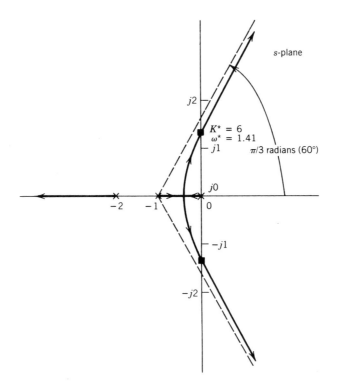

Figure 5.13 Root locus plot for Example 5.11.

EXAMPLE 5.11

We illustrate the calculation of breakaway points by using Rule 6 in this example. Let $GH(s)$ be given by

$$GH(s) = \frac{K}{s(s + 1)(s + 2)} \tag{5.53}$$

where the negative feedback configuration of Figure 5.5 is again applicable. Using Rule 2, we determine that root loci occur on that section of the real axis between $s = 0$ and $s = -1$ and to the left of $s = -2$. Observe the similarities and differences between this case and the one in Example 5.9. There are three poles and no zeros in both cases. Both have a pole at $s = -2$ and a root locus on the section of the real axis to the left of -2. While the other two poles were complex in Example 5.9 and hence required an angle of departure calculation, the $GH(s)$ in (5.53) has poles only on the real axis. As shown in Figure 5.13, the root loci that start at $s = 0$ and $s = -1$ approach each other, form a breakaway point at some point $s = s_B$ (to be determined), and approach asymptotes of $\pi/3$ radians and $-\pi/3$ radians (Rule 4). The intersection of these asymptotes with the real axis occurs at

$$\sigma_I = \frac{\Sigma \text{ Poles} - \Sigma \text{ Zeros}}{p - z} = \frac{(0) + (-1) + (-2)}{3 - 0} = -1 \qquad (5.54)$$

A rough sketch of the root locus in Figure 5.13 can be made without further calculations.

Additional details for a more precise sketch require a calculation of the $j\omega$-axis crossings (Rule 3) and the breakaway point (Rule 6). A $j\omega$-axis crossing is obtained for $K^* = 6$ at $\omega^* = \sqrt{2} \cong 1.41$ radians per second. In applying Rule 6, we first write the characteristic equation as

$$1 + \frac{K}{s(s + 1)(s + 2)} = 0 \qquad (5.55)$$

Then we solve (5.55) for K as a function of s to obtain

$$K(s) = -(s^3 + 3s^2 + 2s) \qquad (5.56)$$

Because $K(s)$ is a relative maximum at the breakaway point ($s = s_B$) somewhere between -1 and 0, we set $dK(s)/ds$ equal to zero to give

$$\left. \frac{dK(s)}{ds} \right|_{s = s_B} = 0 = -(3s^2 + 6s + 2) \qquad (5.57)$$

Solving (5.57) gives

$$s_B = \frac{-6 \pm \sqrt{36 - 4(2)(3)}}{2(3)} = -0.42, -1.58 \qquad (5.58)$$

The value of s_B between -1 and 0 is -0.42. We note that the other solution in (5.58) is a valid breakaway point for the root locus plot obtained when K varies over the range between 0 and $-\infty$. Cases involving negative K are considered in the following section.

EXAMPLE 5.12

To illustrate both breakaway and re-entry points in the same root locus sketch, we consider the linear plant with transfer function given by

$$GH(s) = \frac{Ks(s + 1)}{(s - 1)(s - 2)} \qquad (5.59)$$

with a negative feedback configuration. There are no asymptotes since we have an

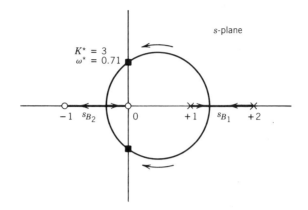

Figure 5.14 Root locus plot for Example 5.12.

equal number of poles and zeros ($p = z = 2$). According to Rule 2, root loci must occur along the real axis between the two poles at $+1$ and $+2$ and between the two zeros at -1 and 0. As shown in Figure 5.14, the two roots for this second-order system start at the poles at $+1$ and $+2$, move toward each other along the real axis, form a breakaway point (s_{B_1}), and proceed along a circular path into the left-half s-plane. The loci form a re-entry point (s_{B_2}) between -1 and 0 on the real axis and then separate and move along the real axis toward the zeros at -1 and 0.

The Routh table can be used to determine the $j\omega$-axis crossing for $K^* = 3$ at $\omega^* = \sqrt{2}/2 \cong 0.71$ radians per second. To determine the breakaway point and re-entry point, we find $K(s)$ from the characteristic equation as

$$K(s) = -\left(\frac{s^2 - 3s + 2}{s^2 + s}\right) \tag{5.60}$$

Setting $dK(s)/ds$ equal to zero yields

$$\frac{dK}{ds}\bigg|_{s=s_B} = -\left[\frac{(s^2 + s)(2s - 3) - (s^2 - 3s + 2)(2s + 1)}{(s^2 + s)^2}\right]\bigg|_{s=s_B} = 0 \tag{5.61}$$

which is satisfied when

$$4s_B^2 - 4s_B - 2 = 0 \tag{5.62}$$

Therefore, the breakaway point (s_{B_1}) is the positive root of (5.62) and the re-entry point (s_{B_2}) is the negative root. We find that $s_{B_1} = 1.37$ and $s_{B_2} = -0.37$, as shown in Figure 5.14.

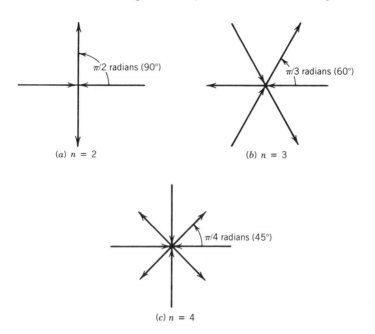

Figure 5.15 Angles of approach and departure for breakaway and re-entry points involving n roots.

In Examples 5.11 and 5.12, we considered only pairs of roots in the formation of breakaway points or re-entry points, that is, two roots coming together from opposite directions. In such cases these two roots depart from the breakaway or re-entry point at angles of $\pi/2$ radians from the directions of approach. In general, when n roots come together to form a breakaway or re-entry point s_B, these n roots depart from s_B at angles of π/n radians from their directions of approach. Figure 5.15 illustrates cases for $n = 2$, 3, and 4 roots.

EXAMPLE 5.13

As a final example in this section, we consider the linear plant transfer function given by

$$GH(s) = \frac{K(s + 1)}{s^2(s + a)} \tag{5.63}$$

where $a > 1$, and the negative feedback configuration of Figure 5.5 is used. We apply Rule 4 to determine that the angles of the asymptotes for $p - z = 3 - 1 = 2$ are $\pi/2$ and $-\pi/2$ radians. The intersection of these asymptotes with the real

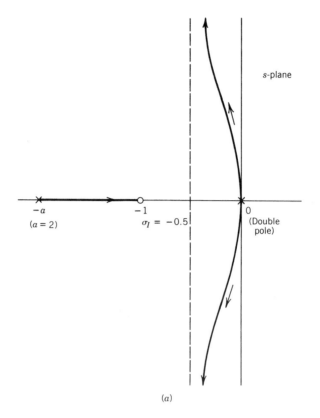

(a)

Figure 5.16 Root locus plots for $a = 2$ and $a = 50$ in Example 5.13.

axis is

$$\sigma_I = \frac{\Sigma \text{ Poles} - \Sigma \text{ Zeros}}{p - z} = \frac{(0) + (0) + (-a) - (-1)}{3 - 1} = \frac{1 - a}{2} \quad (5.64)$$

We use the Routh table to determine that there are no $j\omega$-axis crossings.

The shape of the root locus plots in the *s*-plane depends on the value of a, as shown for the two cases in Figure 5.16. For both cases the loci depart from the double pole at the origin at angles of $\pi/2$ and $-\pi/2$ radians (Rule 5) and then proceed into the second and third quadrants. By Rule 2, that section of the real axis between $-a$ and -1 must contain a segment of the root locus plot. However, the root locus sketch for small values of a (e.g., $a = 2$) has quite a different shape than the sketch for large values of a (e.g., $a = 50$). As shown in Figure 5.16a, for $a = 2$ the root loci move away from the double pole at the origin and proceed smoothly toward asymptotes of $\pi/2$ and $-\pi/2$ radians with $\sigma_I = -1/2$, while the third root moves directly from $-a$ to -1 along the real axis. However, as shown in Figure 5.16b, for $a = 50$, the root loci that start on the poles at the origin move around the zero at -1 in an approximately circular

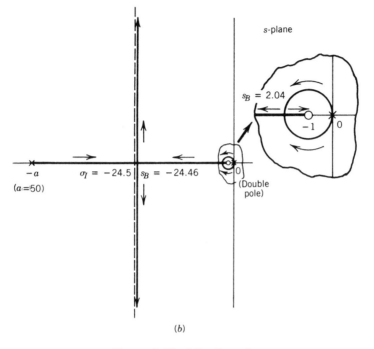

(b)

Figure 5.16 (*Continued*)

contour and form a re-entry point on the real axis to the left of -1. One of these roots then moves to the right along the real axis toward the zero (as $K \to \infty$). The other root moves to the left along the real axis and meets the third root, which started at $-a$, to form a breakaway point. From the breakaway point on the real axis, these two roots move toward asymptotes of $\pi/2$ and $-\pi/2$ radians with $\sigma_I = -24.5$.

We can determine these breakaway and re-entry points for $a = 50$ by using Rule 6. We form $K(s)$ from the characteristic equation as

$$K(s) = -\frac{s^2(s + a)}{s + 1} \tag{5.65}$$

Setting $dK(s)/ds$ equal to zero yields

$$\left.\frac{dK(s)}{ds}\right|_{s=s_B} = 0 = -\left[\frac{(s + 1)(3s^2 + 2as) - s^2(s + a)(1)}{(s + 1)^2}\right]\Bigg|_{s=s_B} \tag{5.66}$$

Therefore, the equation for the breakaway or re-entry points is

$$s_B\left(2s_B^2 + (a + 3)s_B + 2a\right) = 0 \tag{5.67}$$

We see that a breakaway point occurs at $s_B = 0$ (for $K = 0$). Solving the quadratic equation in (5.67) gives

$$s_B = \frac{-(a + 3) \pm \sqrt{(a + 3)^2 - 4(2)(2a)}}{2(2)} \tag{5.68}$$

For $a = 50$, (5.68) yields values of -2.04 and -24.46 for s_B. Thus, the re-entry point in Figure 5.16b is at -2.04 and the breakaway point at -24.46. From (5.38), the corresponding values of K are 191.9 and 651.3, respectively. For $a = 2$, (5.68) gives $s_B = -1.25 \pm j\sqrt{7/16}$, which is not a valid breakaway point because the root locus does not even pass through this point. Although it is possible to obtain breakaway points not on the real axis, such cases are rarely encountered. When these complex breakaway points do occur, we use Rule 6 to determine s_B and then verify by using (5.33) and (5.34) that $s = s_B$ is a point on the root loci.

Let us attempt to explain why large values of a (e.g., $a = 50$) yield a much different shape of root locus plots than small values of a (e.g., $a = 2$). For $a = 50$, we obtain a root locus plot for relatively small values of K which essentially neglects the effect of the pole at $-a$. Because a is so large, the double pole at the origin and the zero at -1 tend to dominate the root locus plot for small K. However, as K becomes larger, the root that started at $-a$ moves into the vicinity of another root which is moving to the left from the re-entry point of the root locus plot. In fact, as $K \to \infty$, these two roots tend to dominate the root locus plot. As an approximation for $K \to \infty$, we may even neglect the zero at -1 and the root (closed-loop pole) approaching it from the re-entry point on its left. In brief, we may think of the root locus plot for large a in Figure 5.16b as the combination of two plots: one for relatively small K which neglects the pole at $-a$ and another for relatively large K which neglects the zero at -1 and the closed-loop pole approaching it from the re-entry point on the left.

For small values of a (e.g., $a = 2$), Figure 5.16a is not easily decomposable into separate parts for small K and large K. We see from (5.64) that the intersection of the asymptotes with the real axis (σ_I) is determined from the location of the open-loop zero at -1 and the open-loop pole at $-a$, that is, $\sigma_I = (1 - a)/2$. This pole-zero pair tends to attract the root loci that start from the double pole at $s = 0$ because the zero (-1) is nearer these poles. Consequently, the root loci move into the second and third quadrants, as shown in Figure 5.16a. If we were to allow a to be 1, then cancellation would occur and the two root loci from the double pole at $s = 0$ would move along asymptotes coincident with the $j\omega$-axis. If the pole at $-a$ were to be placed between -1 and 0, then σ_I would be positive and the root loci from the double pole would move into the first and fourth quadrants and approach the asymptotes there.

Figure 5.17 shows the root locus plot for the case of $a = 9$, an intermediate value between $a = 2$ and $a = 50$. This particular value of a was determined from (5.68) by setting the quantity under the radical to zero. We see that (5.68) yields two complex roots for s_B with $1 < a < 9$, a double root $s_B = -3$ for $a = 9$, and

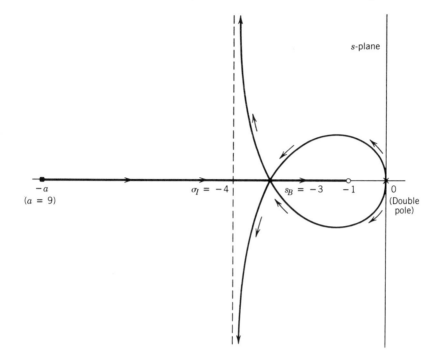

Figure 5.17 Root locus plot for $a = 9$ in Example 5.13.

two real roots for $a > 9$. Therefore, the shapes of all root locus plots for $1 < a < 9$ are similar to the plot in Figure 5.16a, and the shapes of all plots for $a > 9$ are similar to the one in Figure 5.16b. However, the root locus plot for $a = 9$ is quite different from either of the other two plots. The breakaway point at $s_B = -3$ has three roots coming together: the two roots that started at the double pole at the origin and the third root that started at $s = -9$. These roots approach the breakaway point from angles of π, $\pi/3$, and $-\pi/3$ radians. Moreover, these roots depart from this breakaway point at angles of 0, $2\pi/3$, and $-2\pi/3$ radians. These angles correspond to those shown in Figure 5.15b for $n = 3$, though the loci directions are opposite. As $K \to \infty$, the root at 0 radians proceeds along the real axis toward the zero at -1 and the other two roots approach asymptotes of $\pi/2$ and $-\pi/2$ radians, as shown in Figure 5.17.

In summary, we have considered the construction of a root locus plot that requires the calculation of breakaway points (Rule 6) even before the shape of the root locus plot can be determined. Without the results of Rule 6, we would not know whether the plot is similar to the one in Figure 5.16a or that in Figure 5.16b. In the critical case, we might even obtain the root locus plot that has a simultaneous re-entry and breakaway point involving three roots, as shown in Figure 5.17.

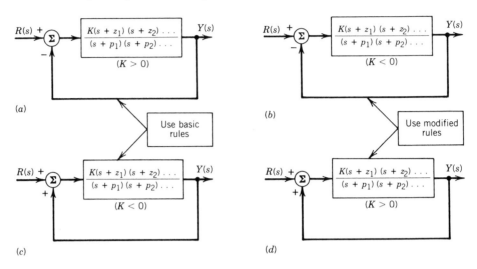

Figure 5.18 The four cases of positive and negative gains and feedback.

5.5 ROOT LOCI FOR NEGATIVE K OR POSITIVE FEEDBACK

The basic root locus construction rules of Table 5.10 are modified in this section to provide s-plane plots as K is varied from 0 to $-\infty$ for negative feedback. These modified rules are also applicable to positive feedback, with K varied from 0 to $+\infty$. The specific changes from the rules in Table 5.10 are that the word "odd" in Rule 2 is replaced with "even" and $(2k + 1)\pi$ is replaced by $2k\pi$ in Rules 4 and 5. Rules 1, 3, and 6 remain the same as before.

5.5.1 Development of Modified Root Locus Rules

Consider the four cases shown in Figure 5.18. Case (a), with $K > 0$ and negative feedback, and Case (c), with $K < 0$ and positive feedback, have the same characteristic equation. To show this equivalence, we express the characteristic equation for Case (c) as

$$1 - \frac{K(s + z_1)(s + z_2) \cdots}{(s + p_1)(s + p_2) \cdots} = 0 \qquad (5.69)$$

Since $K < 0$, we may set $K = -|K|$. Therefore, (5.69) becomes

$$1 - \left(\frac{-|K|(s + z_1)(s + z_2) \cdots}{(s + p_1)(s + p_2) \cdots} \right) = 0$$

$$1 + \frac{|K|(s + z_1)(s + z_2) \cdots}{(s + p_1)(s + p_2) \cdots} = 0 \qquad (5.70)$$

where is the same as (5.30) and (5.31) for Case (a) where $K > 0$. The basic root locus construction rules in Table 5.10 apply to both Cases (a) and (c).

Modified rules are needed for Cases (b) and (d), because the closed-loop system transfer function is different from that of Cases (a) and (c). The characteristic equation for Case (b) is

$$1 + \frac{K(s + z_1)(s + z_2) \cdots}{(s + p_1)(s + p_2) \cdots} = 0 \qquad (5.71)$$

for $K < 0$. Thus, $K = -|K|$ and (5.71) becomes

$$1 - \frac{|K|(s + z_1)(s + z_2) \cdots}{(s + p_1)(s + p_2) \cdots} = 0 \qquad (5.72)$$

which is equivalent to the characteristic equation for Case (d) when $K > 0$. We may express (5.72) as

$$GH(s) = \frac{|K|(s + z_1)(s + z_2) \ldots}{(s + p_1)(s + p_2) \ldots} = +1 \qquad (5.73)$$

from which

$$|GH(s)| = 1 \qquad (5.74)$$

$$\angle GH(s) = 2k\pi \qquad (k = 0, \pm 1, \pm 2, \ldots) \qquad (5.75)$$

Therefore, those basic construction rules that are based on the angle requirement must be changed for Case (b)—negative feedback and $K < 0$—and Case (d)—positive feedback and $K > 0$. In particular, Rule 2 has the word "even" in place of the word "odd" and $2k\pi$ replaces $(2k + 1)\pi$ in Rules 4 and 5. Table 5.16 gives the angles of the asymptotes (θ_A) for several values of ($p - z$), and Table 5.17 shows the modified root locus rules.

TABLE 5.16   Angles of Asymptotes for Modified Rules

$p - z$	Angle of Asymptotes (θ_A)
0	No asymptote
1	0
2	$0, \pi$
3	$0, 2\pi/3, -2\pi/3$
4	$0, \pi/2, \pi, -\pi/2$
5	$0, 2\pi/5, 4\pi/5, -2\pi/5, -4\pi/5$
$\vdots$	$\vdots$

TABLE 5.17   Modified Rules for Root Locus Construction

---

1.   STARTING AND ENDING POINTS

Root locus plots start ($K = 0$) on the open-loop poles and end ($|K| \rightarrow \infty$ or $K \rightarrow -\infty$) on the open-loop zeros.

2.   ROOT LOCUS SEGMENTS ON THE REAL AXIS

Root loci occur on a particular segment of the real axis if and only if there are an even number of total poles and zeros of the open-loop transfer function lying to the right of that segment.

3.   IMAGINARY AXIS INTERSECTIONS

Use the Routh-Hurwitz criterion to determine $j\omega$-axis crossings of the root locus plots. Both the gain K^* and the value of ω^* may be found from the Routh table.

4.   ASYMPTOTES (FOR $p \neq z$)

Root locus plots are asymptotic to straight lines with angles given by

$$\theta_A = \frac{2k\pi}{p - z}$$

as s approaches infinity. These straight lines intersect at a point σ_I on the real axis specified by

$$\sigma_I = \frac{\Sigma \text{ Poles of } GH(s) - \Sigma \text{ Zeros of } GH(s)}{p - z}$$

where p is the number of finite poles and z is the number of finite zeros of $GH(s)$.

5.   ANGLES OF DEPARTURE AND ARRIVAL

Assume a point s_θ arbitrarily near the pole (for departure) or the zero (for arrival) and then apply the fundamental angle relationship

$$\Sigma \text{ Angles of zeros of } GH(s) - \Sigma \text{ Angles of poles of } GH(s) = 2k\pi$$

6.   BREAKAWAY POINTS

Breakaway points may be determined by expressing the characteristic equation for the gain K as a function of s and then solving for the breakaway points s_B from

$$\left. \frac{dK(s)}{ds} \right|_{s = s_B} = 0$$

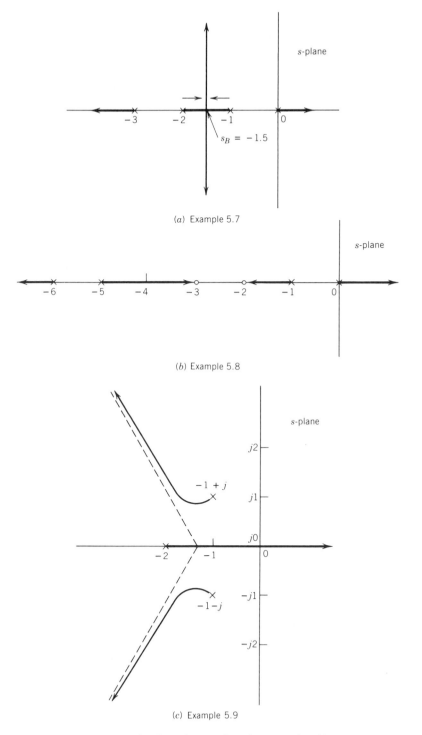

(a) Example 5.7

(b) Example 5.8

(c) Example 5.9

Figure 5.19 Root locus plots for negative *K*.

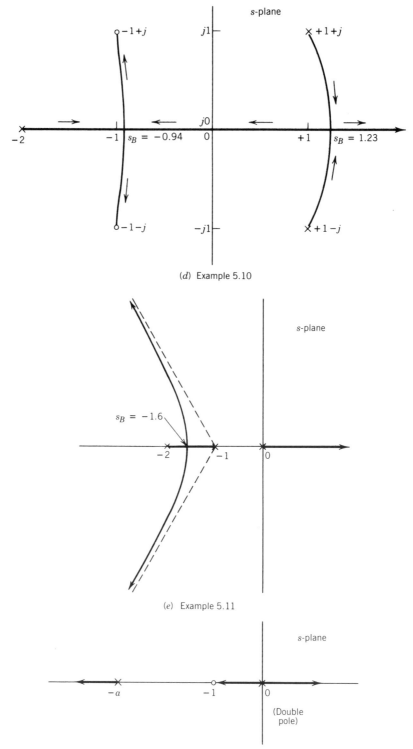

(d) Example 5.10

(e) Example 5.11

(f) Example 5.13

Figure 5.19 (*Continued*)   Root locus plots for negative K.

Figure 5.19 shows sketches of the root locus plots for Examples 5.7 through 5.11 and Example 5.13 for negative K—that is, as K in Figure 5.18b is varied from 0 to $-\infty$ for each example. Example 5.7 in Figure 5.19a has asymptotes at angles of 0, $\pi/2$, π, and $-\pi/2$ radians. Moreover, Rule 6 can be used to compute $s_B = -1.5$. Therefore, since $\sigma_I = -1.5$ also, the two loci at -1 and -2 move together along the real axis, then break away at -1.5, and are coincident with the asymptotes thereafter. Example 5.8 in Figure 5.19b may be sketched by using only Rules 1 and 2. On the other hand, Example 5.9 in Figure 5.19c requires the use of the first five rules. From Table 5.16, θ_A is 0, $2\pi/3$, and $-2\pi/3$ radians. We compute σ_I as

$$\sigma_I = \frac{(-1+j)+(-1-j)+(-2)}{p-z} = -\frac{4}{3} \tag{5.76}$$

The root at -2 moves along the real axis into the right-half s-plane. We determine the value of K for which the closed-loop pole is at the origin either by setting $\rho_{0,1} = K + 4 = 0$ in Table 5.14 or by using s-plane vectors, that is,

$$|K| = \frac{\text{Product of } s\text{-plane vector lengths}}{\text{from poles to origin}}$$

$$= |-1+j| \cdot |-1-j| \cdot |-2| = 4 \tag{5.77}$$

Therefore, we conclude that $K = -4$, since only the negative range of K is being considered here. The angle of departure from the pole at $-1 + j$ is $-3\pi/4$ radians. The difference in this angle and that obtained by using the basic root locus construction rules is π radians.

Figure 5.19d shows the negative-K root locus for Example 5.10. Loci leave the two complex poles and form a re-entry point on the real axis. One closed-loop pole moves to the right along the real axis toward the asymptote of 0 radians, and the other moves to the left to meet the locus coming from the open-loop pole at -2. They break away from the real axis and move toward the two complex open-loop zeros. From Rule 5, the angle of departure from the pole at $+1 + j$ is -1.11 radians $(-63.4°)$, and the angle of arrival at the zero at $-1 - j$ is $-\pi/2$ radians $(-90°)$. Both of these angles are π radians $(180°)$ away from the respective angles obtained from the basic rules for $K > 0$. Figures 5.19e and 5.19f are straightforward to interpret and as such require no further explanation.

EXAMPLE 5.14

Sketch the locus of closed-loop poles in the s-plane as K varies from $-\infty$ to $+\infty$ for the system of Figure 5.20. Provide all details, including exact breakaway points.

We will use the basic root locus rules in Table 5.10 to construct the sketch for $0 < K < +\infty$ and the modified rules in Table 5.17 for $-\infty < K < 0$. In both

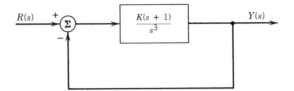

Figure 5.20 System for Example 5.14.

cases, we have the open-loop transfer function given by

$$GH(s) = \frac{K(s + 1)}{s^3} \tag{5.78}$$

An odd number of total poles and zeros lies to the right of the real axis section between -1 and the origin. Therefore, according to the basic rules ($K > 0$), one of the triple poles at the origin moves to the left along the real axis toward the open-loop zero at -1. The other two break away from the axis immediately at the origin, as shown in Figure 5.21a. We wish to determine the angles of departure from this triple pole. Using Rule 5, we have

$$\Sigma\underline{/\text{Zeros of } GH(s)} - \Sigma\underline{/\text{Poles of } GH(s)} = (2k + 1)\pi$$

$$0 - 3\theta_x = (2k + 1)\pi$$

$$\theta_x = \frac{(2k + 1)\pi}{-3} = \frac{\pi}{3}, \pi, -\frac{\pi}{3} \tag{5.79}$$

Thus, the departure angle of π radians already found by Rule 2 has now been reconfirmed. The other two poles depart at angles of $+\pi/3$ and $-\pi/3$ radians. By Rule 4, these two loci approach asymptotes of $+\pi/2$ and $-\pi/2$ radians with the intersection of the asymptotes, with the real axis given by

$$\sigma_I = \frac{0 + 0 + 0 - (-1)}{p - z} = \frac{1}{3 - 1} = +\frac{1}{2} \tag{5.80}$$

The complete sketch for $K > 0$ is shown in Figure 5.21a.

For $K < 0$, we note that the modified Rule 2 requires that an even number of total poles and zeros must be to the right of a permissible section of the real axis. Allowable sections lie to the left of -1 and to the right of the origin. Thus, we expect the triple pole at the origin to break apart, with one pole moving along the positive real axis, and the others to move off the real axis and form a re-entry point onto the real axis to the left of -1. We compute the angle of departure

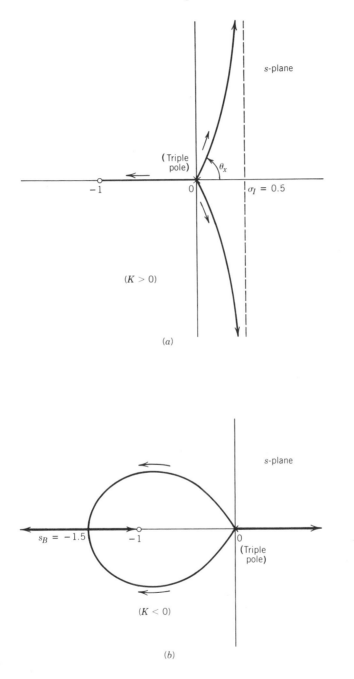

Figure 5.21 Root locus plots for Example 5.14.

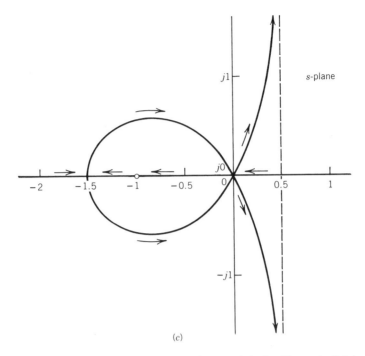

(c)

Figure 5.21 (*Continued*) Root locus plots for Example 5.14.

from the triple pole using the modified Rule 5 as

$$\Sigma \underline{/\text{Zeros of } GH(s)} - \Sigma \underline{/\text{Poles of } GH(s)} = 2k\pi$$

$$0 - 3\theta_x = 2k\pi$$

$$\theta_x = \frac{2k\pi}{-3} = 0, \frac{2\pi}{3}, -\frac{2\pi}{3} \qquad (5.81)$$

We write the characteristic equation as

$$1 + \frac{K(s+1)}{s^3} = 0 \qquad (5.82)$$

Solving for $K(s)$ yields

$$K(s) = \frac{-s^3}{(s+1)} \qquad (5.83)$$

The re-entry point s_B to the left of -1 may be found by Rule 6 as

$$\frac{dK(s)}{ds}\bigg|_{s=s_B} = \frac{d}{ds}\left[\frac{-s^3}{s+1}\right]\bigg|_{s=s_B} = -\left[\frac{(s+1)3s^2 - s^3(1)}{(s+1)^2}\right]\bigg|_{s=s_B} = 0 \quad (5.84)$$

from which $s_B = -1.5$. We note from Table 5.16 that the angles of the asymptotes are 0 and π radians. Therefore, after re-entering onto the real axis at $s_B = -1.5$, one closed-loop pole moves toward infinity along the negative real axis and the other moves toward the open-loop zero at -1. The third closed-loop pole, as we have already noted, moves along the positive real axis toward infinity. The sketch for $K < 0$ is shown in Figure 5.21b.

A composite of the two sketches for $K > 0$ and $K < 0$ is shown in Figure 5.21c. We can trace the movement of the closed-loop poles as K is varied from $-\infty$ to $+\infty$ by reversing the previous direction for the negative K portion of the plot, that is, by starting at $K = -\infty$ and proceeding toward $K = 0$. On reaching $K = 0$, we arrive at the open-loop poles. The remainder of the movement utilizes the basic root locus plot for $K > 0$, which may be traced in the ordinary direction used in its construction. Observe that we started on the left of the open-loop zero at -1 for $K = -\infty$ and arrived back at its right side for $K = +\infty$ with an angle

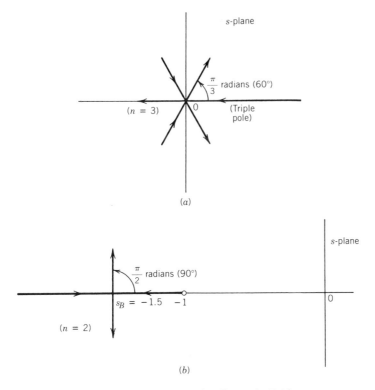

(a)

(b)

Figure 5.22 Breakaway points for Example 5.14.

difference of π radians. This characteristic of an angle difference of π radians also distinguishes the modified rules of Table 5.17 from the basic rules of Table 5.10.

Observe that the tracing of the closed-loop pole movement as K varies from $-\infty$ to $+\infty$ makes apparent the triple re-entry and immediate breakaway point at the origin as K passes from negative to positive values. For the triple pole, the direction of approach toward the origin for negative K is 0, $2\pi/3$, and $-2\pi/3$ radians, as shown in Figure 5.22a. The angles of breakaway (departure) from the origin are $\pi/3$, π, and $-\pi/3$, differing by $\pi/3$ radians (for $n = 3$), as shown in Figure 5.22a. On the other hand, the breakaway point at -1.5 involves only two closed-loop poles and the angular difference in direction of approach and departure in $\pi/2$ radians, as shown in Figure 5.22b.

EXAMPLE 5.15

The closed-loop system in Figure 5.23a represents a simplified model of a large rocket in its vertical position just prior to launch. The supports that held the rocket in place earlier have been removed. The resulting unstable model behaves much as an inverted pendulum, ready to crash to the ground with the slightest disturbance. We sketch the root locus for $K > 0$ (basic rules) in Figure 5.23b and for $K < 0$ (modified rules) in Figure 5.23c. We note that the angles of the asymptotes (θ_A) are $+\pi/2$ and $-\pi/2$ radians for $K > 0$ and 0 and π radians for $K < 0$. In Figure 5.23b, σ_I is -14. Observe that one closed-loop pole is in the right-half s-plane for all K between $-\infty$ and $+\infty$. Thus, we cannot stabilize this unstable system by simple gain adjustment.

The power of the root locus method for design is exhibited by an analysis of the modified feedback strategy of Figure 5.24a. We propose to feed back the output $Y(s)$ through the transfer function $(s + 1)/(s - 1)$ to combine negatively with $R(s)$. The corresponding $GH(s)$ is given by

$$GH(s) = \frac{Ks(s + 1)}{(s - 1)(s - 2)(s + 10)(s + 20)} \tag{5.85}$$

The basic root locus rules are applied for $K > 0$ in Figure 5.24b. The important difference from Figure 5.23b is that the loci starting at the two right-half s-plane open-loop poles at $+1$ and $+2$ come together, break away from the real axis between the two open-loop zeros, and proceed along the real axis toward those two zeros. The closed-loop system is asymptotically stable for $K > K^*$, where K^* and the associated ω^* can be obtained from the Routh array (Rule 3). The root locus plot for $K < 0$ is given in Figure 5.24c. As in Figure 5.23c, the closed-loop system is unstable for all $K < 0$.

We have stabilized the original system in Figure 5.23a by placing a second open-loop pole in the right-half s-plane, as shown in Figure 5.24a, and adjusting the gain K to yield all four closed-loop poles in the left-half s-plane. This

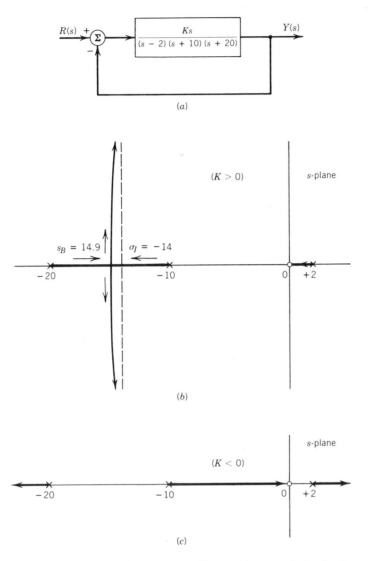

(a)

(b)

(c)

Figure 5.23 Closed-loop system diagram for a rocket prior to launch and associated root locus plots.

solution corresponds to the balancing of an inverted broom on the palm of your hand by moving the extended arm in a direction to restore any imbalance. Implementing this algorithm can prove to be an interesting project for the control engineering student. A "broom" or pole, resting on a small cart, can be constrained by wires so as to allow movement in only one plane. As the broom begins to lean toward one side, the angle and motion are sensed and the cart can be made to accelerate suddenly in the same direction until balance is restored.

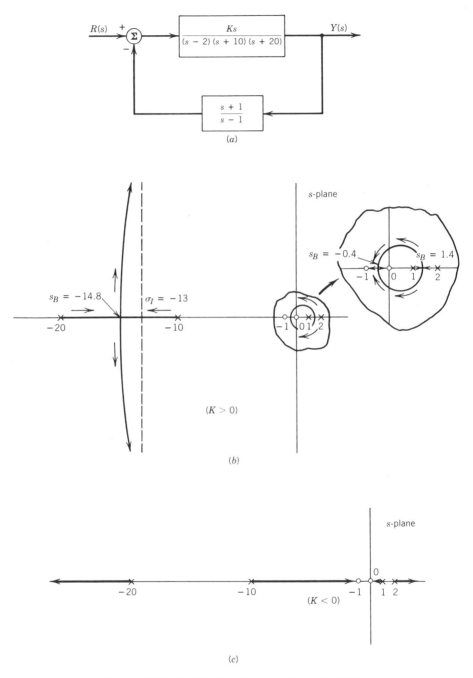

Figure 5.24 Modified design for Example 5.15.

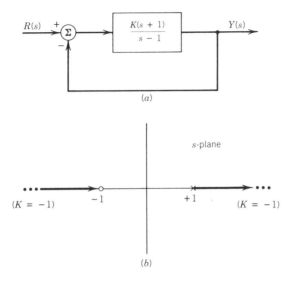

(a)

(b)

Figure 5.25 An illustration of asymptotes for finite *K*.

5.5.2 Asymptotes for Finite *K*

A special situation can occur for negative-*K* root loci when *p* equals *z*. To illustrate this situation, consider the simple case in Figure 5.25a. We may express the closed-loop transfer function as

$$\frac{Y(s)}{R(s)} = \frac{G(s)}{1 + G(s)} = \frac{\dfrac{K(s+1)}{s-1}}{1 + \dfrac{K(s+1)}{s-1}}$$

$$= \frac{K(s+1)}{(K+1)s + (K-1)} \tag{5.86}$$

Therefore, we see that for *K* = −1, *Y*(*s*)/*R*(*s*) becomes

$$\frac{Y(s)}{R(s)} = \frac{-(s+1)}{-2} = \frac{1}{2}(s+1) \tag{5.87}$$

The system order is 1 if *K* ≠ −1 but is reduced to zero for *K* = −1. From (5.86), the characteristic equation is given by

$$(K+1)s + (K-1) = 0 \tag{5.88}$$

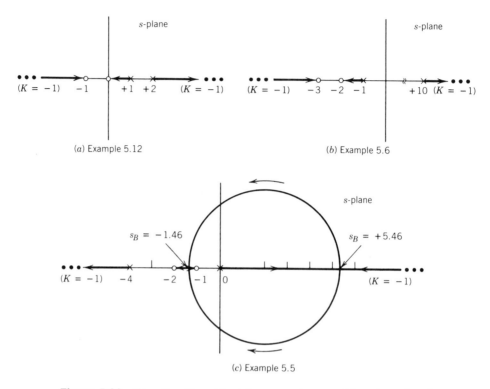

(a) Example 5.12

(b) Example 5.6

(c) Example 5.5

Figure 5.26   Negative-*K* root loci for examples considered previously.

which yields

$$s = \frac{1 - K}{1 + K} \tag{5.89}$$

The root locus for negative K is shown in Figure 5.25b. Rule 2 of the modified rules yields roots on the real axis to the right of the open-loop pole at $+1$ and to the left of the open-loop zero at -1. We see from (5.89) that the closed-loop pole actually moves along the real axis to the right of $+1$ toward infinity and then moves from infinity on the negative real axis toward the zero at -1. One might have previously asked whether closed-loop poles can transfer from the right half to the left half of the s-plane (or vice versa) without crossing the $j\omega$-axis. The system in Figure 5.25a provides a case where this transfer is possible without a finite crossing of the $j\omega$-axis.

Figure 5.26 shows the negative-K root loci for the systems of Examples 5.12, 5.6, and 5.5. The first two of these examples are straightforward applications of Rules 1 and 2. In both cases, $K = -1$ is the value of K for which an infinite closed-loop pole is obtained for finite K. Example 5.5 in Figure 5.26c presents two possible cases. In one case, the locus may move from the right-half s-plane at infinity on the real axis around to the left-half s-plane, then form a breakaway point, move in a circular arc to form a re-entry point between the two open-loop zeros, and finally move along the axis toward the two zeros. Another possibility is for the locus from the left pole to move around to the right-half s-plane, from a breakaway point there, and so forth. The decision between these cases may be determined by computing the location of the breakaway point s_B as

$$\left.\frac{dK(s)}{ds}\right|_{s=s_B} = \left.\frac{d}{ds}\left(\frac{-s(s+4)}{s^2+3s+2}\right)\right|_{s=s_B} = 0$$

$$\therefore -s_B^2 + 4s_B + 8 = 0 \tag{5.90}$$

from which $s_B = 2 \pm \sqrt{12} \cong -1.46$ and $+5.46$. Figure 5.26c shows the desired root locus plot. Observe that both breakaway and re-entry points are given by (5.90).

5.6 ROOT CONTOURS

The parameter to be varied in root locus construction may not necessarily be the open-loop gain K as in all previous examples of this chapter. We show in this section how to vary other parameters that appear linearly in the numerator or denominator of the open-loop transfer function $GH(s)$.

Consider the system of Figure 5.27a with the parameters K and α to be varied. We set α to some desired (nominal) value and proceed to vary K from $-\infty$ to $+\infty$ according to the procedures already established in Tables 5.10 (basic rules for $K > 0$) and 5.17 (modified rules for $K < 0$). Now suppose we want to set K at some nominal value and vary α from $-\infty$ to $+\infty$. We can apply the previous rules if we are able to determine another closed-loop system that has the same

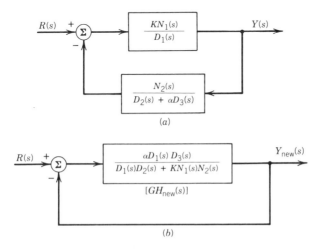

Figure 5.27 Two systems having the same character-istic equation.

characteristic equation as the given system. Moreover, this second system must have the parameter α appearing in the open-loop gain position. The characteristic equation for the system of Figure 5.27a is

$$1 + \left(\frac{KN_1(s)}{D_1(s)}\right)\left(\frac{N_2(s)}{D_2(s) + \alpha D_3(s)}\right) = 0 \tag{5.91}$$

which may be simplified to yield

$$D_1(s)D_2(s) + \alpha D_1(s)D_3(s) + KN_1(s)N_2(s) = 0 \tag{5.92}$$

To place α in the open-loop gain position, we divide (5.92) throughout by the sum of all terms not containing α, to obtain

$$1 + \frac{\alpha D_1(s)D_3(s)}{D_1(s)D_2(s) + KN_1(s)N_2(s)} = 0 \tag{5.93}$$

Thus, we have formed a second system that has an equivalent characteristic equation. The new open-loop transfer function is given by

$$GH_{\text{new}}(s) = \frac{\alpha D_1(s)D_3(s)}{D_1(s)D_2(s) + KN_1(s)N_2(s)} \tag{5.94}$$

A closed-loop diagram for the second system is shown in Figure 5.27b.

The two output responses of the systems are generally different, since the vectors **b** and **c** in their state representation are not the same. The eigenvalues of the matrix A, and hence the stability properties of the systems, are identical.

EXAMPLE 5.16

As a first example, we want to plot the root contour as the damping coefficient ζ for the second-order system in Figure 5.28a varies from $-\infty$ to $+\infty$. As in (5.91), we first form the characteristic equation as

$$1 + \frac{\omega_n^2}{s(s + 2\zeta\omega_n)} = 0 \tag{5.95}$$

which simplifies to yield

$$s^2 + 2\zeta\omega_n s + \omega_n^2 = 0 \tag{5.96}$$

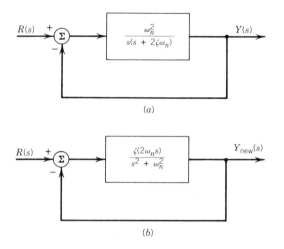

(a)

(b)

Figure 5.28 Two second-order systems having the same characteristic equation.

Dividing (5.96) throughout by the sum of all terms not containing ζ gives

$$1 + (\zeta)\frac{2\omega_n s}{s^2 + \omega_n^2} = 0 \tag{5.97}$$

Therefore, the new open-loop transfer function with ζ in the forward gain position is

$$GH_{new}(s) = \frac{\zeta(2\omega_n s)}{s^2 + \omega_n^2} \tag{5.98}$$

This $GH_{new}(s)$ is used in Figure 5.28b to yield a new closed-loop system, which has the same stability properties (characteristic equation) as the one in Figure 5.28a.

Root locus plots for $\zeta > 0$ (basic rules) and $\zeta < 0$ (modified rules) are shown in Figures 5.29a and 5.29b, respectively, and the combined plot is shown in Figure 5.29c. For the combined plot, the closed-loop poles for $\zeta = -\infty$ are at the open-loop zero at the origin and at $+\infty$ on the positive real axis. As ζ increases, these roots move together to form a breakaway point at $s_B = \omega_n$ for $\zeta = -1$ (Rule 6). The loci then move in a perfect circle around the open-loop zero at the origin, produce a $j\omega$-axis crossing (Rule 3) for $\zeta^* = 0$ and $\omega^* = \omega_n$, and form a re-entry point on the real axis at $s_B = -\omega_n$ for $\zeta = +1$ (Rule 6). One root then proceeds along the real axis toward the zero at the origin and the other moves to the left along the real axis to the asymptote of π radians (Rules 2 and 4 of basic rules). The closed-loop system is unstable for $\zeta < 0$, marginally stable for $\zeta = 0$, and asymptotically stable for $\zeta > 0$. Recall that in Section 4.2 we further classified the asymptotically stable time response as underdamped for $0 < \zeta < 1$, critically damped for $\zeta = 1$, and overdamped for $\zeta > 1$.

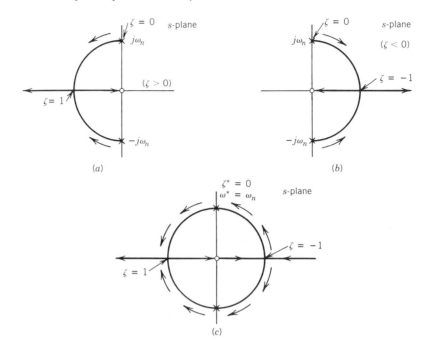

Figure 5.29   Root contour plots for Example 5.16.

EXAMPLE 5.17

As a preview of series compensation design by root locus procedures, consider the negative unity-feedback system shown in Figure 5.30a. A phase-lead series compensation network has been inserted into the loop with the linear plant to yield a combined transfer function given by

$$G_c G_p(s) = \left(\frac{s + z}{s + az} \right) \left(\frac{K}{s(s + 2)} \right) \tag{5.99}$$

where the compensation zero is placed at $-z$ and the pole at $-az$ for $a > 1$. While a design procedure described fully in Chapter 7 determines values of a and z to satisfy prespecified design criteria, our analysis here focuses on the construction of root contours for $K = 40$ as a varies from 1 to 6 and z from 2 to 6.

Figure 5.30b shows the root locus plot without compensation—that is, $G_c(s) = 1$—as K varies from 0 to $+\infty$. The points on the loci where $K = 40$ are determined as $s = -1 \pm j\sqrt{39}$. Root contours for $K = 40$ are determined in Figure 5.30c as a varies and in Figure 5.30d as z varies. These contours are

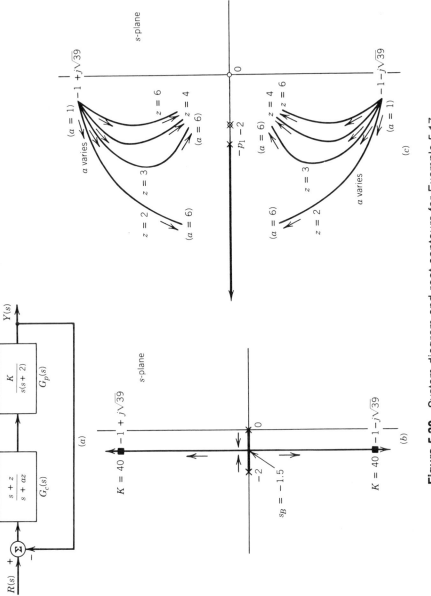

Figure 5.30 System diagram and root contours for Example 5.17.

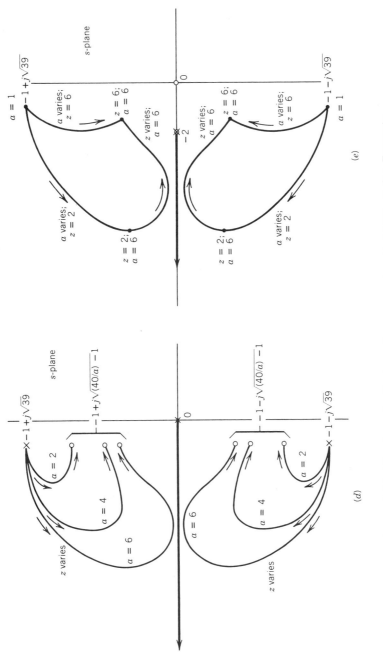

Figure 5.30 (*Continued*) System diagram and root contours for Example 5.17.

obtained by rearranging the characteristic equation for Figure 5.30a to obtain

$$1 + a\left[\frac{zs(s + 2)}{s^2(s + 2) + 40(s + z)}\right] = 0 \qquad (5.100)$$

and

$$1 + z\left[\frac{a(s^2 + 2s + 40/a)}{s^2(s + 2) + 40s}\right] = 0 \qquad (5.101)$$

The basic root locus construction rules in Table 5.10 are then applied to (5.100) and (5.101) to obtain the desired plots. The combined root contour curves in Figure 5.30e identify the region in the s-plane corresponding to these variations in both a and z.

SUMMARY

The stability of linear control systems has been examined in this chapter by considering s-plane and time-domain techniques. We began by stating the definitions of stability in the time domain and then relating these definitions to procedures in the s-plane. We showed that the Routh-Hurwitz criterion provides a necessary and sufficient condition to determine whether a given linear, time-invariant feedback system is stable. All closed-loop poles must lie in the left-half s-plane for asymptotic stability, but some may appear as single poles on the $j\omega$-axis for stability (but not asymptotic stability). Construction rules for root locus were presented in tables to aid in determining closed-loop pole locations in the s-plane as the loop gain K varies from $-\infty$ to $+\infty$. Root contours were sketched for the variation of other parameters by first deriving new closed-loop systems having the same characteristic equations and then applying these root locus construction rules.

REFERENCES

1. R. W. Brockett. *Finite-Dimensional Linear Systems.* New York: Wiley, 1970.
2. J. E. Gibson. *Nonlinear Automatic Control.* New York: McGraw-Hill, 1963, Chapter 8.
3. J. C. Hsu and A. U. Meyer. *Modern Control Principles and Applications.* New York: McGraw-Hill, 1968, Chapter 11.
4. R. E. Kalman and J. E. Bertram. "Control System Analysis and Design Via the Second Method of Liapunov." *Journal of Basic Engineering, Transactions of the ASME* (D), Vol. 82, (1960), pp. 371–400.

5. J. LaSalle and S. Lefschetz. *Stability by Liapunov's Direct Method*. New York: Academic Press, 1961.

6. K. S. Narendra and J. H. Taylor. *Frequency Domain Criteria for Absolute Stability*. New York: Academic Press, 1973.

7. Z. V. Rekasius. "Lagrange Stability of Nonlinear Feedback Systems." *IEEE Transactions on Automatic Control*, Vol. AC-8, No. 2 (April 1963), pp. 160–163.

8. J. L. Willems. *Stability Theory of Dynamical Systems*. New York: Wiley, 1970.

9. A. Hurwitz. "Über die Bedingungen, unter welchen eine Gleichung nur Wurzeln mit negativen reellen Teilen besitzt." *Mathematische Annalen*, Vol. 46 (1895), pp. 273–284.

10. E. J. Routh. *A Treatise on the Dynamics of a System of Rigid Bodies*. London: Macmillan, 1877.

11. J. J. D'Azzo and C. H. Houpis. *Linear Control System Analysis and Design: Conventional and Modern*, 2nd ed. New York: McGraw-Hill, 1981.

12. W. R. Evans. *Control System Dynamics*. New York: McGraw-Hill, 1954.

13. B. C. Kuo. *Automatic Control Systems*, 4th ed. Englewood Cliffs, N.J.: Prentice-Hall, 1982.

PROBLEMS

(§5.1) **5.1** The poles and zeros of four transfer functions are plotted in the *s*-plane in Figure 5.31. Which of the systems are asymptotically stable, which are marginally stable, and which are unstable?

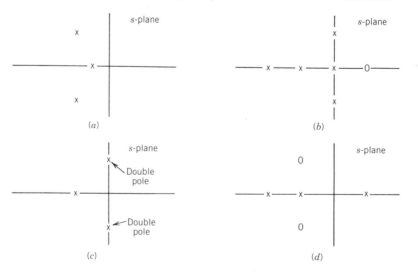

Figure 5.31 Four pole/zero configurations in the *s*-plane for Problem 5.1.

(§5.1) **5.2** Let the impulse response function $g(t)$ for a linear time-invariant system be given by

$$g(t) = e^{-t} - e^{-2t}$$

    a. Prove that the linear system satisfies Equation (5.3).

    b. Determine the system transfer function, locate the poles in the s-plane, and verify asymptotic stability.

    c. Express the system in direct phase variable form, determine the eigenvalues of the matrix A, and prove that the system is asymptotically stable from inspection of these eigenvalues.

(§5.2) **5.3** For the following characteristic polynomials for closed-loop systems, construct the Routh table and then apply the Routh-Hurwitz criterion to determine the number of closed-loop poles in the right-half s-plane, the number in the left-half s-plane, and the number on the $j\omega$-axis. In each case, specify whether the closed-loop system is asymptotically stable, marginally stable, or unstable.

    a. $f(s) = s^4 + 3s^3 + 2s^2 + s + 5$.
    b. $f(s) = s^5 + 2s^4 + s^3 + 2s^2 + s + 1$.
    c. $f(s) = s^5 + 2s^4 + 2s^3 + 4s^2 + 2s + 4$.
    d. $f(s) = s^5 + 2s^4 + 2s^3 + 4s^2 + s + 2$.

(§5.2) **5.4** For the system given in Figure 5.32, how many open-loop poles are in the right-half plane? Determine the range of K such that the closed-loop system is asymptotically stable.

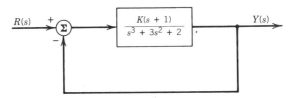

Figure 5.32   System diagram for Problem 5.4.

(§5.2) **5.5** Consider the system in Figure 5.33.

    a. Is the open-loop plant stable? Given your reason.

    b. For $K = 4$, is the closed-loop system stable?

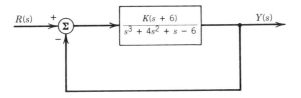

Figure 5.33   System diagram for Problem 5.5.

    c. Determine the entire range of K for which the closed-loop system is asymptotically stable.

(§5.2) **5.6** Use the Routh-Hurwitz criterion to determine the range of K for asymptotic stability for the following open-loop systems when placed in a negative unity-feedback configuration.

    a. $GH(s) = \dfrac{K(s+1)^2}{s(s-2)(s+3)}$

    b. $GH(s) = \dfrac{K(s+2)}{(s+1)^4}$

(§5.2) **5.7** Find the range of positive K for which the system of Figure 5.34 is stable.

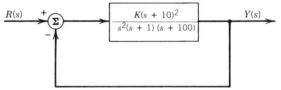

Figure 5.34 System diagram for Problem 5.7.

(§5.2) **5.8** Sometimes it is possible to stabilize an unstable open-loop plant by using negative feedback.

    a. Find the range of K for which the closed-loop system of Figure 5.35 is stable.

    b. Determine that value of K for which closed-loop poles are located on the $j\omega$-axis in the s-plane, and specify their exact location on the $j\omega$-axis.

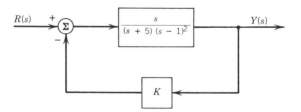

Figure 5.35 System diagram for Problem 5.8.

(§5.2) **5.9** In the first quadrant only of the K versus T plane ($K > 0, T > 0$), indicate the region of stability for the system given in Figure 5.36.

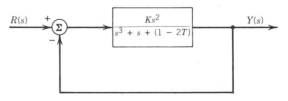

Figure 5.36 System diagram for Problem 5.9.

(§5.3) **5.10** Provide rough sketches of the root locus plots for $K > 0$ for systems having the following open-loop transfer functions. Assume negative-unity feedback in each case. Do not solve explicitly for angles of departure or arrival or for breakaway points (Rules 5 and 6). However, you are asked to indicate $j\omega$-axis crossings and the angles and real-axis intersections of the asymptotes (Rules 3 and 4).

    a. $GH(s) = \dfrac{K}{s(s + 1)(s + 5)}$

    b. $GH(s) = \dfrac{K(s + 1)}{s(s - 2)}$

    c. $GH(s) = \dfrac{K(s - 1)}{(s + 1)(s + 2)(s - 6)}$

(§5.3) **5.11** Sketch the root locus for the system of Figure 5.37, and specify the number of closed-loop poles in the right-half s-plane as K varies from 0 to ∞. (Do not find exact locations of breakaway points, if any.)

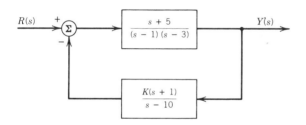

Figure 5.37 System diagram for Problem 5.11.

(§5.3) **5.12** Give a rough sketch of the root locus, that is, the locations of the closed-loop poles, in the s-plane as K varies from 0 to $+\infty$. You are not to find breakaway points, angles of departure or arrival, or $j\omega$-axis crossings. Only the general shape is required for each root locus plot. Assume negative feedback for each system.

    a. $G(s)H(s) = \dfrac{K(s + 2)(s + 3)}{(s + 4)(s^2 - 2s + 2)}$

    b. $G(s)H(s) = \dfrac{Ks}{(s + 3)(s^2 + 2s + 2)}$

    c. $G(s)H(s) = \dfrac{4K}{s^2(s + 1)^2(s + 2)}$

    d. $G(s)H(s) = \dfrac{K(s^2 + 2s + 2)}{s}$

(§5.3) **5.13** Sketch the root locus of the closed-loop poles in the s-plane for the system of Figure 5.38. Determine all $j\omega$-axis crossings, but do not specify exact breakaway points. K varies from 0 to $+\infty$.

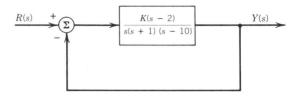

Figure 5.38 System diagram for Problem 5.13.

(§§5.3, **5.14** For the open-loop transfer functions given below, sketch the locus
5.4) of the closed-loop poles as K varies from 0 to $+\infty$. Assume
negative-unity feedback. Give all pertinent information, except the
exact location of breakaway points, if any.

   a. $G(s) = \dfrac{K(s + 1)(s + 2)(s + 3)}{(s + 4)(s + 5)(s + 6)(s + 7)}$

   b. $G(s) = \dfrac{K(s + 1)}{s(s - 1)(s + 3)}$

(§§5.3, **5.15** **a.** Sketch the root locus for the system in Figure 5.39 for $K > 0$.
5.4)   **b.** Determine the angle of arrival of the locus at the zero at
$s = +j$.
    c. Show that the root locus passes through the point $s = +1.0 + j0$, and calculate the value of K for which it does.

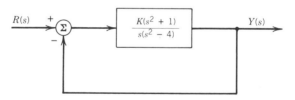

Figure 5.39 System diagram for Problem 5.15.

(§§5.3, **5.16** Consider the feedback control system in Figure 5.40. Determine the
5.4) root locus plot in the *s*-plane of the closed-loop poles as the gain K
varies from 0 to $+\infty$. Determine the location of all breakaway
points (if any) and any angles of departure or arrival (if appropriate). Determine all $j\omega$-axis crossings (if any), and specify the
value(s) of K^* and ω^* for which these occur. In other words,

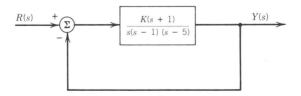

Figure 5.40 System diagram for Problem 5.16.

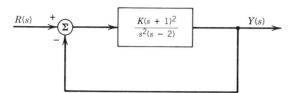

Figure 5.41 System diagram for Problem 5.17.

provide these and all other pertinent details associated with your root locus plot.

(§§5.3, **5.17** Investigate the stability of the closed-loop system in Figure 5.41 for
5.4) positive K by a detailed application of the root locus method. You may omit the calculation of any breakaway locations.

(§§5.3, **5.18** Give a detailed plot of the root locus for the systems shown in
5.4) Figure 5.42, and specify the number of closed-loop poles in the right-half plane as K varies from 0 to $+\infty$. Find all breakaways, angles of departure or arrival, and $j\omega$-axis crossings.

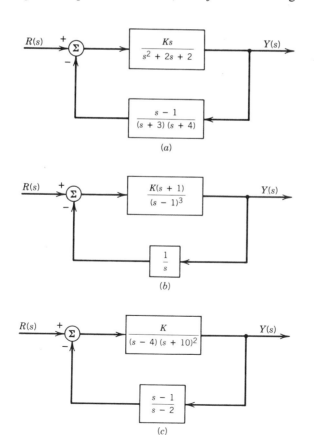

Figure 5.42 Closed-loop system diagrams for Problem 5.18.

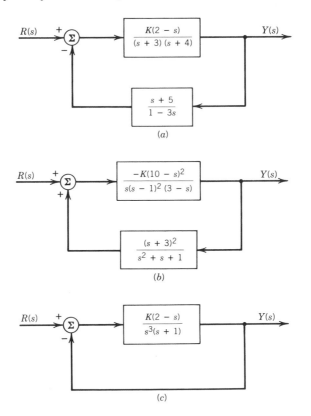

(a)

(b)

(c)

Figure 5.43 Closed-loop system diagrams for Problem 5.19.

(§§5.3, **5.19** Express $GH(s)$ for the systems of Figure 5.43 in "standard" form
5.4, so that the root loci may be plotted. In each case, determine
5.5) whether the basic rules or the modified rules apply when the given
gain K varies from 0 to $+\infty$. Do not plot the root loci.

(§§5.3, **5.20** For the system of Figure 5.44,
5.4,
5.5) **a.** Plot the root loci as K varies from 0 to $+\infty$. Specify $j\omega$-axis
crossings and ranges of K for stability, but do not solve
explicitly for breakaway points.
b. Repeat these instructions for the case where K varies from 0 to
$-\infty$.

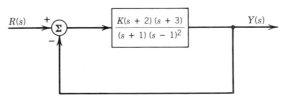

Figure 5.44 System diagram for Problem 5.20.

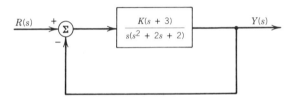

Figure 5.45 System diagram for Problem 5.21.

(§§5.3, **5.21** Consider the closed-loop system in Figure 5.45.
5.4,
5.5)
    a. Sketch the root locus in the s-plane as K varies from 0 to $+\infty$. Do not solve for breakaway points, if any occur.
    b. Specify those values of positive K for which closed-loop poles are on the $j\omega$-axis.
    c. What is the frequency of sustained oscillation (ω^*) when K is adjusted to that value found in Part (b)?
    d. Determine the angle at which the root locus departs from the pole at $-1 + j$.
    e. Sketch the locus of closed-loop poles in the s-plane as K varies from 0 to $-\infty$ (modified rules). Do not solve for breakaway points, if any occur.

(§§5.3, **5.22** Sketch the locus of the roots of the characteristic equation as K
5.4,
5.5)
varies from 0 to $+\infty$ for the six $GH(s)$ functions that follow. In each case, determine the range of positive K for stability. Find all breakaway points, angles of departure or arrival, and all other pertinent details. Assume negative feedback, as in Figure 5.5.

    a. $GH(s) = \dfrac{K(s + 1)}{s^2(s + 2)(s + 4)}$

    b. $GH(s) = \dfrac{K(s + 1)}{s^2(s - 2)}$

    c. $GH(s) = \dfrac{K(s + 10)^2}{s^2(s + 1)(s + 100)}$

    d. $GH(s) = \dfrac{-K}{s(s^2 - 2s + 2)}$

    e. $GH(s) = \dfrac{K(s + 1)^2}{(s - 1)^2(s + 4)}$

    f. $GH(s) = \dfrac{Ks^2(s + 1)}{(s + 2)(s - 1)(s - 2)}$

(§§5.3, **5.23** Sketch the root locus plot for each system in Figure 5.46. Specify
5.4,
5.5)
ranges of $K > 0$ for stability, but do not find exact locations of breakaways, if any.

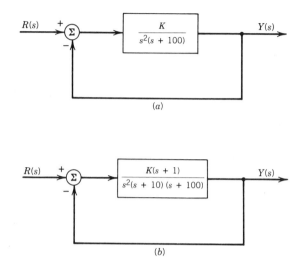

(a)

(b)

Figure 5.46 System diagrams for Problem 5.23.

(§§5.3, **5.24** Determine the range of positive K for stability for the system of
5.4, Figure 5.47. Sketch the root locus plot for $0 < K < \infty$.
5.5)

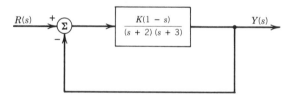

Figure 5.47 System diagram for Problem 5.24.

(§§5.3, **5.25** For the system in Figure 5.48, give a rough sketch (shape only; no
5.4, details required) of the locations of the poles of $Y(s)/R(s)$ as the
5.5) parameter A is varied from 0 to $+\infty$.

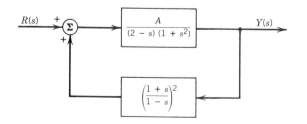

Figure 5.48 System diagram for Problem 5.25.

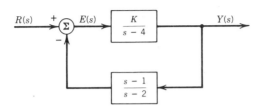

Figure 5.49 System diagram for Problem 5.26.

(§§5.3, **5.26** Consider the closed-loop control system shown in Figure 5.49.
5.4,   Sketch the s-plane locations of the closed-loop poles (i.e., root locus)
5.5)   as K varies from $-\infty$ to $+\infty$. Note that both basic and modified
rules are needed. Determine all $j\omega$-axis crossings and breakaway
points.

(§§5.3, **5.27** For the system shown in Figure 5.50, provide a root locus sketch of
5.4,   the closed-loop pole locations in the s-plane as
5.5)
a. K is varied from 0 to $+\infty$.
b. K is varied from 0 to $-\infty$.
In both parts, determine all $j\omega$-axis crossings, but do not specify
exact breakaway points.

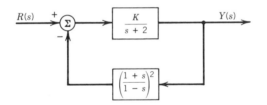

Figure 5.50 System diagram for Problem 5.27.

(§§5.3, **5.28** Plot the root locus of the closed-loop poles in the s-plane as K
5.4,   varies from $-\infty$ to $+\infty$ for the $GH(s)$ that follows. Provide all
5.5)   details, such as $j\omega$-axis crossings, angles of departure, and breakaways.

$$GH(s) = \frac{Ks(s+1)}{(s-1)(s-2)(s+3)(s+4)(s+5)}$$

(§§5.3, **5.29** For the system of Figure 5.51, use the root locus method to plot the
5.4,   closed-loop pole locations in the s-plane as K varies from $-\infty$ to
5.5) $+\infty$. Supply all pertinent information.

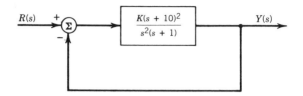

Figure 5.51   System diagram for Problem 5.29.

(§§5.3, **5.30**   Consider the closed-loop control system shown in Figure 5.52.
5.4,   Sketch the *s*-plane locations of the closed-loop poles (i.e., root locus)
5.5)   as *K* varies from $-\infty$ to $+\infty$. Note that both basic and modified
             rules are needed. Determine all $j\omega$-axis crossings and breakaway
             points.

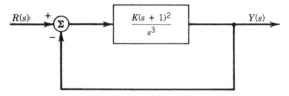

Figure 5.52   System diagram for Problem 5.30.

(§5.6) **5.31**   Sketch the root contour in the *s*-plane for the system of Figure 5.53
             as *T* varies from 0 to $+\infty$, with $K = +3$.

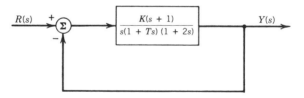

Figure 5.53   System diagram for Problem 5.31.

(§5.6) **5.32**   Consider the system in Figure 5.54. Plot the location of the closed-
             loop poles as *b* varies from $-\infty$ to $+\infty$. Provide all pertinent
             information on your plot.

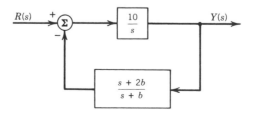

Figure 5.54   System diagram for Problem
5.32.

6

Stability Analysis in the Frequency Domain

A sinusoidal steady-state output is obtained when an asymptotically stable linear system is subjected to a sinusoidal input. The linear system generally modifies the amplitude and phase of the entering sinusoid; the frequency ω is unchanged. We investigate the stability of closed-loop linear systems in this chapter by using frequency response methods. Initially, we define frequency response performance specifications and then apply these to underdamped second-order systems. We identify four kinds of frequency response plots and present contour mapping as a preliminary step in developing the Nyquist criterion. Several examples of this criterion are presented, together with detailed comparisons with the root locus method. In fact, we combine our knowledge of these two powerful procedures to determine $j\omega$-axis crossings (root locus) and phase crossover frequencies (Nyquist criterion) in the simplest way. We complete the chapter by considering relative stability and the closed-loop frequency response.

6.1 FREQUENCY RESPONSE PERFORMANCE SPECIFICATIONS

Performance specifications relative to frequency response curves include bandwidth, peak magnitude, gain margin, and phase margin [1–4]. The closed-loop system bandwidth and the closed-loop system peak magnitude are usually of interest in feedback analysis and design. These specifications are indicated on the closed-loop magnitude and phase curves of Figure 6.1a. Gain margin and phase margin specifications pertain to the open-loop plant frequency response curves, as shown in Figure 6.1b. We require that this open-loop plant be placed in a negative unity-feedback configuration. Each of these four frequency response specifications is defined more precisely below.

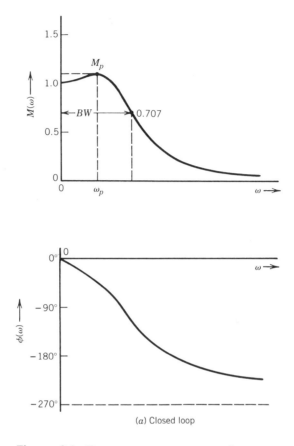

Figure 6.1 Frequency response performance specifications.

6.1.1 Bandwidth

The closed-loop system bandwidth (BW) is defined as that value of frequency at which the magnitude of the closed-loop system frequency response is reduced to $1/\sqrt{2}$ of its low frequency value.[1] Recall that we calculated this closed-loop system bandwidth in the gain-versus-bandwidth discussion of Section 3.2. Whether a large or small bandwidth is required depends on the particular application. In some cases, a large bandwidth is needed to enable the system to respond properly for a wide frequency range of input signals and control commands. In other cases, a small bandwidth is needed to provide attenuation of high-frequency noise on the input signal.

---

[1]Alternate definitions of bandwidth are sometimes used in other applications. For example, the bandwidth of tuned circuits is often defined as the frequency range between those points on the magnitude curve that are located at $1/\sqrt{2}$ of the peak value.

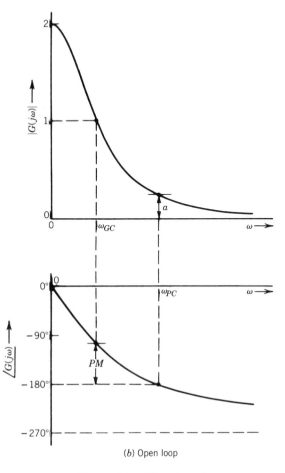

(b) Open loop

Figure 6.1 (*Continued*)

6.1.2 Peak Magnitude

The closed-loop system peak magnitude (M_p) is a measure of the relative stability of the system. Values of 1.2 to 1.5 are usually not considered excessive for most control system applications, and these "rule-of-thumb" values are often used for design purposes.

6.1.3 Gain Margin

The gain margin is a performance specification that is based on the open-loop response but that provides analysis and design information about the closed-loop system response. It is assumed that the open-loop plant is placed in a negative unity-feedback configuration. The gain margin (GM) is defined as

$$GM = \frac{\text{Maximum allowable open-loop gain for closed-loop system stability}}{\text{Actual open-loop gain}}$$

$$(6.1)$$

For example, if the open-loop gain K is 10 and the closed-loop system would become marginally stable for $K = K^* = 60$, then the gain margin is $60/10 = 6$. A gain margin of 4 or greater is an arbitrary "rule-of-thumb" in many design applications, although lower values are sometimes used.

Observe that the definition of gain margin in (6.1) is not restricted to a frequency response interpretation. In fact, we could apply the Routh-Hurwitz criterion (Rule 3 of the root locus construction rules) to determine K^*. However, we show later in this chapter that, for a stable open-loop plant, the closed-loop system is marginally stable if and only if the open-loop frequency response magnitude curve is unity at the same frequency where the phase is 180°. We define a phase crossover frequency (ω_{PC}) as that ω at which the phase is 180°. Figure 6.1*b* shows open-loop plant response curves with a magnitude of $a < 1$ at $\omega = \omega_{PC}$. Since the actual open-loop gain could be increased by a factor of $1/a$ to reach unity, we say that the gain margin is $1/a$ for this case. Careful reflection on these procedures should reveal that this result is consistent with (6.1).

6.1.4 Phase Margin

In accordance with the above discussion, we identify the gain crossover frequency (ω_{GC}) as that ω at which the open-loop frequency response magnitude is unity. Figure 6.1*b* shows a typical case. We define the phase margin (PM) as

$$PM = 180° - \left| \angle G(j\omega_{GC}) \right| \qquad (6.2)$$

where $\angle G(j\omega_{GC})$ is expressed as a negative angle, usually in the third quadrant. As another measure of relative stability, the phase margin indicates the amount of additional phase allowable in a stable open-loop plant frequency response before the point of marginal stability for the closed-loop system is reached.

Four performance specifications have been defined in this section for evaluating system behavior in the frequency domain. This overview of analysis and design objectives is intended to structure our thinking toward first guaranteeing closed-loop system stability and then ensuring that a desired degree of relative stability is maintained. These performance measures are applied to second-order under-damped systems in Section 6.2, with a more extensive explanation of their significance in general design cases provided in Section 6.6.

6.2 UNDERDAMPED SECOND-ORDER SYSTEMS

This section applies the performance specifications defined in Section 6.1 to the underdamped system ($0 < \zeta < 1$) of Figure 6.2. Typical open-loop and closed-loop frequency response curves for this system are shown in Figure 6.3. Closed-loop system bandwidth and peak magnitude are calculated by using the closed-loop frequency response curves, while gain and phase margin computations are performed by referring to the open-loop frequency response curves.

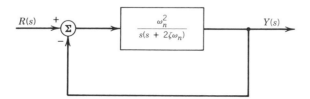

Figure 6.2 An underdamped second-order system.

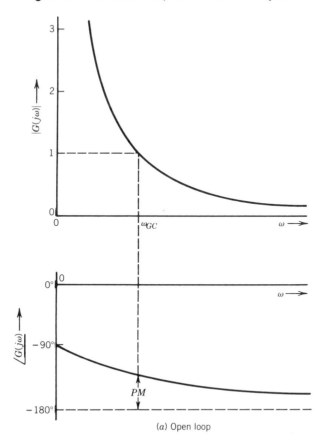

(a) Open loop

Figure 6.3 Open-loop and closed-loop frequency response curves for the system of Figure 6.2.

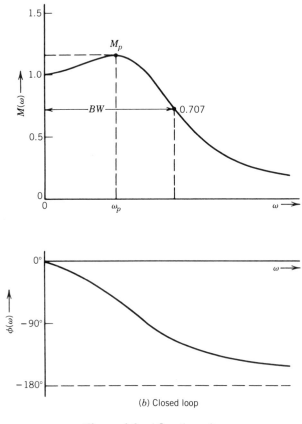

(b) Closed loop

Figure 6.3 (*Continued*)

6.2.1 Open-Loop and Closed-Loop Frequency Responses

Setting $s = j\omega$ for the open-loop transfer function $G(s)$ gives

$$G(j\omega) = \frac{\omega_n^2}{(j\omega)(j\omega + 2\zeta\omega_n)} = \frac{\omega_n/2\zeta}{(j\omega)(1 + j\omega/2\zeta\omega_n)} \qquad (6.3)$$

The corresponding open-loop magnitude and phase relationships are

$$|G(j\omega)| = \frac{\omega_n/2\zeta}{\omega\sqrt{1 + (\omega/2\zeta\omega_n)^2}}$$

$$\underline{/G(j\omega)} = -90° - \tan^{-1}\left(\frac{\omega}{2\zeta\omega_n}\right) \qquad (6.4)$$

Solving for the closed-loop transfer function gives

$$\frac{Y(s)}{R(s)} = \frac{\omega_n^2}{s^2 + 2\zeta\omega_n s + \omega_n^2} \tag{6.5}$$

Setting $s = j\omega$, the closed-loop magnitude and phase expressions are

$$M(\omega) = \left| \frac{Y(j\omega)}{R(j\omega)} \right| = \frac{1}{\sqrt{\left(1 - \frac{\omega^2}{\omega_n^2}\right)^2 + \left(\frac{2\zeta\omega}{\omega_n}\right)^2}}$$

$$\phi(\omega) = \angle \frac{Y(j\omega)}{R(j\omega)} = -\tan^{-1}\left[\frac{\frac{2\zeta\omega}{\omega_n}}{1 - \frac{\omega^2}{\omega_n^2}}\right] \tag{6.6}$$

6.2.1.1 Bandwidth

We determine the closed-loop system bandwidth (BW) as that frequency at which $M(\omega)$ in (6.6) is equal to $1/\sqrt{2}$. Performing the indicated operation yields

$$BW = \omega_n\sqrt{1 - 2\zeta^2 + \sqrt{4\zeta^4 - 4\zeta^2 + 2}} \tag{6.7}$$

A curve of the normalized closed-loop bandwidth (BW/ω_n) versus ζ is shown in Figure 6.4.

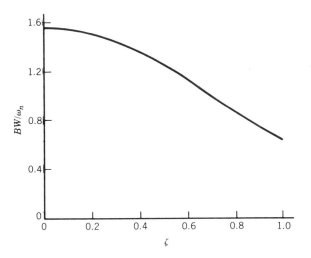

Figure 6.4 Normalized closed-loop bandwidth (BW/ω_n) versus ζ for the system of Figure 6.2.

6.2.1.2 Peak Magnitude

For the closed-loop curve in Figure 6.3b, we wish to find M_p and the frequency (ω_p) at which the peak occurs. Using $M(\omega)$ from (6.6), we set $dM(\omega)/d\omega$ equal to zero for $\omega = \omega_p$ and simplify to obtain

$$\omega_p = \omega_n\sqrt{1 - 2\zeta^2} \qquad (6.8)$$

where $0 < \zeta < 1/\sqrt{2}$. Substituting (6.8) into $M(\omega)$ in (6.6) yields

$$M_p = \frac{1}{2\zeta\sqrt{1 - \zeta^2}} \qquad (6.9)$$

Plots of M_p and (ω_p/ω_n) versus ζ for the closed-loop system are given in Figure 6.5.

6.2.1.3 Gain Margin

Both the gain margin (GM) and the phase margin (PM) for the second-order system in Figure 6.2 are measured with respect to the open-loop frequency response in Figure 6.3a. Since the open-loop phase $\angle G(j\omega)$ in (6.4) varies only over the range from $-90°$ to $-180°$, the gain margin is infinite. As $\omega \rightarrow \infty$, $\angle G(j\omega)$ approaches $-180°$, and thus the phase crossover point occurs at $\omega = \infty$ with a magnitude of zero. Consequently, GM is infinite for this case.

6.2.1.4 Phase Margin

The phase margin (PM) is measured at the gain crossover frequency ω_{GC} defined as that frequency at which the magnitude of the open-loop frequency response function is unity. Setting $|G(j\omega_{GC})| = 1$ from (6.4), we obtain

$$\omega_{GC} = \omega_n\sqrt{\sqrt{4\zeta^4 + 1} - 2\zeta^2} \qquad (6.10)$$

From (6.4), the corresponding open-loop phase at the gain crossover frequency is given by

$$\angle G(j\omega_{GC}) = -90° - \tan^{-1}\left[\frac{1}{2\zeta}\sqrt{\sqrt{4\zeta^4 + 1} - 2\zeta^2}\right] \qquad (6.11)$$

The phase margin may then be expressed from (6.2) as

$$PM = 180° - \left|\angle G(j\omega_{GC})\right|$$

$$= 90° - \tan^{-1}\left[\frac{1}{2\zeta}\sqrt{\sqrt{4\zeta^4 + 1} - 2\zeta^2}\right] \qquad (6.12)$$

Curves of phase margin and (ω_{GC}/ω_n) versus ζ are shown in Figure 6.6.

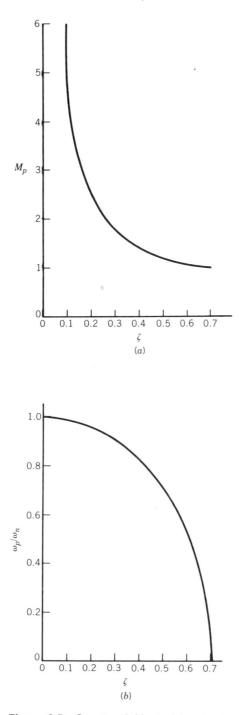

Figure 6.5 Curves of M_p and (ω_p / ω_n) versus ζ for the closed-loop frequency response in Figure 6.3b.

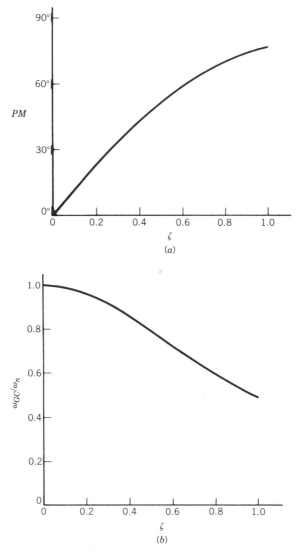

Figure 6.6 Curves of phase margin (*PM*) and normalized gain crossover frequency (ω_{GC} / ω_n) versus ζ for the system of Figure 6.2.

EXAMPLE 6.1

Determine the frequency response specifications identified above if $\omega_n = 10$ radians/second and $\zeta = 0.1$ for the system of Figure 6.2.

By direct evaluation from (6.7), the closed-loop bandwidth (*BW*) is 15.5 radians/second. From (6.8) and (6.9), the closed-loop peak magnitude (M_p) is

5.0, and this peak value occurs at a frequency (ω_p) of 9.9 radians/second. The phase margin (PM) in (6.12) is 11.4°, occurring at the gain crossover frequency (ω_{GC}) of 9.9 radians/second, and the gain margin is infinite.

6.3 FREQUENCY RESPONSE PLOTS

Four types of open-loop frequency response plots are described and compared in this section. The first of these is the linear plot used in the previous two sections. We next introduce Bode diagrams and log-log plots, which are both based on the properties of logarithms. These curves are useful for quick frequency response sketches or for more detailed analysis and design. The fourth type of frequency response representation is the polar plot, which is used in applying the Nyquist criterion described later in the chapter.

Frequency response curves of magnitude and phase versus frequency may be plotted directly on linear scales. Although an advantage is the directness of this procedure, a major disadvantage usually identified for linear plots is the cumbersome computations. However, the use of calculators or personal computers has tended to reduce this objection to some extent for individual point calculations. Nevertheless, the shapes of curves for linear plots are not as easily determined as those for Bode diagrams or log-log plots.

Bode diagrams are plotted on semi-log paper with linear vertical scales and logarithmic horizontal scales. The logarithm to the base 10 of the open-loop frequency response function magnitude is computed as a first step. Magnitude values in decibels, defined as $20 \log_{10} |G(j\omega)|$, are then entered on the vertical scale. The logarithmic property allows us to form the resultant Bode magnitude curve by summing the contributions due to individual factors. The Bode phase curve has a linear vertical scale in degrees and a logarithmic horizontal curve for frequency. The phase curve is usually sketched beneath the magnitude curve with frequency values aligned. Magnitude and phase curves for four types of factors (constant gains, linear factors, poles or zeros at the origin, and quadratic factors) are sketched on semi-log paper in Figure 6.7.

We begin the construction of Bode magnitude and phase curves by expressing the open-loop frequency response function in the "standard" form of (3.16) with $s = j\omega$, as shown in Figure 6.7a. For a constant gain factor of the form K in Figure 6.7b, the Bode magnitude is $20 \log_{10} |K|$, and the phase is 0° for $K > 0$ and $\pm 180°$ for $K < 0$. The Bode magnitude curve in Figure 6.7c, for a linear factor of the form $(1 \pm j\omega T)$, has a low-frequency asymptote of zero decibels and a straight-line high-frequency asymptote that intersects the horizontal axis at the corner frequency $\omega = 1/T$. Alternatively, we may express the linear factor as $(1 \pm j\omega/\omega_n)$, where ω_n is the corner frequency. The actual magnitude curve differs from the asymptotes by approximately three decibels at the corner frequency, and this difference is smaller for all other values of ω. When a linear

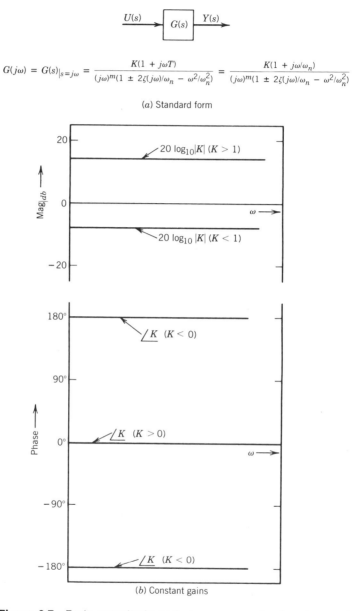

$$G(j\omega) = G(s)|_{s=j\omega} = \frac{K(1 + j\omega T)}{(j\omega)^m(1 \pm 2\zeta(j\omega)/\omega_n - \omega^2/\omega_n^2)} = \frac{K(1 + j\omega/\omega_n)}{(j\omega)^m(1 \pm 2\zeta(j\omega)/\omega_n - \omega^2/\omega_n^2)}$$

(a) Standard form

(b) Constant gains

Figure 6.7 Bode magnitude and phase curves for four types of factors.

factor appears in the numerator of the open-loop frequency response function, we form a straight-line high-frequency asymptote that has a slope of $+20$ decibels per decade. This asymptote has a slope of -20 db/decade when a linear factor appears in the denominator. The phase is $+\tan^{-1}(\omega T)$ both for $(1 + j\omega T)$ in the numerator and for $(1 - j\omega T)$ in the denominator; it is $-\tan^{-1}(\omega T)$ for $(1 - j\omega T)$ in the numerator and for $(1 + j\omega T)$ in the denominator.

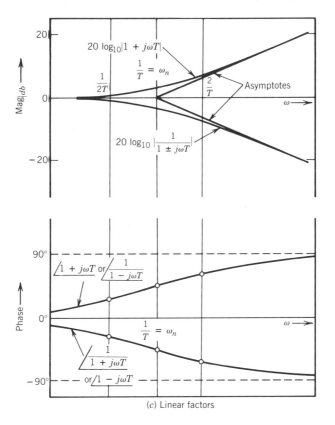

(c) Linear factors

Figure 6.7 (*Continued*)

Bode magnitude curves in Figure 6.7*d* for poles or zeros at the origin, that is, $(j\omega)^m$, are straight lines with slopes of $-20m$ db/decade, where m is the number of poles (for positive m) or the number of zeros (for negative m) at the origin. The corresponding phase is a constant of $(-90°)m$ for all ω. Finally, quadratic factors of the form $(1 \pm 2\zeta(j\omega)/\omega_n - \omega^2/\omega_n^2)$ are considered in Figure 6.7*e*. While different curves are obtained for different values of the damping coefficient ζ, observe that low-frequency asymptotes of zero decibels and high-frequency asymptotes of ± 40 db/decade are obtained for all Bode magnitude curves. Moreover, Bode phase curves are given by $\pm \tan^{-1}[(2\zeta\omega/\omega_n)/(1 - \omega^2/\omega_n^2)]$ and vary between 0° and 180° or between 0° and $-180°$, depending on whether the term appears in the numerator or denominator and whether a plus or minus sign appears before the term involving ζ. The distinction can be discerned in a manner analogous to the linear factor cases. For convenience, only a denominator term with a plus sign is considered in Figure 6.7*e*. Observe that a quadratic factor can be expressed as the product of two linear factors for $\zeta > 1$.

Log-log plots, as the name implies, require logarithmic scales on both vertical and horizontal axes. The advantage in plotting magnitude curves is identical to that described for Bode diagrams—that is, contributions of individual factors are

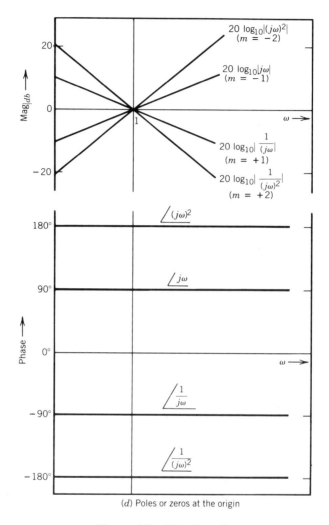

(d) Poles or zeros at the origin

Figure 6.7   (*Continued*)

summed to form the resultant curve. A disadvantage is that the phase curve requires a linear vertical scale. Consequently, sketching the phase curve on the log-log paper beneath the magnitude curve presents a minor problem usually resolved by modifying the vertical log scale to yield a linear scale.

Polar plots for open-loop plants are necessary for understanding and applying the Nyquist criterion for stability analysis in the frequency domain. Figure 6.8 shows a typical polar plot. The gain crossover frequency (ω_{GC}) occurs at a phase of approximately $-110°$, and the phase crossover frequency (ω_{GC}) occurs at a magnitude of a. Therefore, we may determine the phase margin as $70°$ and the gain margin as $1/a$. The polar plot is plotted more easily by first sketching the corresponding Bode diagrams.

In summary, Bode plots are recommended over log-log plots for magnitude curves because of the presence of a linear vertical scale, which is also needed for

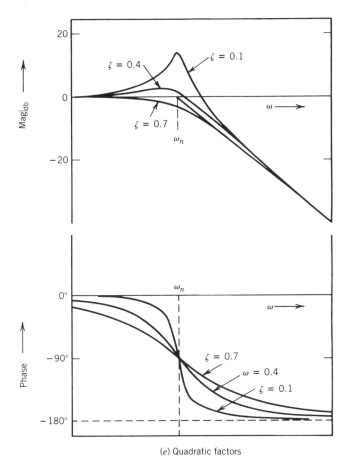

(e) Quadratic factors

Figure 6.7 (*Continued*)

phase plots. Both of these plots are preferred over direct linear plots that can be determined point-by-point by using electronic calculators but that are more difficult to sketch without a large number of these computations. Easily sketched from Bode diagrams, polar plots are needed to apply the Nyquist criterion described in Section 6.5.

EXAMPLE 6.2

Sketch linear plots, Bode diagrams, log-log plots, and polar plots for each of the open-loop transfer functions given below for $s = j\omega$.

$$\text{a.} \quad G(s) = \frac{1000(s + 1)}{(s + 2)(s + 10)(s + 20)}$$

$$\text{b.} \quad G(s) = \frac{100(s - 1)}{s(s + 2)^2}$$

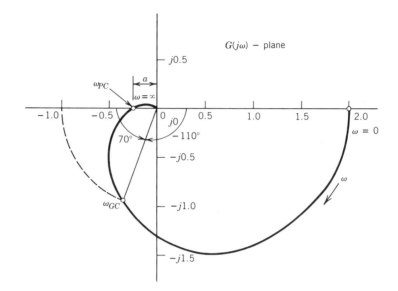

Figure 6.8   A typical polar plot showing gain and phase margins.

To obtain the frequency response function for (a), we substitute $s = j\omega$ to form

$$G(j\omega) = \frac{1000(j\omega + 1)}{(j\omega + 2)(j\omega + 10)(j\omega + 20)} \tag{6.13}$$

The magnitude for the linear plot is given by

$$|G(j\omega)| = \frac{1000\sqrt{\omega^2 + 1}}{\sqrt{\omega^2 + 2^2}\sqrt{\omega^2 + 10^2}\sqrt{\omega^2 + 20^2}} \tag{6.14}$$

and the phase by

$$\angle G(j\omega) = \tan^{-1}(\omega) - \tan^{-1}\left(\frac{\omega}{2}\right)$$

$$- \tan^{-1}\left(\frac{\omega}{10}\right) - \tan^{-1}\left(\frac{\omega}{20}\right) \tag{6.15}$$

These curves are plotted in Figure 6.9a. We discuss procedures for determining the peak magnitude, gain margin, and phase margin in Section 6.6.

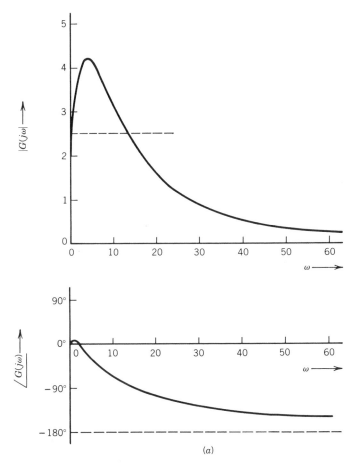

Figure 6.9   Frequency response plots for Example 6.2(a).

To obtain the Bode diagrams and log-log plots, we first express (6.13) in the "standard" form

$$G(j\omega) = \frac{\left(\dfrac{1000}{2(10)(20)}\right)(1 + j\omega)}{\left(1 + \dfrac{j\omega}{2}\right)\left(1 + \dfrac{j\omega}{10}\right)\left(1 + \dfrac{j\omega}{20}\right)}$$

$$= \frac{2.5(1 + j\omega)}{\left(1 + \dfrac{j\omega}{2}\right)\left(1 + \dfrac{j\omega}{10}\right)\left(1 + \dfrac{j\omega}{20}\right)} \qquad (6.16)$$

The Bode and log-log magnitude plots differ only by a scale factor and hence may

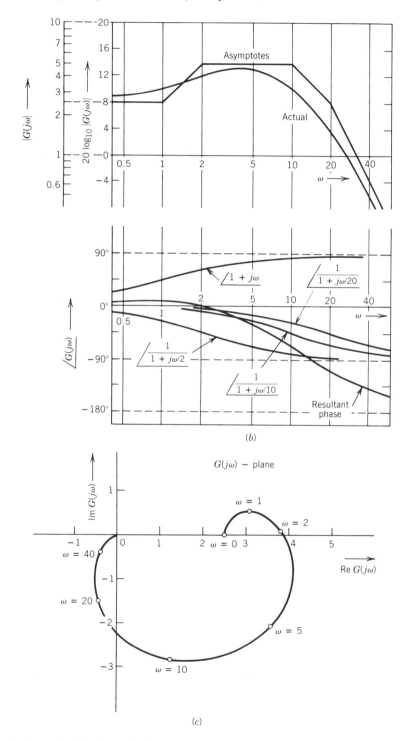

Figure 6.9 (*Continued*) Frequency response plots for Example 6.2(a).

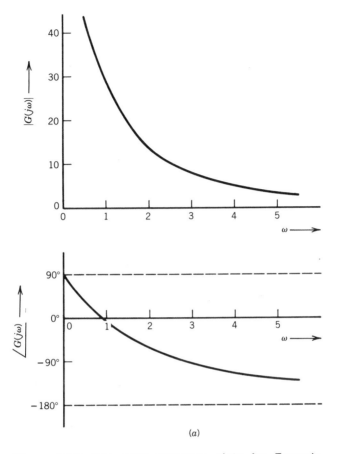

Figure 6.10 Frequency response plots for Example 6.2(b).

be plotted as the same curve with two different scales, as shown in Figure 6.9b. The corresponding phase diagram is identical for Bode and log-log plots.

The polar plot for (a) in Figure 6.9c starts ($\omega = 0$) at a magnitude of 2.5 and phase of 0°. We use the Bode plots of Figure 6.9b to help us sketch this polar plot. As ω increases above zero, the phase goes positive briefly and the magnitude increases, resulting in a plot that is in the first quadrant of the $G(j\omega)$-plane. For larger values of ω, the phase returns to 0° at a magnitude somewhat larger than 2.5 and then continues to decrease toward $-180°$ as $\omega \to \infty$. The Bode magnitude plot decreases steadily, thus indicating a zero magnitude as $\omega \to \infty$. This frequency response is typical for an open-loop plant that passes signals having frequencies below 20 radians/second and attenuates those with higher frequencies.

The required frequency response plots for (b) are given in Figure 6.10. We say that a stable plant transfer is "minimum phase" if its zeros are located in the left-half s-plane. The transfer function for (a) is minimum phase. On the other hand, we classify a stable plant transfer function as "nonminimum phase" if one

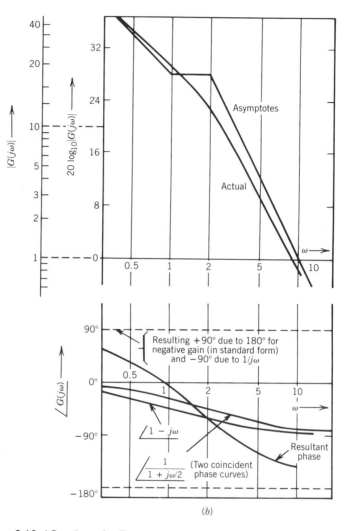

Figure 6.10 (*Continued*) Frequency response plots for Example 6.2(b).

or more of its zeros are located in the right-half *s*-plane. The case in (b) is an example of a nonminimum phase transfer function.

6.4 CONTOUR MAPPINGS

Mappings of a function $F(s)$ from the *s*-plane to the $F(s)$-plane are considered in this section as the point *s* moves around a closed contour in the *s*-plane. We show that the net number of encirclements (N) of the origin in the $F(s)$-plane is equal

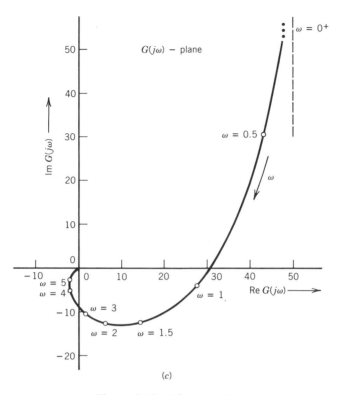

Figure 6.10 (*Continued*)

to the number of zeros (Z) minus the number of poles (P) contained within the s-plane closed contour. This principle, based on the net change of angle of $F(s)$, is central to the development of the major frequency-domain stability criterion, which is presented in the following section.

Consider the function $F(s)$ given by

$$F(s) = \frac{K(s + z_1)(s + z_2)\ldots(s + z_m)}{(s + p_1)(s + p_2)\ldots(s + p_n)} \tag{6.17}$$

The locations of the zeros and poles of $F(s)$ are shown in Figure 6.11. Let us arbitrarily select a closed contour in the s-plane that encircles only some of the zeros and poles of $F(s)$ and does not pass through any. We begin at Point P_1 and evaluate $F(s)|_{s=P_1}$ by using s-plane vectors. This value of $F(P_1)$ is then plotted in the $F(s)$-plane. We continue by calculating $F(P_i)$ for other points P_i along a closed contour in the s-plane and by plotting the results in the $F(s)$-plane.

Let us examine the properties of the net angle change for the $F(s)$-plane contour. We see from Figure 6.12a that if we have encircled only a single zero (and no poles) in proceeding around a closed contour in the s-plane, then the net change in angle of $F(s)$ is 2π radians. This net angle change of 2π radians for $F(s)$ corresponds to a single encirclement of the origin in the $F(s)$-plane.

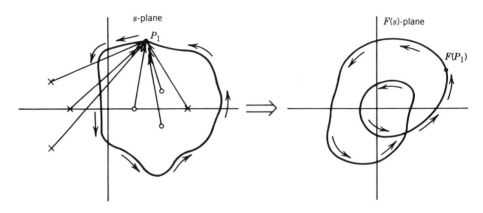

Figure 6.11 Mapping from the s-plane into the $F(s)$-plane for the function in (6.17).

Moreover, the direction of this encirclement is the same as the direction of movement around the closed contour in the s-plane. For example, if the s-plane contour is counterclockwise about a single zero, then the corresponding contour in the $F(s)$-plane is also counterclockwise once about the origin. Observe from Figure 6.12b that zeros of $F(s)$ lying outside the given contour in the s-plane do not contribute to the net angle change for $F(s)$ as the contour is traversed.

We may express the angle of $F(s)$ in (6.17) as

$$\underline{/F(s)} = \sum_{i=1}^{m} \underline{/(s + z_i)} - \sum_{i=1}^{n} \underline{/(s + p_i)} \tag{6.18}$$

As we observed in Figure 6.12a, the net change in $\underline{/(s + z_i)}$ as a closed contour in the s-plane is traversed is 2π radians if z_i is inside the closed contour and 0 radians if z_i is outside. We refrain from constructing a contour that passes

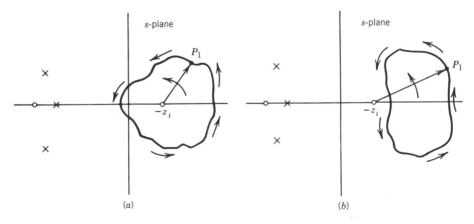

Figure 6.12 Determining net angle changes of $F(s)$ due to inside and outside zero placements.

through any zero (or pole) of $F(s)$. If there are Z zeros inside the closed contour in the s-plane, then the net contribution to the angle change in $F(s)$ is $2\pi Z$ radians. Similarly, if there are P poles inside the closed s-plane contour, the net contribution to the angle change in $F(s)$ is $-2\pi P$ radians. Note that the negative sign is obtained because the net angle change due to poles—that is, $\underline{/(s + p_i)}$—is subtracted in (6.18). Thus, (6.18) becomes

$$\underline{/F(s)}\Big|_{\substack{\text{Due to closed} \\ \text{s-plane contour}}} = 2\pi Z - 2\pi P \tag{6.19}$$

Since the net change in the angle of $F(s)$ may itself be written as 2π times the net number of encirclements (N) of the origin in the $F(s)$-plane, we have $2\pi N = 2\pi Z - 2\pi P$. Dividing by 2π yields

$$N = Z - P \tag{6.20}$$

which is the main result of this section. In words, (6.20) states that the net number of encirclements of the origin by the conformal mapping of this closed s-plane contour into the $F(s)$-plane is equal to the number of zeros minus the number of poles of $F(s)$ inside the closed s-plane contour.

EXAMPLE 6.3

Sketch the $F(s)$-plane contour obtained by mapping the function

$$F(s) = \frac{2(s^2 + 4)}{s^2(s^2 - 4)} \tag{6.21}$$

for the closed s-plane contour in Figure 6.13a. We use the s-plane vector concept to calculate $F(P_i)$ for the points P_1 through P_9. The results are plotted in the $F(s)$-plane in Figure 6.13b. Because of the symmetry involved both in $F(s)$ and in the given s-plane contour, the mapping obtained from points P_5 through P_9 is identical to that obtained from points P_1 through P_5. Since $F(P_5) = F(P_1)$ and $P_9 = P_1$, the $F(s)$-plane contour for points P_5 through P_9 is superimposed over the same contour for points P_1 through P_5. The $F(s)$-plane contour encircles the origin twice in the opposite direction from the direction of traversing the s-plane contour. Whereas the s-plane contour is counterclockwise, the $F(s)$-plane contour encircles the origin twice in the clockwise direction. We note that there are no zeros $(Z = 0)$ and two poles $(P = 2)$ inside the s-plane contour, which yields a value of $N = -2$ from (6.20). This result is consistent with Figure 6.13b.

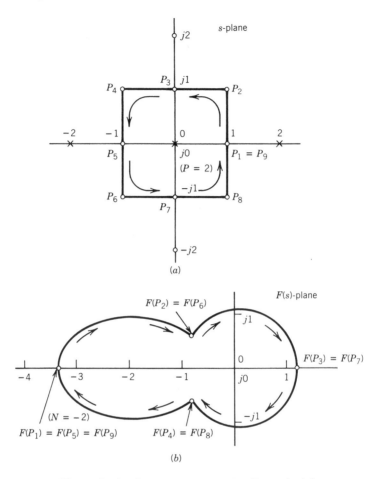

Figure 6.13 Contour mapping for Example 6.3.

6.5 THE NYQUIST CRITERION

We use the result in (6.20) in this section to develop the primary frequency response stability criterion for linear, time-invariant feedback systems [5–8]. In particular, we select the closed contour in the s-plane as the entire right-half plane and let the function $F(s)$ be the characteristic polynomial for a closed-loop system. We show that encircling the $-1 + j0$ point in the $GH(s)$-plane is equivalent to encircling the origin in the $F(s)$-plane, since $F(s) = 1 + GH(s)$. Therefore, a polar plot in the $GH(s)$-plane with $s = j\omega$ is constructed first, and the net number of encirclements (N) of the $-1 + j0$ point is determined. We show that P is the number of open-loop poles in the right-half s-plane. Thus, Z may be determined from (6.20) as the sum of N and P, where Z is the number of zeros (or roots) of the characteristic equation, or equivalently the number of closed-loop poles, in the right-half s-plane. If Z is zero, there are no closed-loop poles in the right-half s-plane, and the system is stable. If Z is greater than zero,

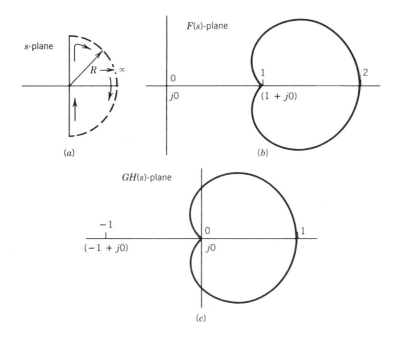

Figure 6.14   Definition of planes for contour mapping.

the system is unstable. Contour modifications in the s-plane are discussed for open-loop poles on the $j\omega$-axis, and several examples are worked to illustrate these details in a variety of situations.

To apply the results of Section 6.4 for the development of the Nyquist criterion as summarized above, we select the entire right-half s-plane as the closed contour of interest. Moreover, we investigate the conformal mapping from this s-plane contour to the $F(s)$-plane defined by choosing

$$F(s) = 1 + GH(s) \tag{6.22}$$

The right-half s-plane contour is indicated in Figure 6.14a, the $F(s)$-plane in 6.14b, and the corresponding $GH(s)$-plane in 6.14c. The mapping for a typical stable second-order plant transfer function is shown as an example. Observe from (6.22) that $F(s) = 0$ implies $GH(s) = -1$. Therefore, encircling the origin in the $F(s)$-plane is equivalent to encircling the -1 (or $-1 + j0$) point in the $GH(s)$-plane. Hereafter, we will consider only the mapping from the s-plane in Figure 6.14a to the $GH(s)$-plane in Figure 6.14c. Note that there are no encirclements ($N = 0$) of the -1 point in the $GH(s)$-plane for the example. As shown in Figure 6.14a, the s-plane contour in general consists of a portion from the origin along the $j\omega$-axis toward $+j\infty$, an infinite semicircle from $+90°$ to $-90°$ containing the first and fourth quadrants, and a portion along the $-j\omega$-axis from $-j\infty$ to the origin. The contour mapping into the $GH(s)$-plane corresponding to these parts of the s-plane contour is composed of the polar plot $GH(j\omega)$ for the first portion and a mirror reflection of this polar plot about the real axis in the $GH(s)$-plane for the portion between $-j\infty$ and the origin along the $-j\omega$-axis. The contribution of the infinite semicircle is simply a redirection of the $GH(s)$-plane contour, provided the order of the numerator of $GH(s)$ is not greater than the order of the denominator.

Suppose $GH(s)$ has the form

$$GH(s) = \frac{K(s + \alpha_1)(s + \alpha_2)\ldots(s + \alpha_z)}{(s + \beta_1)(s + \beta_2)\ldots(s + \beta_p)} \tag{6.23}$$

Thus, $F(s)$ may be expressed as

$$F(s) = 1 + GH(s)$$

$$= 1 + \frac{K(s + \alpha_1)\ldots(s + \alpha_z)}{(s + \beta_1)\ldots(s + \beta_p)}$$

$$= \frac{(s + \beta_1)(s + \beta_2)\ldots(s + \beta_p) + K(s + \alpha_1)\ldots(s + \alpha_z)}{(s + \beta_1)(s + \beta_2)\ldots(s + \beta_p)} \tag{6.24}$$

We see that the poles of $F(s)$ are the poles of the open-loop transfer function $GH(s)$. The number of these poles inside the right-half s-plane closed contour is designated as P. Moreover, the zeros of $F(s)$ in (6.24), that is, the roots of the characteristic equation, are the closed-loop poles. We denote the number of these zeros inside the right-half s-plane as Z, where Z is to be determined by applying the Nyquist criterion. We further note that the open-loop zeros, that is, the zeros of $GH(s)$, affect the shape of the $GH(s)$-plane plot but do not otherwise enter into the criterion equations. These observations are summarized in Table 6.1.

In brief, we first determine P directly from the open-loop $GH(s)$, then find N from the $GH(j\omega)$-plane plot, and finally determine Z from $Z = N + P$. The closed-loop system is stable if and only if $Z = 0$. The steps of this algorithm are listed in Table 6.2. The value of P may be found quite easily since $GH(s)$ is assumed to be in factored form and the open-loop poles can be located in the s-plane directly. Next, we express $GH(j\omega)$ in standard form and sketch the Bode magnitude and phase diagrams. The polar plot is then sketched from these Bode

TABLE 6.1   Definitions for the Nyquist Criterion

Notation	Relation to $F(s)$ and $GH(s)$	Relation to Criterion Result
P	Poles of $F(s)$ in right-half s-plane	Open-loop poles, that is, poles of $GH(s)$ in right-half s-plane
Z	Zeros of $F(s)$ in right-half s-plane	Closed-loop poles in right-half s-plane
α_i	Zeros of $GH(s)$	Affect shape of $GH(s)$-plane plot

TABLE 6.2   Steps in Applying the Nyquist Criterion

---

1. Determine the number of open-loop poles in the right-half s-plane (P) by inspecting the factored open-loop transfer function $GH(s)$.

2. Sketch Bode magnitude and phase diagrams of the open-loop transfer function $GH(j\omega)$.

3. Sketch the polar plot in the $GH(j\omega)$-plane and complete the $GH(s)$-plane contour closure resulting from the selection of the entire right-half s-plane as the contour to be mapped into the $GH(s)$-plane.

4. Determine the net number of encirclements (N) of the $-1 + j0$ point, where N is positive if the encirclements are in the same direction as the s-plane contour and negative if the direction is opposite.

5. Compute $Z = N + P$, where Z is the number of right-half s-plane closed-loop poles. The closed-loop system is stable if and only if Z is zero. If Z is negative, an error in finding P or N (probably N) has been made!

---

diagrams for $s = j\omega$ as ω goes from 0 to $+\infty$. As noted above, the corresponding polar plot for negative ω, that is, for $s = j\omega$ as ω varies from $-\infty$ to 0, is obtained by constructing the mirror reflection about the real axis in the $GH(s)$-plane.

The net number of encirclements (N) of the $-1 + j0$ point may vary as K becomes larger or smaller. We determine the ranges of K by first calculating the phase crossover frequency (ω_{PC}) from

$$\text{Im } GH(j\omega)|_{\omega = \omega_{PC}} = 0 \qquad (6.25)$$

At $\omega = \omega_{PC}$ we compute the value of gain K^* for which the $GH(j\omega)$-plane plot passes through the $-1 + j0$ point by setting

$$\text{Re } GH(j\omega_{PC})|_{K = K^*} = -1 \qquad (6.26)$$

Ranges of K versus N are then tabulated, and Z is determined from $Z = N + P$, where P was determined in the first step of the algorithm.

EXAMPLE 6.4

Apply the Nyquist criterion steps identified in Table 6.2 to determine as a function of positive K the stability of the closed-loop system formed by feeding back negatively the output of a combined plant/controller transfer function given by

$$GH(s) = \frac{K}{(s + 1)(s + 2)(s + 3)} \qquad (6.27)$$

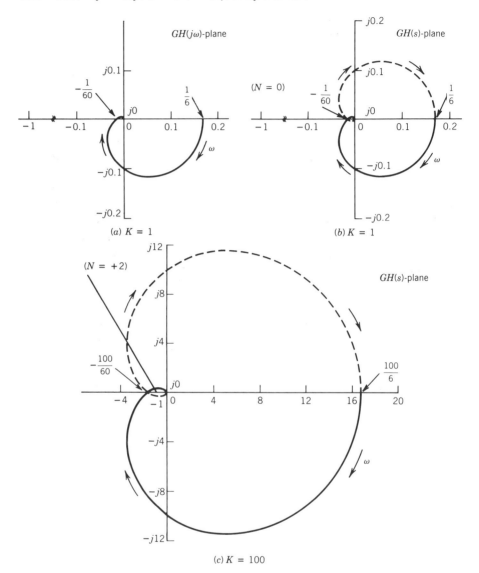

Figure 6.15   Polar plots for the application of the Nyquist criterion to Example 6.4.

Figure 6.15a shows a polar plot for $GH(j\omega)$ for $K = 1$. This plot corresponds to the conformal mapping of $GH(s)$ for that section of the s-plane contour along the positive imaginary axis—that is, for $s = j\omega$ with $0 < \omega < +\infty$. Therefore, we arbitrarily begin traversing the s-plane contour at the origin of the s-plane and consider all points along the positive imaginary axis. Next, we continue by traversing the s-plane contour clockwise along the infinite semicircle $Re^{j\theta}$, where $R \to \infty$ and θ varies from $+90°$ to $-90°$. Since $R \to \infty$, then $GH(Re^{j\theta})$ approaches zero, regardless of the value of θ. Consequently, the conformal mapping of the infinite semicircular arc from the s-plane into the $GH(s)$-plane results in a redirection of the $GH(s)$-plane contour with an infinitesimal magnitude. The traversing of the negative imaginary axis in the s-plane, that is, $s = j\omega$ for $-\infty < \omega < 0$, completes the closure of the s-plane contour. This section of the s-plane contour yields a $GH(s)$-plane contour that is the mirror reflection about the real axis of the corresponding section for $s = j\omega$ with $0 < \omega < +\infty$. The result of traversing the entire s-plane closed contour is shown in the $GH(s)$-plane plot of Figure 6.15b for $K = 1$.

We have arbitrarily chosen to proceed around the right-half s-plane in a clockwise direction beginning at the origin. This choice coincides with traversing the positive imaginary axis from $\omega = 0$ to $\omega = +\infty$, the convention used in constructing first the Bode diagrams and then the polar plot of $GH(j\omega)$. However, once the polar plot and its mirror reflection about the real axis in the $GH(s)$-plane has been constructed, the direction of traversal in the s-plane is arbitrary.[2] If a clockwise direction in the s-plane is selected and the corresponding mapping in the $GH(s)$-plane encircles the -1 point in a clockwise direction n times, then N is positive. If the corresponding encirclements of the -1 point are in a counterclockwise direction, then N is negative. Reversing the direction selected in the s-plane (counterclockwise) simply reverses the direction of encirclements of the -1 point in the $GH(s)$-plane, yielding a positive value of N for counterclockwise encirclements and a negative value for clockwise encirclements. Whichever direction is selected, N is positive if the encirclements in the $GH(s)$-plane are in the same direction as the s-plane contour and negative if the direction is opposite. This simple statement, appearing in Step 4 of Table 6.2, circumvents continual references to clockwise and counterclockwise arguments.

In applying the Nyquist criterion to the closed-loop system defined by the $GH(s)$ in (6.27), we first determine the number of open-loop poles in the right-half s-plane (P) as zero. This is Step 1 of the procedure itemized in Table 6.2. From Bode diagrams (Step 2), we have already sketched the polar plot in the $GH(j\omega)$-plane and have completed the closure (Step 3) in Figure 6.15b for $K = 1$. We see that there are no encirclements of the -1 point $(N = 0)$ for this case. Therefore, $Z = N + P = 0$ for $K = 1$, and the closed-loop system is stable for this value of K.

Figure 6.15c shows the same $GH(s)$-plane sketch as in Figure 6.15b, except it is for a value of $K = 100$ and, consequently, is 100 times larger. The scale has also been reduced in Figure 6.15c for convenience. Tracing along the $GH(s)$-plane contour corresponding to the clockwise direction arbitrarily selected for the s-plane contour around the right-half plane yields two net encirclements of the -1 point: $N = 2$ for $K = 100$ (Step 4). Extending a ray from the -1 point outward in any direction toward infinity, as shown in Figure 6.15c, allows us to determine the value of N easily. Therefore, we can now use Step 5 in Table 6.2 to

---

[2] Mathematicians generally insist on traversing the s-plane contour in a counterclockwise direction because a region is said to be "enclosed" by a closed contour if that region lies on the left as the contour is traversed. Many control textbooks conform to this convention and require counterclockwise encirclements in the $GH(s)$-plane for positive N and clockwise encirclements for negative N.

compute $Z = N + P = 2 + 0 = 2$ for $K = 100$. Observe that the value of P remains the same throughout the application of the criterion; only N (and hence Z) may change as K varies.

We have now shown that $Z = 0$ for $K = 1$ and $Z = 2$ for $K = 100$. Reviewing the steps for each of these two values of K reveals that $Z = 0$ for all K for which the polar plot of $GH(j\omega)$ crosses the negative real axis in the $GH(j\omega)$-plane (Figures 6.15b and 6.15c) at values to the right of the -1 point. This crossing of the negative real axis occurs at the phase crossover frequency ω_{PC}, which may be determined from (6.25). To determine the imaginary part of $GH(j\omega)$, we first expand and rationalize the denominator of that function, that is,

$$GH(j\omega) = \frac{K}{(j\omega + 1)(j\omega + 2)(j\omega + 3)} = \frac{K}{(j\omega)^3 + 6(j\omega)^2 + 11(j\omega) + 6}$$

$$= \frac{K}{(6 - 6\omega^2) + j\omega(11 - \omega^2)} \left[\frac{(6 - 6\omega^2) - j\omega(11 - \omega^2)}{(6 - 6\omega^2) - j\omega(11 - \omega^2)} \right]$$

$$= \frac{K\left[(6 - 6\omega^2) - j\omega(11 - \omega^2)\right]}{(6 - 6\omega^2)^2 + \omega^2(11 - \omega^2)^2} \tag{6.28}$$

Setting the imaginary part of $GH(j\omega)$ in (6.28) equal to zero yields

$$\operatorname{Im} GH(j\omega) = \frac{K\left[-\omega(11 - \omega^2)\right]}{(6 - 6\omega^2)^2 + \omega^2(11 - \omega^2)^2} = 0 \tag{6.29}$$

from which ω equals either zero or $\sqrt{11}$ radians per second. The value of $\omega = 0$ identifies a positive real axis crossing; the value of ω we seek is the phase crossover frequency $\omega_{PC} = \sqrt{11}$. Observe that adjusting the value of K does not alter this frequency at which the angle of $GH(j\omega)$ is $-180°$. In fact, we see from the Bode diagrams that adjusting K has no effect at all on $\big/ GH(j\omega)$, only on the magnitude of $GH(j\omega)$.

To determine that value of K, denoted as K^*, for which the $GH(j\omega)$-plane plot passes through the -1 point, we use (6.26) and (6.28) to obtain

$$\operatorname{Re} GH(j\omega_{PC})\big|_{K=K^*} = \frac{K^*(6 - 6\omega^2)}{(6 - 6\omega^2)^2 + \omega^2(11 - \omega^2)^2}\bigg|_{\omega=\sqrt{11}}$$

$$= \frac{K^*\left(6 - 6(\sqrt{11})^2\right)}{\left(6 - 6(\sqrt{11})^2\right)^2 + (\sqrt{11})^2\left[11 - (\sqrt{11})^2\right]^2}$$

$$= \frac{-K^*}{60} = -1 \tag{6.30}$$

Therefore, we find that $K^* = 60$. These results for Example 6.4 are summarized in Table 6.3. Observe that results from the Routh-Hurwitz criterion for the same system in Example 5.4 are in agreement with this solution.

TABLE 6.3 Application of Nyquist Criterion to Example 6.4

Range of K	P	N	Z	Stability
$0 < K < 60$	0	0	0	Stable
$60 < K < +\infty$	0	2	2	Unstable

EXAMPLE 6.5

Apply the Nyquist criterion to determine the stability of a closed-loop system having $GH(s)$ given by

$$GH(s) = \frac{K(s-1)}{(s-2)(s-4)} \tag{6.31}$$

where the output of $GH(s)$ is fed back negatively to form the closed-loop system.

We begin by observing that there are two (open-loop) poles of $GH(s)$ in the right-half s-plane, that is, $P = 2$. Next we set $s = j\omega$ and express $GH(j\omega)$ in the standard form for sketching Bode diagrams as

$$GH(j\omega) = \frac{(-K/8)(1 - j\omega)}{(1 - j\omega/2)(1 - j\omega/4)} \tag{6.32}$$

The Bode magnitude and phase diagrams for $GH(j\omega)$ are sketched in Figure 6.16 for $K = 1$ with the corresponding polar plot shown in Figure 6.17. The conformal mapping of the closed contour around the entire right-half s-plane is obtained by performing a mirror reflection of the polar plot for $0 < \omega < \infty$ about the real axis in the $GH(s)$-plane; this resulting plot is shown in Figure 6.17a also. We see that there are no encirclements of the -1 point for $K = 1$, that is, $N = 0$, and consequently $Z = N + P = 0 + 2 = 2$. In other words, the unstable open-loop plant remains unstable when connected in a closed-loop system with the gain K set equal to 1. However, we observe that one or more encirclements do occur when K is adjusted to sufficiently larger values.

To determine the number of closed-loop poles in the right-half s-plane (Z) for larger values of K, we first express $GH(j\omega)$ in rectangular form as

$$GH(j\omega) = \frac{K(j\omega - 1)}{(j\omega - 2)(j\omega - 4)} = \frac{K(-1 + j\omega)}{(j\omega)^2 - 6(j\omega) + 8}$$

$$= \frac{K(-1 + j\omega)}{(8 - \omega^2) - j6\omega}\left(\frac{(8 - \omega^2) + j6\omega}{(8 - \omega^2) + j6\omega}\right)$$

$$= \frac{K\left[(-8 - 5\omega^2) + j\omega(2 - \omega^2)\right]}{(8 - \omega^2)^2 + (6\omega)^2} \tag{6.33}$$

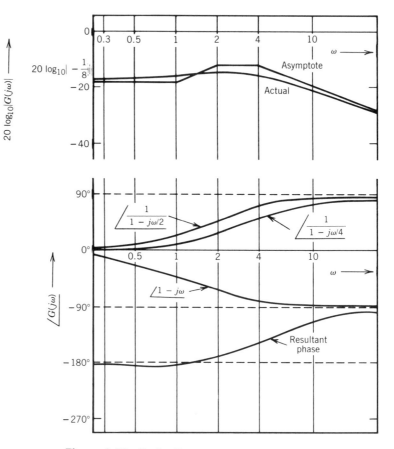

Figure 6.16 Bode diagrams for Example 6.5.

Using (6.25), we can solve for ω_{PC} from

$$\text{Im } GH(j\omega)\big|_{\omega=\omega_{PC}} = \frac{K\left[\omega(2-\omega^2)\right]}{(8-\omega^2)^2 + (6\omega)^2}\bigg|_{\omega=\omega_{PC}} = 0 \qquad (6.34)$$

to obtain $\omega_{PC} = 0$ and $\omega_{PC} = \sqrt{2}$. Substituting these values of ω_{PC} into (6.26) for the $GH(j\omega)$ in (6.33) yields

$$\text{Re } GH(j\omega_{PC})\big|_{K=K^*} = \frac{K^*\left(-8 - 5\omega_{PC}^2\right)}{\left(8 - \omega_{PC}^2\right)^2 + \left(6\omega_{PC}\right)^2} = -1 \qquad (6.35)$$

from which $K^* = 8$ for $\omega_{PC} = 0$ and $K^* = 6$ for $\omega_{PC} = \sqrt{2}$. Therefore, we see that the value of N changes for $K = 6$ and for $K = 8$. Figures 6.17b and 6.17c show $GH(s)$-plane plots for $K = 7$ and $K = 10$, respectively, from which we can determine N as -2 for $K = 7$ and $N = -1$ for $K = 10$. The minus sign is correct

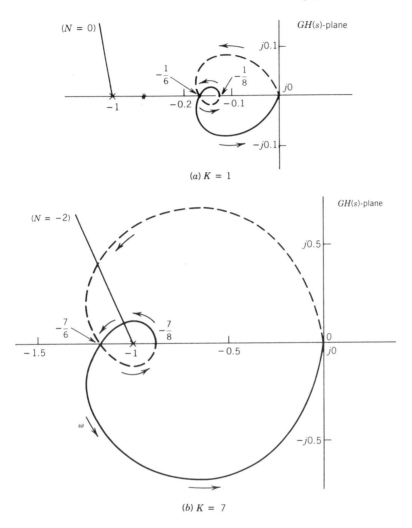

Figure 6.17 *GH(s)*-plane plots for Example 6.5.

for N in both cases because encirclements about the -1 point occur in a direction opposite to that of the direction that the contour was traversed in the s-plane. Therefore, $Z = N + P = -2 + 2 = 0$ for $6 < K < 8$, and $Z = -1 + 2 = 1$ for $8 < K < +\infty$. These results are summarized in Table 6.4.

6.5.1 Open-Loop Poles on the $j\omega$-Axis

The s-plane contour must be modified when poles of $GH(s)$ occur on the $j\omega$-axis. Consider the case of a single open-loop pole at the origin, as shown in Figure 6.18a. Let the s-plane contour pass around this pole in an infinitesimal semicircle

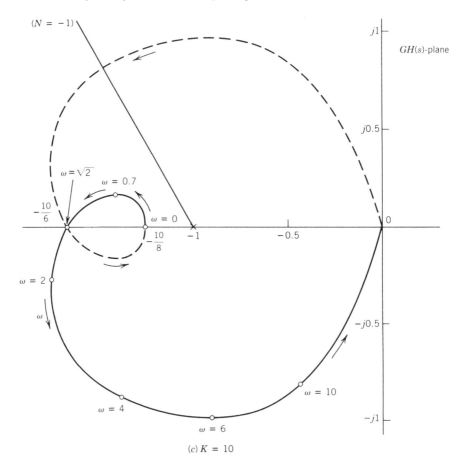

(c) $K = 10$

Figure 6.17 (*Continued*) $GH(s)$-plane plots for Example 6.5.

TABLE 6.4   Application of Nyquist Criterion to Example 6.5

Range of K	P	N	Z	Stability
$0 < K < 6$	2	0	2	Unstable
$6 < K < 8$	2	-2	0	Stable
$8 < K < +\infty$	2	-1	1	Unstable

that excludes the pole from the interior of the closed contour. The remainder of the contour is unchanged from the previous discussion. We examine only that part of the semicircle in the first quadrant and obtain the result for the fourth quadrant by a mirror reflection about the real axis in the $GH(s)$-plane. In the first quadrant, we have

$$s = \varepsilon e^{j\theta} \tag{6.36}$$

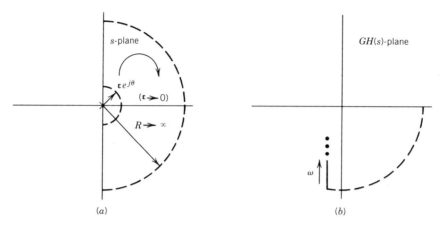

Figure 6.18 Contour mapping for an open-loop pole at the origin.

where ε is arbitrarily small ($\varepsilon > 0$) and θ varies from $0°$ to $90°$. Let the $GH(s)$ function under consideration have the form

$$GH(s) = \frac{K(1 + s/\alpha_1)\ldots(1 + s/\alpha_z)}{s(1 + s/\beta_1)\ldots(1 + s/\beta_p)} \tag{6.37}$$

Thus, for $s = \varepsilon e^{j\theta}$, $GH(s)$ becomes

$$GH(s)\big|_{s=\varepsilon e^{j\theta}} = \frac{K\left(1 + \dfrac{\varepsilon e^{j\theta}}{\alpha_1}\right)\ldots\left(1 + \dfrac{\varepsilon e^{j\theta}}{\alpha_z}\right)}{\varepsilon e^{j\theta}\left(1 + \dfrac{\varepsilon e^{j\theta}}{\beta_2}\right)\ldots\left(1 + \dfrac{\varepsilon e^{j\theta}}{\beta_p}\right)}$$

$$\cong \frac{K}{\varepsilon e^{j\theta}} = \frac{K}{\varepsilon}e^{-j\theta} \tag{6.38}$$

Since ε is arbitrarily small, the magnitude of $GH(s)$ for $s = \varepsilon e^{j\theta}$ in (6.38) becomes infinite. Moreover, since θ varies between $0°$ and $90°$, the angle of $GH(s)$ in (6.38), that is, $-\theta$, varies between $0°$ and $-90°$. Figure 6.18b shows the $GH(s)$-plane mapping for the part of the s-plane contour in (6.38) when K is positive. We show in Example 6.6 below that the real part of $GH(j\omega)$ is some finite nonzero value as $\omega \to 0$. Consequently, the small finite offset in Figure 6.18b from the negative imaginary axis is indeed correct. Note that the angle of $GH(j\omega)$ as $\omega \to 0$ is $-90°$ even with this small offset, since the magnitude is infinite.

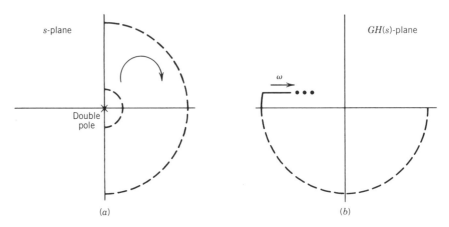

Figure 6.19 Contour mapping for a double open-loop pole at the origin in the s-plane.

For double open-loop poles at the origin in the s-plane, we have

$$GH(s) = \frac{K\left(1 + \dfrac{\varepsilon e^{j\theta}}{\alpha_1}\right)\cdots\left(1 + \dfrac{\varepsilon e^{j\theta}}{\alpha_z}\right)}{\left(\varepsilon e^{j\theta}\right)^2\left(1 + \dfrac{\varepsilon e^{j\theta}}{\beta_3}\right)\cdots\left(1 + \dfrac{\varepsilon e^{j\theta}}{\beta_p}\right)}$$

$$\cong \frac{K}{\left(\varepsilon e^{j\theta}\right)^2} = \frac{K}{\varepsilon^2}e^{-j2\theta} \tag{6.39}$$

As shown in Figure 6.19, the $GH(s)$-plane contour moves at an infinite magnitude from the positive real axis through $-180°$ (clockwise through $180°$), since $\underline{/GH(s)} = -2\theta$ and θ varies, as before, from $0°$ to $+90°$ (counterclockwise) around the pole at the origin in the s-plane. In general, if there are m poles at the origin in the s-plane, then the same rules apply, except that the $GH(s)$-plane contour varies through m quadrants, that is, m times $-90°$, at an infinite clockwise direction from its starting angle.

EXAMPLE 6.6

Consider a negative feedback system with open-loop transfer function given by

$$GH(s) = \frac{K}{s(s + 1)} \tag{6.40}$$

The Bode diagrams and polar plot for $\varepsilon < \omega < +\infty$ are sketched in Figure 6.20a

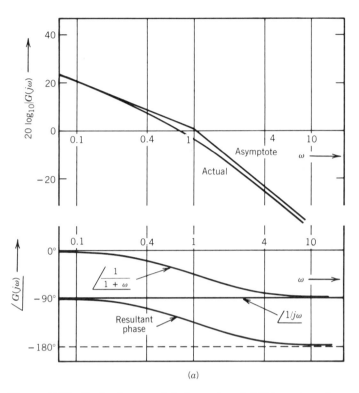

(a)

Figure 6.20 Bode diagram (a), the polar plot (b), and contour mapping (c) for Example 6.6.

for $K = 1$. To determine the real part of $GH(j\omega)$ as $\omega \to 0$, we write

$$\text{Re}\{GH(j\omega)\} = \text{Re}\left[\frac{K}{j\omega(1+j\omega)}\left(\frac{1-j\omega}{1-j\omega}\right)\right]$$

$$= \text{Re}\left[\frac{-jK(1-j\omega)}{\omega(1+\omega^2)}\right] = \frac{-K}{1+\omega^2} \qquad (6.41)$$

Therefore, the real part of $GH(j\omega)$ as $\omega \to 0$ is $-K$, as indicated in Figure 6.20b for $K = 1$.

We continue by mapping the infinitesimal s-plane contour $s = \varepsilon e^{j\theta}$ for $0° \leq \theta \leq 90°$ into the $GH(s)$-plane, as shown in Figure 6.20c. Then we use the polar plot of Figure 6.20b to map the s-plane contour along the positive imaginary axis from $\omega = +\varepsilon$ to $\omega = +\infty$. The infinite circular arc in the first quadrant of the s-plane yields only a redirection of the $GH(s)$-plane mapping, since the magnitude of $GH(s)$ for that arc is zero. We reflect the mapping just described for all parts of the s-plane contour in the first quadrant to yield the remainder of the s-plane mapping for the fourth quadrant. Applying the Nyquist criterion yields

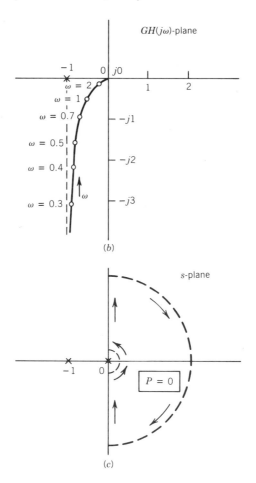

Figure 6.20 (*Continued*)

$Z = N + P = 0 + 0 = 0$, since no net encirclements occur ($N = 0$) for any positive K and the open-loop plant is stable ($P = 0$).

EXAMPLE 6.7

As a second example in this subsection, consider two open-loop transfer functions given by

$$\text{a.} \quad GH(s) = \frac{K}{s^2(s + 1)}$$

$$\text{b.} \quad GH(s) = \frac{K(s + 1)}{s^2}$$

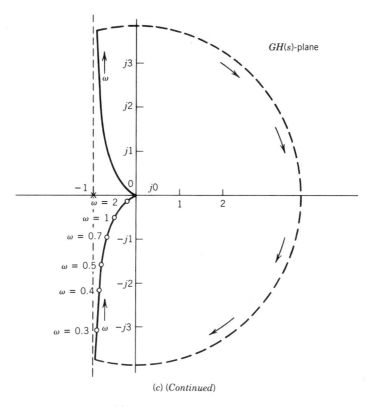

(c) (*Continued*)

Figure 6.20 (*Continued*)

Closed-loop systems are formed by feeding back the outputs of these $GH(s)$ functions negatively.

Polar plots in the $GH(s)$-plane are shown for these two cases in Figure 6.21. Both $GH(s)$ functions contain a double pole at the origin in the s-plane and, consequently, an infinitesimal semicircular arc ($s = \varepsilon e^{j\theta}$) is selected to bypass these poles, excluding them from the interior of the closed contour in the s-plane. Therefore, $P = 0$ for both cases. Considering only that part of the infinitesimal arc in the first quadrant, where $0 \leq \theta \leq 90°$, we obtain a $GH(s)$-plane mapping as an infinite arc that begins at $0°$ and varies through $-180°$ — that is, a $180°$ clockwise rotation. Figure 6.21a shows the resulting plot in the $GH(s)$-plane for (a), which has the $(s + 1)$ factor appearing in the denominator of $GH(s)$, thus contributing a phase component that varies as an arc tangent curve from $0°$ to $-90°$. The resultant phase curve for $GH(j\omega)$ varies from $-180°$ to $-270°$ as shown. For all values of ω (and especially for $\omega \to 0$), the polar plot lies in the second quadrant of the $GH(j\omega)$-plane. On the other hand, the polar plot for the $GH(s)$ function in (b) lies entirely in the third quadrant. Since the $(s + 1)$ factor appears in the numerator for (b), we have a corresponding phase component that varies as an arc tangent curve from $0°$ to $+90°$. Figure 6.21b shows the resulting polar plot for the combined $GH(j\omega)$ function.

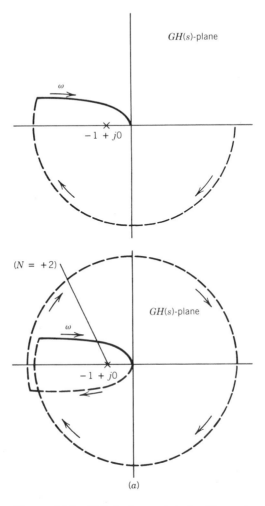

Figure 6.21 $GH(s)$-plane plots for Example 6.7.

Figure 6.21 also shows complete $GH(s)$-plane plots consisting of the polar plots just described and their mirror reflections about the positive real axes. We see that $N = +2$ for (a) and $N = 0$ for (b). Therefore, for all positive K the values of Z determined from $N + P$, where $P = 0$ in both cases, are 2 and 0, respectively.

6.5.2 Applications Involving Negative K

The Nyquist criterion has been applied to cases thus far having variations in K from 0 to $+\infty$. Bode diagrams and the resulting polar plots have been constructed for $K = 1$, and the complete $GH(s)$-plane mappings have been formed by combining polar plots and infinite contours for the first quadrant of the s-plane with their mirror reflections about the real axis in the $GH(s)$-plane. When

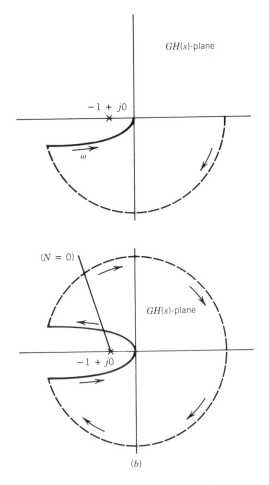

Figure 6.21   (*Continued*)

a finite crossing of the negative real axis occurs at some phase crossover frequency ω_{PC}, we obtain different values of N (and hence Z) by varying K over its positive ranges.

We now extend these results to negative K ranges by first rotating the $GH(s)$-plane plots 180° to account for the additional negative sign on K and then applying the Nyquist criterion as before. Ranges of negative K yielding different values of N (and Z) are possible when the rotated $GH(s)$-plane plots have finite negative real axis crossings. Alternatively, we could also perform the same analysis on the original $GH(s)$-plane plots before the 180° rotation by identifying N as the net number of encirclements of the $+1$ point in the $GH(s)$-plane. This alternate viewpoint corresponds to positive K and positive feedback that, as indicated in the early part of Section 5.5 on negative K root locus, is equivalent to negative K and negative feedback.

Figures 6.22a through 6.22e show the $GH(s)$-plane plots for negative K (normalized to $K = -1$) for the $GH(s)$ transfer functions considered in Exam-

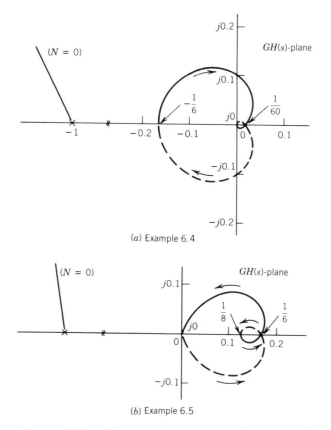

(a) Example 6.4

(b) Example 6.5

Figure 6.22 $GH(s)$-plane plots for Example 6.4 through 6.7 with negative K.

ples 6.4 through 6.7. For Figure 6.22b, we see that $N = 0$ for all $K < 0$. However, since $P = 2$ for Example 6.5, we obtain $Z = N + P = 2$. For the $GH(s)$ function in Example 6.6 (Figure 6.22c), we have $N = +1$ and $Z = N + P = 1 + 0$ for $K < 0$. We also obtain $N = +1$ and $Z = 1$ for $K < 0$ for both $GH(s)$ functions of Example 6.7 in Figures 6.22d and 6.22e. Example 6.4 in Figure 6.22a yields a finite crossing of the negative real axis after the $GH(s)$-plane plot has been rotated 180° to account for the negative sign on K. This finite crossing occurs at $\omega_{PC} = 0$. Therefore, for Example 6.4 we obtain $N = 0$ ($Z = 0$) for $-6 < K < 0$ and $N = +1$ ($Z = 1$) for $K < -6$.

EXAMPLE 6.8

Consider the negative feedback system having the open-loop transfer function given by

$$GH(s) = \frac{K(s + 1)^2}{s^3(s + 6)^2} \tag{6.42}$$

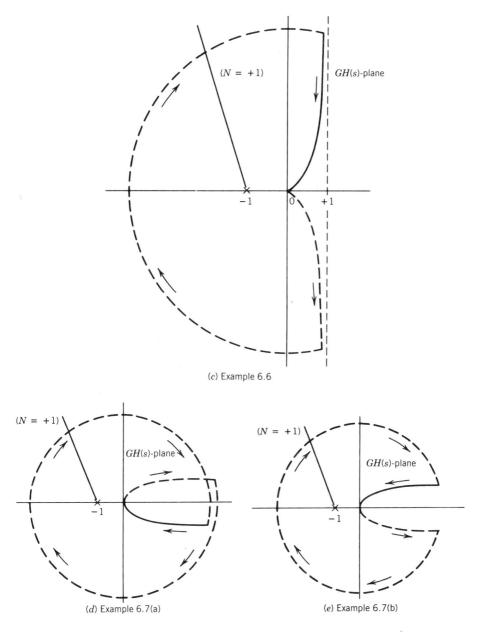

(c) Example 6.6

(d) Example 6.7(a)

(e) Example 6.7(b)

Figure 6.22 (*Continued*)

Bode diagrams and the polar plot of $GH(j\omega)$ for $K = 1$ are shown in Figure 6.23a, with the complete $GH(s)$-plane mapping of the right-half s-plane shown in Figure 6.23b. The two phase crossover frequencies are obtained from (6.25) as 2 and 3 radians per second. Adjusting the gain K in (6.42) to values of 64 and 121.5 for these two frequencies, respectively, places the $GH(s)$-plane plot on the -1 point. Therefore, we identify $N = +2$ both for $0 < K < 64$ and for $121.5 <$

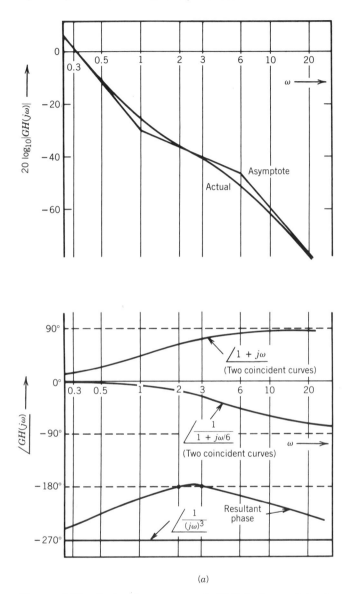

(a)

Figure 6.23 Bode diagrams and $GH(s)$-plane plots for Example 6.8.

$K < +\infty$ and $N = 0$ for $64 < K < 121.5$. Moreover, we see from Figure 6.23c that $N = 1$ for all $K < 0$. In all of these cases, $P = 0$ and $Z = N$. These findings are summarized in Table 6.5. We say that the closed-loop system is "conditionally stable" because the system is stable for $64 < K < 121.5$, but adjusting the gain below or above this range results in an unstable system.

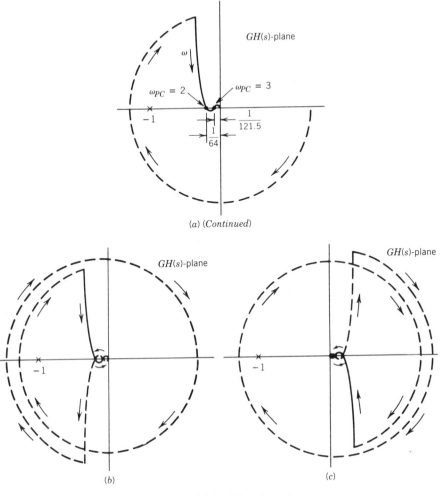

(a) (Continued)

(b)

(c)

Figure 6.23   (Continued)

TABLE 6.5   Application of Nyquist Criterion to Example 6.8

Range of K	P	N	Z	Stability
$0 < K < 64$	0	+2	2	Unstable
$64 < K < 121.5$	0	0	0	Stable
$121.5 < K < +\infty$	0	+2	2	Unstable
$-\infty < K < 0$	0	+1	1	Unstable

6.6 RELATIVE STABILITY

This section applies three of the frequency response specifications defined in Section 6.1 to linear time-invariant systems of order n. These performance specifications are peak magnitude, phase margin, and gain margin. We illustrated

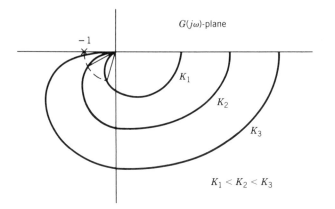

Figure 6.24   Polar plots for the system in (6.43).

their application in Section 6.2 for underdamped second-order systems. We continue to investigate these performance specifications here as measures of relative stability. Initially, we show that feedback systems having acceptable gain margins and phase margins, determined from the open-loop frequency response, may yet yield excessive values of peak magnitude for the closed-loop system frequency response. Contours of constant values of closed-loop frequency magnitude (M) are derived and plotted as circles in the $G(j\omega)$-plane. Finally, we present algorithms based on Newton's method for calculating quite accurately phase margin and gain margin.

6.6.1 Measures of Relative Stability

Consider the $G(j\omega)$-plane plots shown in Figure 6.24 for three values of positive K for the second-order system with open-loop transfer function given by

$$G(s) = \frac{K}{\left(1 + \dfrac{s}{\beta_1}\right)\left(1 + \dfrac{s}{\beta_2}\right)} \tag{6.43}$$

where the output of $G(s)$ is fed back negatively to form the closed-loop system. Note that the gain margin is infinite, regardless of the value of K, because the negative real axis in Figure 6.24 is intersected by the $G(j\omega)$-plane polar plot only at the origin (as $\omega \to \infty$). Therefore, on the basis of the gain margin specification alone, the system in (6.43) appears to behave in an acceptable manner. Furthermore, we see that, as K increases, the phase margin becomes smaller. For a sufficiently large K, a preset phase margin specification—for example, $PM \geq 30°$ —would be violated. On the other hand, Figure 6.25 shows a different $G(j\omega)$-plane plot for which the phase margin specification is satisfied but the gain margin is too small. Thus, we apparently need both phase margin and gain margin

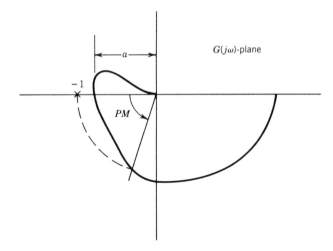

Figure 6.25 A polar plot illustrating an acceptable phase margin and unacceptable gain margin.

specifications to help ensure an acceptable system performance in the frequency domain.

Let us reappraise the meaning of these two performance specifications. To satisfy gain margin and phase margin requirements, we specify only that the open-loop plant frequency response characteristics obey preset conditions at exactly two frequencies—that is, at the phase crossover (ω_{PC}) and gain crossover (ω_{GC}). While it is true because of the construction of the $G(j\omega)$ polar plot that the magnitude and phase at neighboring frequencies are also constrained to some extent, we observe that it is possible to satisfy both gain margin and phase margin specifications and yet have an unacceptable closed-loop frequency response. If the $G(j\omega)$-plane plot passes through the $-1 + j0$ point, the closed-loop system has s-plane poles on the $j\omega$-axis. Consequently, it follows that if the $G(j\omega)$-plane plot approaches very near to the -1 point, the closed-loop system becomes less stable. A third measure of relative stability is the peak magnitude of the closed-loop frequency response (M_p). We show in the following subsection that M_p approaches infinity as the $G(j\omega)$-plane plot approaches the -1 point. For completeness, we illustrate in Figure 6.26 a $G(j\omega)$-plane plot for which the gain margin and phase margin are both acceptable and yet M_p is excessive. Additional restrictions beyond these three performance specifications may need to be placed on the frequency response to achieve a desired closed-loop system bandwidth, for example, or to obtain a sufficiently sharp cutoff.

6.6.2 Contours of Constant *M*

Let the frequency response for an open-loop plant be expressed as

$$G(j\omega) = X(\omega) + jY(\omega) \tag{6.44}$$

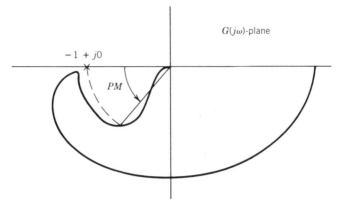

Figure 6.26 A polar plot for which M_p is excessive and gain and phase margins are acceptable.

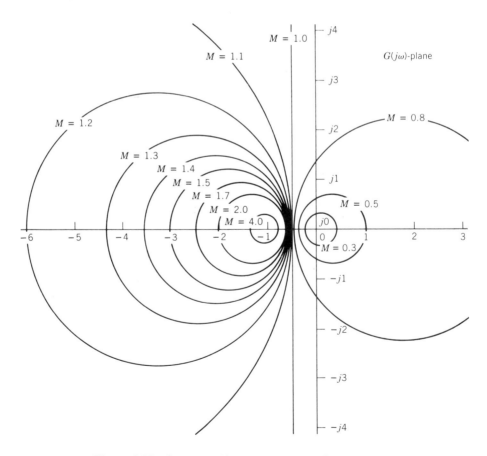

Figure 6.27 Constant-M contours in the $G(j\omega)$-plane.

where $X(\omega) = \operatorname{Re} G(j\omega)$ and $Y(\omega) = \operatorname{Im} G(j\omega)$. For a negative unity-feedback system with the plant frequency response function $G(j\omega)$, we have

$$M(\omega)\underline{/\phi(\omega)} = \frac{G(j\omega)}{1 + G(j\omega)} \tag{6.45}$$

where $M(\omega)$ is the magnitude and $\phi(\omega)$ is the phase of the closed-loop frequency response. Therefore, we may express the closed-loop magnitude $M(\omega)$ as

$$M(\omega) = \frac{|X + jY|}{|1 + X + jY|} = \frac{\sqrt{X^2 + Y^2}}{\sqrt{(1 + X)^2 + Y^2}} \tag{6.46}$$

We wish to derive curves of constant M and plot these contours in the $G(j\omega)$-plane. Setting $M(\omega) = M$ (a constant), squaring both sides of (6.46), and rearranging yields

$$\left(X + \frac{M^2}{M^2 - 1}\right)^2 + Y^2 = \left(\frac{M}{M^2 - 1}\right)^2 \tag{6.47}$$

Equation (6.47) indicates that constant-M contours are circles in the Y versus X plane, that is, $G(j\omega)$-plane, with centers at $X = -M^2/(M^2 - 1)$, $Y = 0$, and radii r of $|M/(M^2 - 1)|$. These curves are shown in Figure 6.27.

EXAMPLE 6.9

Use the constant-M circles in the $G(j\omega)$-plane to determine a linear magnitude plot of the closed-loop frequency response from the open-loop polar plot of

$$G(j\omega) = \frac{5}{(1 + j\omega)\left(1 + \dfrac{j\omega}{2}\right)\left(1 + \dfrac{j\omega}{3}\right)} \tag{6.48}$$

Figure 6.28a shows the desired polar plot and its intersections with several constant-M circles. These values of M are plotted versus the specific intersection frequencies (ω) in Figure 6.28b. The derivation of the curve of constant closed-loop phase in the $G(j\omega)$-plane is postponed until Section 6.8, which also describes a procedure for using a Nichols chart to transform directly from open-loop to closed-loop Bode diagrams for unity feedback systems.

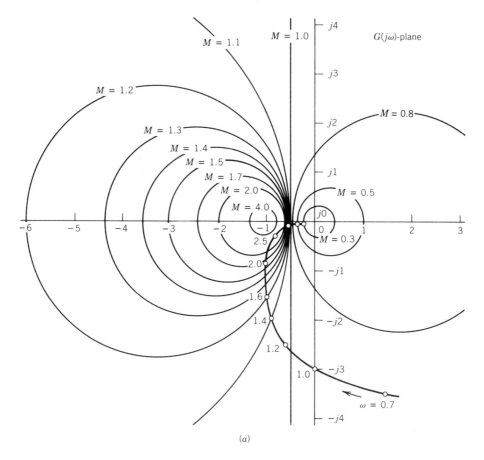

Figure 6.28   Use of constant-M circles to obtain closed-loop frequency response magnitude plots.

6.6.3 Calculation of Gain Margin

We may use Newton's method to calculate the phase crossover frequency (ω_{PC}) according to the recursive formulas

$$\Delta\omega_k = -\left[\frac{\theta(\omega)_k - [-180°]}{\dfrac{d\theta}{d\omega}\Big|_{\omega=\omega_k}}\right]\left(\frac{\pi}{180°}\right)$$

$$\omega_{k+1} = \omega_k + \Delta\omega_k \tag{6.49}$$

where $\theta(\omega)$ is the open-loop phase as a function of ω. For simple linear factors in

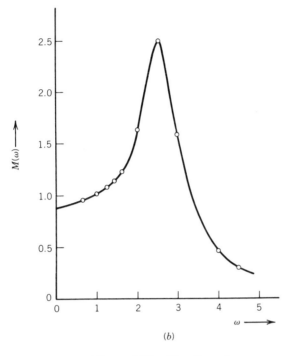

(b)

Figure 6.28 (*Continued*)

$G(j\omega)$, we may write

$$\theta(\omega_k) = \sum_{i=1}^{z} \tan^{-1}\left(\frac{\omega_k}{\alpha_i}\right) - \sum_{i=1}^{p} \tan^{-1}\left(\frac{\omega_k}{\beta_i}\right) \tag{6.50}$$

where $-\alpha_i$ are open-loop zeros and $-\beta_i$ are open-loop poles. In this case, the derivative $d\theta/d\omega_k$ becomes

$$\frac{d\theta}{d\omega_k} = \sum_{i=1}^{z} \frac{\dfrac{1}{\alpha_i}}{1 + \left(\dfrac{\omega_k}{\alpha_i}\right)^2} - \sum_{i=1}^{p} \frac{\dfrac{1}{\beta_i}}{1 + \left(\dfrac{\omega_k}{\beta_i}\right)^2} \tag{6.51}$$

Although more complicated terms are obtained for quadratic factors, such factors occur far less often in practice than the linear factors to which (6.50) and (6.51) apply.

Convergence is assured, in general, only for initial guesses sufficiently near the final solution. Once ω_{PC} has been determined by the recursive application of

TABLE 6.6   Numerical Results for Example 6.10

Iteration	ω_k	$\theta(\omega_k)$	$d\theta(\omega_k)/d\omega_k$
1	3.00	$-172.9°$	-0.42
2	3.30	$-179.6°$	-0.37
3	3.32	$-180.0°$	-0.37

(6.49), the magnitude of $G(j\omega)$ is computed and the gain margin is given by

$$GM = \frac{1}{|G(j\omega_{PC})|} \tag{6.52}$$

as indicated earlier in Section 6.1.

EXAMPLE 6.10

Determine the phase crossover frequency and gain margin for the open-loop frequency response in (6.48) of Example 6.9.

Table 6.6 shows the results of those calculations. The Routh table may be used to verify that $\omega_{PC} = \sqrt{11} \cong 3.32$. The resulting gain margin is 2.0.

6.6.4 Calculation of Phase Margin

In calculating the gain crossover frequency (ω_{GC}), we must solve the equation $|G(j\omega_{GC})| = 1$. It is somewhat more convenient instead to solve the equivalent equation

$$f(\omega) = \ln|G(j\omega)| = 0 \tag{6.53}$$

We may use Newton's formula recursively as

$$\Delta\omega_k = -\frac{f(\omega_k)}{f'(\omega_k)}$$

$$\omega_{k+1} = \omega_k + \Delta\omega_k \tag{6.54}$$

For simple linear factors, we have

$$f(\omega_k) = \frac{1}{2}\sum_{i=1}^{z}\ln\left(\omega_k^2 + \alpha_i^2\right) - \frac{1}{2}\sum_{i=1}^{p}\ln\left(\omega_k^2 + \beta_i^2\right) \tag{6.55}$$

where $-\alpha_i$ are open-loop zeros and $-\beta_i$ are open-loop poles as in (6.50). The corresponding derivative for $f(\omega_k)$ in (6.55) is

$$f'(\omega_k) = \sum_{i=1}^{z} \frac{\omega_k}{\omega_k^2 + \alpha_i^2} - \sum_{i=1}^{p} \frac{\omega_k}{\omega_k^2 + \beta_i^2} \qquad (6.56)$$

Thus, we have relatively simple expressions for $f(\omega_k)$ and $f'(\omega_k)$ for the case when only linear factors are present.

EXAMPLE 6.11

Use Newton's method to calculate ω_{GC} and PM for the negative unity-feedback system with the open-loop frequency response function given by

$$G(j\omega) = \frac{50}{(j\omega + 1)(j\omega + 2)(j\omega + 3)} \qquad (6.57)$$

Using (6.55) and (6.56), we form

$$f(\omega_k) = \ln 50 - \frac{1}{2}\left\{\ln\left(\omega_k^2 + 1\right) + \ln\left(\omega_k^2 + 2^2\right)\right.$$

$$\left. + \ln\left(\omega_k^2 + 3^2\right)\right\}$$

$$f'(\omega_k) = -\omega_k\left[\frac{1}{\omega_k^2 + 1} + \frac{1}{\omega_k^2 + 2^2} + \frac{1}{\omega_k^2 + 3^2}\right] \qquad (6.58)$$

Applying (6.58) recursively yields $\omega_{GC} = 3.05$ radians per second. Using this value of ω in (6.57) gives $G(j3.05) = 1.0\underline{/-174.01°}$. Thus, the phase margin (PM) is

$$PM = 180° - |\theta(\omega_{GC})| = 5.99° \qquad (6.59)$$

6.6.5 Calculation of Peak Magnitude

The closed-loop magnitude for a negative unity-feedback system with a frequency response function $G(j\omega)$ is given by

$$M(\omega) = \left|\frac{G(j\omega)}{1 + G(j\omega)}\right| \qquad (6.60)$$

We want to determine the peak value of $M(\omega)$ by solving for the ω such that $dM(\omega)/d\omega = 0$. To avoid the square root associated with the magnitude operation, we choose instead to set the derivative of $M^2(\omega)$ equal to zero. Let $G(j\omega)$ have the form $N(j\omega)/D(j\omega)$, and let these numerator and denominator functions for $G(j\omega)$ be expressed in terms of their real and imaginary parts as

$$N(j\omega) = X_N(\omega) + jY_N(\omega)$$

$$D(j\omega) = X_D(\omega) + jY_D(\omega) \tag{6.61}$$

Using (6.61) in (6.60) and setting $dM^2(\omega)/d\omega = 0$ yields

$$\frac{d}{d\omega}\left[\frac{X_N^2 + Y_N^2}{(X_N + X_D)^2 + (Y_N + Y_D)^2}\right] = 0 \tag{6.62}$$

The indicated derivative in (6.62) is equal to zero if

$$\left[(X_N + X_D)^2 + (Y_N + Y_D)^2\right]\frac{d}{d\omega}\left[X_N^2 + Y_N^2\right]$$

$$= (X_N^2 + Y_N^2)\frac{d}{d\omega}\left[(X_N + X_D)^2 + (Y_N + Y_D)^2\right] \tag{6.63}$$

For the special case when $N(j\omega)$ is a constant, that is, not a function of ω, (6.63) becomes

$$(X_N + X_D)\frac{d}{d\omega}[X_N + X_D] + (Y_N + Y_D)\frac{d}{d\omega}[Y_N + Y_D] = 0 \tag{6.64}$$

The solution of (6.63), or of (6.64) for the special case, yields the value of frequency (ω_p) at which the closed-loop peak magnitude (M_p) occurs.

EXAMPLE 6.12

Determine the peak value of the closed-loop magnitude for the system of Figure 6.29 by using (6.64) and compare with the solution obtained by (6.9) of Section 6.2.

We identify real and imaginary parts of the numerator and denominator as $X_N(\omega) = 100$, $Y_N(\omega) = 0$, $X_D(\omega) = -\omega^2$, and $Y_D(\omega) = 4\omega$. Therefore, (6.64)

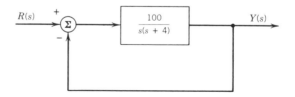

Figure 6.29 The closed-loop system of Example 6.12.

gives

$$(100 - \omega^2) \frac{d}{d\omega} (100 - \omega^2) + (0 + 4\omega) \frac{d}{d\omega} (0 + 4\omega) = 0 \quad (6.65)$$

which simplifies to yield

$$(100 - \omega^2)(-2\omega) + 16\omega = 0 \quad (6.66)$$

Therefore, (6.66) gives the frequency (ω_p) for peak magnitude as $\omega_p = \sqrt{92} \cong 9.59$. We may then form M_p from

$$M_p = \left| \frac{100}{100 - \omega_p^2 + j4\omega_p} \right| = \frac{100}{\sqrt{(100 - 92)^2 + 16(92)}} \cong 2.55 \quad (6.67)$$

Since the given system is a second-order one, we may compute ω_n and then use (6.8) and (6.9) to yield ω_p and M_p as

$$\omega_n = \sqrt{100} = 10$$

$$\zeta = 4/(2\omega_n) = 4/[2(10)] = 0.2$$

$$\omega_p = \omega_n\sqrt{1 - 2\zeta^2} = 10\sqrt{0.92} \cong 9.59$$

$$M_p = \frac{1}{2\zeta\sqrt{1 - \zeta^2}} = \frac{1}{2(0.2)\sqrt{0.96}} \cong 2.55 \quad (6.68)$$

which agrees with the results in (6.66) and (6.67). However, it is important to reiterate that (6.64) is applicable for unity-feedback systems of any order n, while (6.8) and (6.9) are valid only for second-order underdamped systems of the form shown in Figure 6.2.

6.7 COMBINED ROOT LOCUS AND NYQUIST APPROACHES

We view the root locus method of Chapter 5 and the Nyquist criterion of this chapter in a combined context in this section. It is evident that the $j\omega$-axis crossing (Rule 3) of the root locus method (ω^*) corresponds to the phase crossover frequency (ω_{PC}) of the Nyquist criterion. We itemize four procedures for finding ω^* (or ω_{PC}) based on the Routh table, the polar plot, graphical methods with Bode diagrams or s-plane vectors, and Newton's method. Thus, we may select the easiest of these possibilities in a particular stability analysis

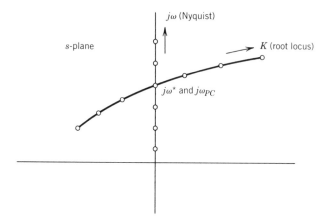

Figure 6.30 Nyquist and root locus interrelations.

problem and then use the result in applying both root locus and Nyquist techniques.

Both root locus and Nyquist approaches for stability analysis utilize the s-plane extensively. The root locus method plots the closed-loop pole locations directly in the s-plane as K varies, and the Nyquist criterion relies on a conformal mapping from the s-plane to the $GH(s)$-plane. Figure 6.30 shows a typical $j\omega$-axis crossing in the s-plane for root locus. The angle of $GH(s)$ is 180° at all points on the root locus as it moves from the left-half s-plane to the $j\omega$-axis at $s = j\omega^*$ and into the right-half plane. We may determine ω^* by the Routh table, which is Rule 3 of the root locus construction rules. Alternatively, we may investigate stability by mapping a closed contour around the entire right-half s-plane into the $GH(s)$-plane. A key aspect of this Nyquist approach is the $GH(s)$-plane plot for all $s = j\omega$. Consequently, we note that the angle of $GH(j\omega)$ is 180° at that frequency ω^* where the s-plane contour intersects the root locus plot. In terms of the Nyquist criterion, this frequency is interpreted as the phase crossover frequency (ω_{PC}). In brief, the angle is always 180° on the root locus plot but the value of s is purely imaginary only on the $j\omega$-axis, whereas this section of the s-plane contour is always purely imaginary for the Nyquist criterion but the angle is 180° only for $\omega = \omega_{PC}$. Thus, we see the correspondence between the two procedures and the usefulness of applying both approaches simultaneously.

We have described four procedures for finding the phase crossover frequency (ω_{PC}). These methods are listed in Table 6.7 as a convenient reference. Generally, we would seek to apply one of the first two methods initially. The graphical methods listed third in the table yield only approximate solutions and, hence, are useful primarily in establishing a starting guess for Newton's method. We recommend this fourth method only when both of the first two prove to be too cumbersome. For example, if Methods 1 and 2 result in a high-order polynomial which is to be solved by Newton's method, it is more straightforward to use Newton's method directly on the phase expression.

TABLE 6.7   Methods for Calculating the Phase Crossover Frequency

Method	Procedure
1. Routh table	Use the s^2 row of the table to compute ω^*.
2. Set $\text{Im}\,[\,GH(j\omega)]\big|_{\omega=\omega_{PC}} = 0$	Solve for ω_{PC} directly from the polar plot.
3. Graphical	Use the Bode diagrams or s-plane vectors to determine the frequency at which the phase is 180°.
4. Newton's method	Use the phase expression and its derivative in Newton's recursive formulas.

EXAMPLE 6.13

Use the methods of Table 6.7 to determine the phase crossover frequency (ω_{PC}) for the system of Figure 6.31, and apply the root locus and Nyquist techniques as K varies from $-\infty$ to $+\infty$.
  We form the characteristic equation as

$$1 + \frac{K(s + 6)^2}{s(s + 1)^2} = 0 \tag{6.69}$$

We simplify (6.69) to yield

$$s^3 + (K + 2)s^2 + (12K + 1)s + 36K = 0 \tag{6.70}$$

which is placed in the Routh array of Table 6.8. The entry $\rho_{1,1}$ is given by

$$\rho_{1,1} = \frac{(K + 2)(12K + 1) - 36K}{K + 2} \tag{6.71}$$

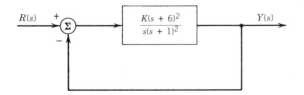

Figure 6.31 The system of Example 6.13.

TABLE 6.8   Routh Array for Example 6.13

s^3	1	$12K + 1$
s^2	$K + 2$	$36K$
s^1	$\rho_{1,1}$	
s^0	$36K$	

which is zero for K^* equal to $1/4$ and $2/3$. By using the s^2 row of the Routh table, we obtain

$$\omega^* = \sqrt{\frac{36K^*}{K^* + 2}} \tag{6.72}$$

Thus, ω^* has the two values of 2 and 3 from the Routh table.

We next use Method 2 in Table 6.7 to yield

$$\operatorname{Im} GH(j\omega)\big|_{\omega=\omega_{PC}} = \operatorname{Im}\left[\frac{K(36 - \omega^2 + j12\omega)}{j\omega(1 - \omega^2 + j2\omega)}\right]\Bigg|_{\omega=\omega_{PC}} = 0 \tag{6.73}$$

which gives

$$\omega_{PC}^4 - 13\omega_{PC}^2 + 36 = 0 \tag{6.74}$$

Solving (6.74) yields the same two values of ω^* obtained earlier by using (6.72).

Using Method 3 of Table 6.7 yields the Bode diagrams in Figure 6.32. The phase crossover frequencies can be determined approximately as 2 and 3.

In applying Method 4, we use s-plane vectors to form

$$\angle G(j\omega_{PC}) = 2\angle(j\omega_{PC} + 6) - 2\angle(j\omega_{PC} + 1) - 90° = 180°$$

$$= 2\tan^{-1}(\omega_{PC}/6) - 2\tan^{-1}(\omega_{PC}) - 90° = 180° \tag{6.75}$$

We may rearrange (6.75) and use Newton's method with appropriate starting values to obtain $\omega_{PC} = 2$ and 3. However, (6.75) may also be solved directly by forming $\tan(\alpha - \beta)$ as

$$\tan(\alpha - \beta) = \frac{\tan\alpha - \tan\beta}{1 + \tan\alpha\tan\beta}$$

$$= \frac{(\omega_{PC}/6) - (\omega_{PC})}{1 + (\omega_{PC}/6)(\omega_{PC})} = \tan 135° = -1 \tag{6.76}$$

which gives

$$\omega_{PC}^2 - 5\omega_{PC} + 6 = 0 \tag{6.77}$$

The solutions to (6.77) are $\omega_{PC} = 2$ and 3 as before.

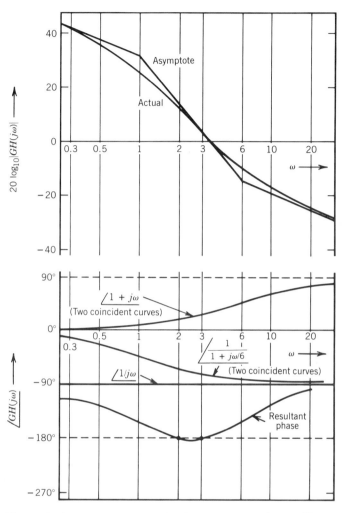

Figure 6.32   Bode diagrams for the open-loop plant in Figure 6.31.

Root locus and Nyquist plots are shown in Figure 6.33 for both positive and negative ranges of K. Table 6.9 summarizes these stability results for the Nyquist criterion.

6.8 CLOSED-LOOP FREQUENCY RESPONSE

We examine the closed-loop frequency response in this section by three methods: (1) contours of constant magnitude and constant phase in the $G(j\omega)$-plane, (2) the Nichols chart, and (3) direct formulas for the magnitude and phase curves as

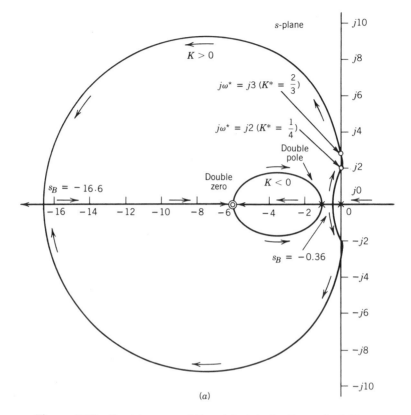

(a)

Figure 6.33 Root locus and Nyquist plots for Example 6.13.

functions of ω. We also determine the sensitivities of these closed-loop system magnitude and phase curves to plant parameter variations.

6.8.1 Constant *M* and *N* Contours

Contours of constant magnitude (M) for the closed-loop frequency response were determined in Section 6.6 as circles in the $G(j\omega)$-plane. To review the development there, we recall that the plant frequency response function $G(j\omega)$ was first expressed in terms of its real part $X(\omega)$ and its imaginary part $Y(\omega)$ as

$$G(j\omega) = X(\omega) + jY(\omega) \tag{6.44}$$

For a negative unity-feedback system, the closed-loop system magnitude $M(\omega)$ and phase $\phi(\omega)$ could then be written as

$$M(\omega)\underline{/\phi(\omega)} = \frac{G(j\omega)}{1 + G(j\omega)} \tag{6.45}$$

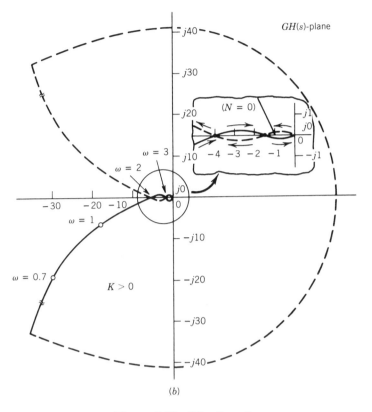

Figure 6.33 (Continued)

Using (6.44) in (6.45) yielded an expression for $M(\omega)$ as

$$M(\omega) = \frac{|X + jY|}{|1 + X + jY|} = \frac{\sqrt{X^2 + Y^2}}{\sqrt{(1 + X)^2 + Y^2}} \qquad (6.46)$$

Setting $M(\omega)$ as a constant M and simplifying (6.46) according to (6.47) yielded the equation of circles in the $G(j\omega)$-plane with centers at $X = -M^2/(M^2 - 1)$, $Y = 0$, and radii r of $|M/(M^2 - 1)|$. Figure 6.27 shows some typical constant-M circles. These contours were useful in Section 6.6 for determining from the polar plot of a given $G(j\omega)$ the peak value of $M(\omega)$, that is, M_p, as an important measure of relative stability.

Contours of constant phase $\phi(\omega)$ can be determined by a similar procedure. Substituting (6.44) into (6.45) and solving for $\phi(\omega)$ yields

$$\phi(\omega) = \tan^{-1}(Y/X) - \tan^{-1}[Y/(1 + X)] \qquad (6.78)$$

Taking the tangent of both sides of (6.78), letting $\alpha = \tan^{-1}(Y/X)$ and $\beta =$

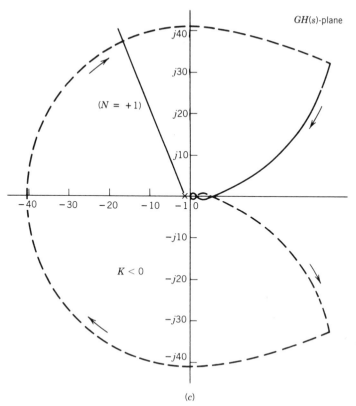

(c)

Figure 6.33 (*Continued*) Root locus and Nyquist plots for Example 6.13.

$\tan^{-1}[Y/(1 + X)]$, and using the two-angle tangent formula gives

$$\tan[\phi(\omega)] = \tan(\alpha - \beta) = \frac{\tan\alpha - \tan\beta}{1 + \tan\alpha\tan\beta}$$

$$= \frac{(Y/X) - [Y/(1 + X)]}{1 + (Y/X)[Y/(1 + X)]} = \frac{Y}{X^2 + X + Y^2} \qquad (6.79)$$

Setting $\tan[\phi(\omega)]$ in (6.79) equal to some constant N and simplifying yields

$$\left(X + \frac{1}{2}\right)^2 + \left(Y - \frac{1}{2N}\right)^2 = \frac{1}{4}\left(1 + \frac{1}{N^2}\right) \qquad (6.80)$$

TABLE 6.9 Application of the Nyquist Criterion to Example 6.13

Range of K	P	N	Z	Stability
$0 < K < 1/4$	0	0	0	Stable
$1/4 < K < 2/3$	0	+2	2	Unstable
$2/3 < K < +\infty$	0	0	0	Stable
$-\infty < K < 0$	0	+1	1	Unstable

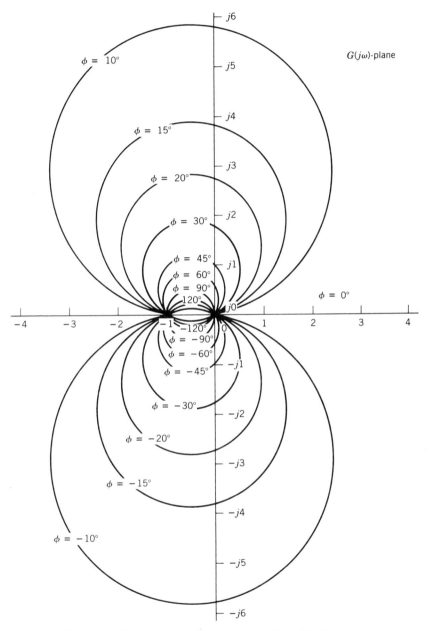

Figure 6.34   Constant-N contours in the $G(j\omega)$-plane.

which enables us to identify constant-N contours as circles with centers at $X = -1/2$, $Y = 1/2N$, and radii r of $(1/2)\sqrt{1 + (1/N^2)}$. Figure 6.34 shows some typical constant-N circles in the $G(j\omega)$-plane. Therefore, curves of $M(\omega)$ and $\phi(\omega)$ versus ω can be obtained by observing intersections of the $G(j\omega)$-plane plot with each of several constant-M and constant-N circles at particular frequencies.

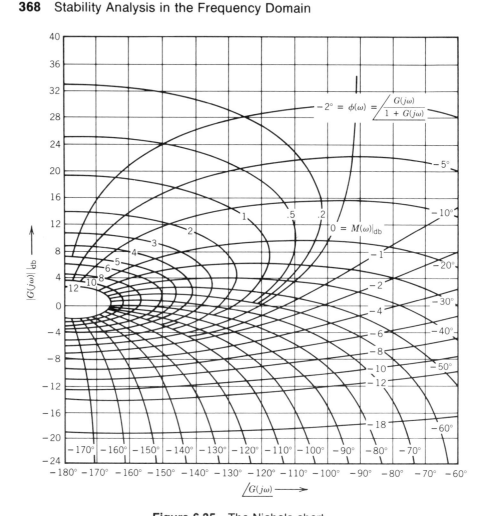

Figure 6.35   The Nichols chart.

6.8.2 The Nichols Chart

While general shapes of polar plots of $G(j\omega)$ can be determined easily, accurate plots require somewhat more careful calculations. Consequently, the use of constant-M and constant-N circles to determine $M(\omega)$ and $\phi(\omega)$ often yields only approximate results. The Nichols chart shown in Figure 6.35 provides a means of obtaining closed-loop frequency response curves by using information from Bode diagrams directly. These Bode diagrams presumably can be constructed fairly accurately. The Bode magnitude plot yields values for the vertical axis of the Nichols chart and the Bode phase curve gives corresponding values for the horizontal axis. Therefore, a plot on the Nichols chart has frequency as the varying parameter along the curve. The constant-M and constant-N circles in the $G(j\omega)$-plane are transferred to the modified contours shown on the Nichols chart of Figure 6.35. As before, the closed-loop frequency response curves can be obtained by noting the intersections of the plot with these contours at particular frequencies. A minor difference from the constant M and N contours is that the closed-loop magnitude plot obtained from the Nichols chart is expressed in decibels. In brief, we avoid the tedious construction of a polar plot in using the Nichols chart to determine closed-loop frequency response curves by transferring information directly from Bode diagrams of the open-loop frequency response.

6.8.3 Direct Formulas

We can obtain exact expressions from (6.45) for $M(\omega)$ and $\phi(\omega)$ in terms of the magnitude and phase of the open-loop frequency response as

$$M(\omega) = \frac{R(\omega)}{\sqrt{1 + R^2(\omega) + 2R(\omega)\cos\theta(\omega)}}$$

$$\phi(\omega) = \theta(\omega) - \tan^{-1}\left(\frac{R(\omega)\sin\theta(\omega)}{1 + R(\omega)\cos\theta(\omega)}\right) \qquad (6.81)$$

where $R(\omega) = |G(j\omega)|$ and $\theta(\omega) = \underline{/G(j\omega)}$. The availability of calculators, personal computers, and larger computers makes these direct formulas especially useful for today's control engineer. These formulas yield highly accurate results, depending on the preciseness with which both $R(\omega)$ and $\theta(\omega)$ are known. Alternatively, we may calculate the closed-loop frequency response function and then determine its magnitude $M(\omega)$ and phase $\phi(\omega)$ directly. The sensitivities of $M(\omega)$ and $\phi(\omega)$ to plant parameter variations are determined later in this section.

EXAMPLE 6.14

Determine the closed-loop frequency response magnitude and phase for the negative unity-feedback system with the plant frequency response function $G(j\omega)$ given by

$$G(j\omega) = \frac{1}{j\omega(j\omega + 1)} \qquad (6.82)$$

Figure 6.36a shows the $G(j\omega)$-plane plot for (6.82) and its intersections with constant-M and constant-N circles. Curves of $M(\omega)$ and $\phi(\omega)$ are shown in Figure 6.36b for these intersection frequencies. Figure 6.37a gives the Bode diagrams for $G(j\omega)$, Figure 6.37b plots this information on the Nichols chart, and Figure 6.37c provides Bode diagrams for the closed-loop frequency response.

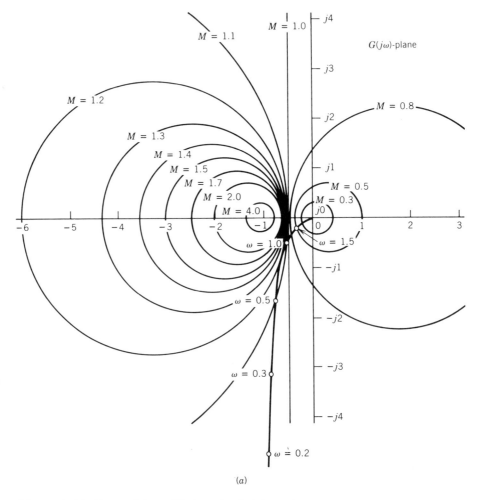

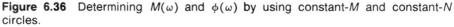

(a)

Figure 6.36 Determining $M(\omega)$ and $\phi(\omega)$ by using constant-M and constant-N circles.

In using the direct formulas of (6.81), we first calculate

$$R(\omega) = |G(j\omega)| = \frac{1}{\omega\sqrt{\omega^2 + 1}}$$

$$\theta(\omega) = \underline{/G(j\omega)} = -90° - \tan^{-1}(\omega) \tag{6.83}$$

Using these results in (6.81) yields the $M(\omega)$ and $\phi(\omega)$ plots with equally spaced frequencies given in Figure 6.38. Alternately, we can determine $M(\omega)$ and $\phi(\omega)$ as the magnitude and phase of the closed-loop frequency response function given

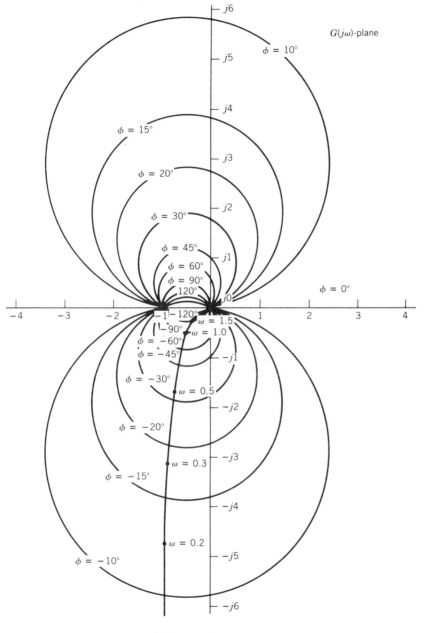

(a) (*Continued*)

Figure 6.36   (*Continued*)

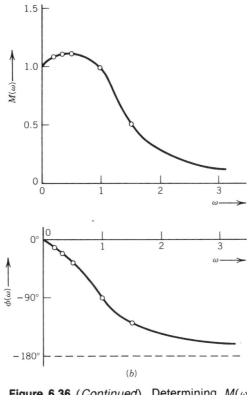

Figure 6.36 (*Continued*) Determining $M(\omega)$ and $\phi(\omega)$ by using constant-M and constant-N circles.

by (6.45). These calculations yield

$$M(\omega) = \frac{1}{\sqrt{(1 - \omega^2)^2 + \omega^2}} = \frac{1}{\sqrt{1 - \omega^2 + \omega^4}}$$

$$\phi(\omega) = -\tan^{-1}\left(\frac{\omega}{1 - \omega^2}\right) \tag{6.84}$$

The student is asked in Problem 6.32 to verify that using (6.83) in (6.81) yields (6.84) for all ω.

6.8.4 Sensitivities to Plant Parameter Variations

Two kinds of errors sometimes associated with closed-loop frequency response curves are: (1) graphical errors, and (2) errors due to an imprecise knowledge of $G(j\omega)$. The first error can occur when using either the constant-M and constant-N

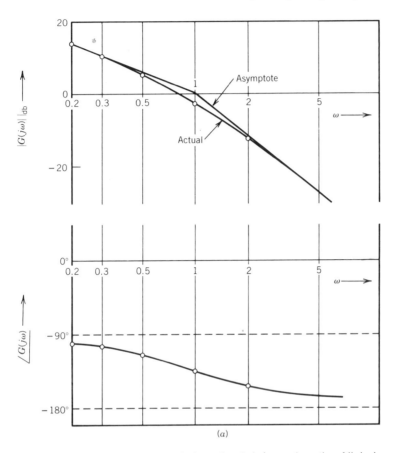

Figure 6.37 Determining $M(\omega)$ and $\phi(\omega)$ by using the Nichols chart.

contours or the Nichols chart in converting from open-loop to closed-loop frequency responses. However, all three methods of this section are subject to errors due to inaccuracies in $G(j\omega)$. We now examine the sensitivities of the closed-loop frequency response magnitude and phase to plant parameter variations.

As discussed in Section 3.4 in introducing the sensitivity concept, we may approximate incremental errors in $M(\omega)$ and $\phi(\omega)$ resulting from small variations in a single plant parameter α as

$$\Delta M(\omega) = \left.\frac{\partial M(\omega, \alpha)}{\partial \alpha}\right|_N \Delta \alpha$$

$$\Delta\phi(\omega) = \left.\frac{\partial \phi(\omega, \alpha)}{\partial \alpha}\right|_N \Delta \alpha \qquad (6.85)$$

where $\Delta\alpha$ is the incremental variation about some nominal value α_N and the

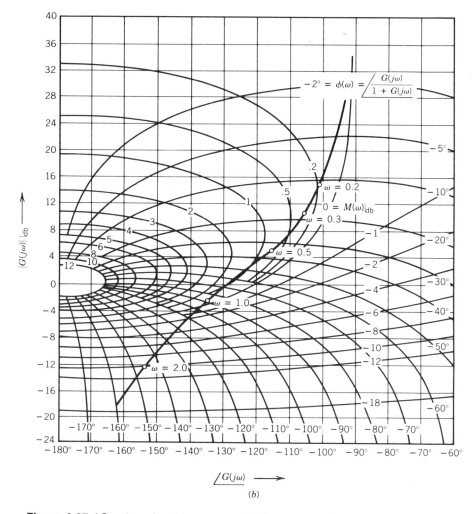

(b)

Figure 6.37 (*Continued*) Determining $M(\omega)$ and $\phi(\omega)$ by using Nichols chart.

notation $(\cdot)|_N$ indicates that the derivatives in (6.85) are evaluated at the nominal condition. By using (6.81) to evaluate the partial derivatives in (6.85), we obtain

$$\frac{\partial M(\omega, \alpha)}{\partial \alpha} = \frac{\partial M(\omega, \alpha)}{\partial R}\left(\frac{\partial R}{\partial \alpha}\right) + \frac{\partial M(\omega, \alpha)}{\partial \theta}\left(\frac{\partial \theta}{\partial \alpha}\right)$$

$$= \frac{1 + R(\omega)\cos\theta(\omega)}{\left[1 + R^2(\omega) + 2R(\omega)\cos\theta(\omega)\right]^{3/2}}\left(\frac{\partial R}{\partial \alpha}\right)$$

$$+ \frac{R^2(\omega)\sin\theta(\omega)}{\left[1 + R^2(\omega) + 2R(\omega)\cos\theta(\omega)\right]^{3/2}}\left(\frac{\partial \theta}{\partial \alpha}\right)$$

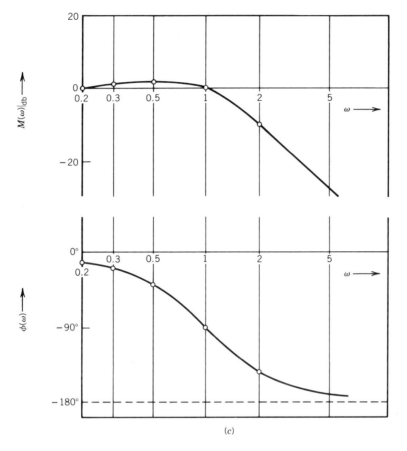

(c)

Figure 6.37 (*Continued*)

$$\frac{\partial \phi(\omega, \alpha)}{\partial \alpha} = \frac{\partial \phi(\omega, \alpha)}{\partial R}\left(\frac{\partial R}{\partial \alpha}\right) + \frac{\partial \phi(\omega, \alpha)}{\partial \theta}\left(\frac{\partial \theta}{\partial \alpha}\right)$$

$$= \frac{-\sin\theta(\omega)}{1 + R^2(\omega) + 2R(\omega)\cos\theta(\omega)}\left(\frac{\partial R}{\partial \alpha}\right)$$

$$+ \frac{1 + R(\omega)\cos\theta(\omega)}{1 + R^2(\omega) + 2R(\omega)\cos\theta(\omega)}\left(\frac{\partial \theta}{\partial \alpha}\right) \qquad (6.86)$$

Alternatively, we may either perform the operations in (6.85) directly on $M(\omega, \alpha)$ and $\phi(\omega, \alpha)$ in (6.45) or determine numerical values for these partial derivatives

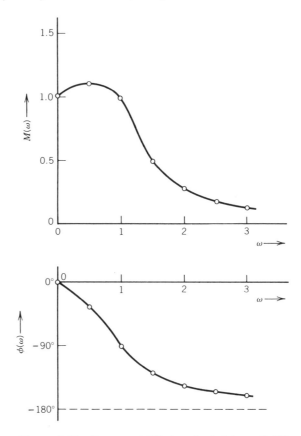

Figure 6.38 Curves of $M(\omega)$ and $\phi(\omega)$ from (6.83).

at a particular frequency by using the incremental relationships

$$\frac{\partial M(\omega, \alpha)}{\partial \alpha} \cong \frac{M(\omega, \alpha + \Delta\alpha) - M(\omega, \alpha)}{\Delta\alpha}$$

$$\frac{\partial \phi(\omega, \alpha)}{\partial \alpha} \cong \frac{\phi(\omega, \alpha + \Delta\alpha) - \phi(\omega, \alpha)}{\Delta\alpha} \tag{6.87}$$

where $\Delta\alpha$ is chosen to be sufficiently small.

EXAMPLE 6.15

Determine the partial derivatives in (6.86) and (6.87) for the plant frequency response function given by

$$G(j\omega) = \frac{1}{j\omega(j\omega + \alpha)} \tag{6.88}$$

TABLE 6.10   Numerical Results for Example 6.15

ω	$M(\omega)$	$\dfrac{\partial M(\omega, \alpha)}{\partial \alpha}\bigg|_N$	$\phi(\omega)$	$\dfrac{\partial \phi(\omega, \alpha)}{\partial \alpha}\bigg|_N^a$
0.0	1.00	0.00	0.0°	0.00
0.5	1.11	− 0.34	− 33.7°	− 0.46
1.0	1.00	− 1.00	− 90.0°	0.00
1.5	0.51	− 0.30	− 129.8°	+ 0.49
2.0	0.28	− 0.09	− 146.3°	+ 0.46
2.5	0.17	− 0.03	− 154.5°	+ 0.38
3.0	0.12	− 0.02	− 159.4°	+ 0.35

[a] These values are yet to be multiplied by 57.3 to convert them from radians to degrees before using in (6.85).

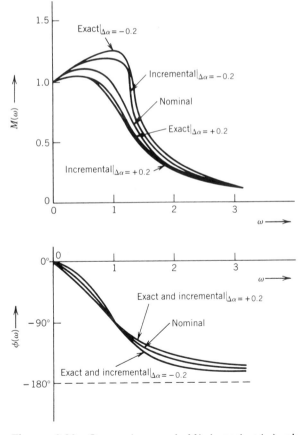

Figure 6.39 Comparisons of $M(\omega)$ and $\phi(\omega)$ obtained by incremental and exact methods for $\Delta\alpha = \pm 0.2$.

where $\alpha_N = 1$. Provide curves that indicate the extent of incremental variations about α_N.

Table 6.10 gives the results of the required calculations, and Figure 6.39 compares these approximate variations with exact calculations for the closed-loop frequency response curves. We see that the incremental method of calculating $M(\omega)$ for $\Delta\alpha = \pm 0.2$ yields values that differ from the exact values by less than 5%, whereas the incremental and exact results for $\phi(\omega)$ are indistinguishable from the curves shown.

SUMMARY

The Nyquist criterion was presented in this chapter for the stability analysis of linear time-invariant feedback systems in the frequency domain. Mapping the s-plane contour which encloses the entire right-half plane into the $GH(s)$-plane was accomplished by first sketching Bode magnitude and phase diagrams for the open-loop frequency response function $GH(j\omega)$. The required polar plot can then be constructed for $s = j\omega$ and the $GH(s)$-plane contour completed by performing a mirror reflection about the real axis, corresponding to $s = -j\omega$. Cases involving poles on the $j\omega$-axis were considered by selecting an infinitesimal semicircle around each such pole, excluding the pole from the interior of the s-plane contour being mapped into the $GH(s)$-plane. Comparisons were made between Nyquist criterion results and those obtained by using the root locus technique of Chapter 5.

Relative stability concepts in the frequency domain involved the closed-loop system peak magnitude of the frequency response (M_p) and margins for open-loop system gain (GM) and phase (PM). Numerical methods were presented for calculating M_p, GM, and PM. The closed-loop system frequency response was obtained for negative unity-feedback systems by constant-M and constant-N circles in the $G(j\omega)$-plane, the Nichols chart, and direct formulas.

REFERENCES

1. F. Jay, editor-in-chief. *IEEE Standard Dictionary of Electrical and Electronics Terms*, 2nd ed. New York: The Institute of Electrical and Electronics Engineers, 1977.

2. J. L. Melsa and D. G. Schultz. *Linear Control Systems*. New York: McGraw-Hill, 1969.

3. J. J. D'Azzo and C. H. Houpis. *Linear Control System Analysis and Design: Conventional and Modern*, 2nd ed. New York: McGraw-Hill, 1981.

4. J. E. Gibson and F. B. Tuteur. *Control System Components*. New York: McGraw-Hill, 1958.

5. H. Nyquist. "Regeneration Theory." *Bell System Technical Journal*, Vol. 11 (January 1932), pp. 126–147.
6. I. M. Horowitz. *Synthesis of Feedback Systems*. New York: Academic Press, 1963.
7. H. M. James, N. B. Nichols, and R. S. Phillips. *Theory of Servomechanisms*. New York: McGraw-Hill, 1947.
8. B. C. Kuo. *Automatic Control Systems*, 4th ed. Englewood Cliffs, N.J.: Prentice-Hall, 1982.

PROBLEMS

(§6.1) **6.1** Determine the closed-loop system bandwidth (BW) and peak magnitude for the closed-loop frequency response curves (magnitude only) shown in Figure 6.40.

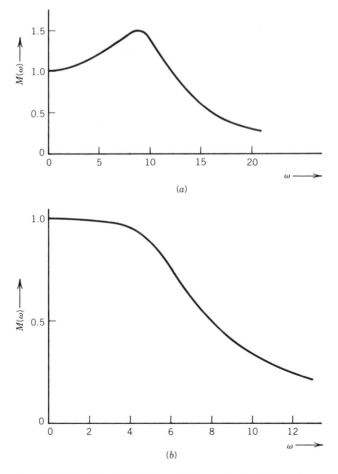

Figure 6.40 The closed-loop frequency response magnitude curves for Problem 6.1.

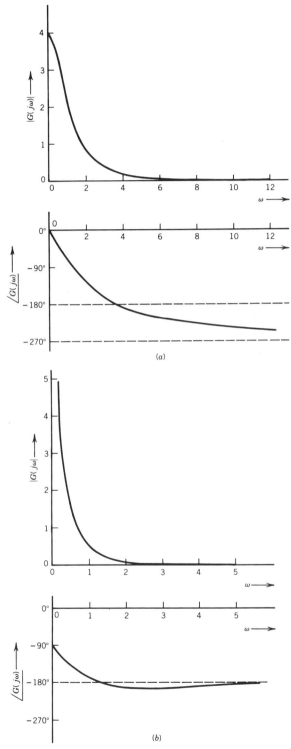

Figure 6.41 The open-loop frequency response curves for Problem 6.2.

(§6.1) **6.2** Determine the gain margin (GM) and phase margin (PM) for each of the two systems having the open-loop frequency response curves for magnitude and phase shown in Figure 6.41. Assume negative unity feedback.

(§6.2) **6.3** Determine the closed-loop bandwidth, closed-loop peak magnitude, gain margin, and phase margin for each of the following systems. The given open-loop transfer functions are fed back negatively in each case to form the closed-loop systems.

a. $G(s) = \dfrac{4}{s(s + 1)}$

b. $G(s) = \dfrac{10}{s(s + 2)}$

c. $G(s) = \dfrac{100}{s(s + 1)}$

(§6.2) **6.4** Suppose we want to ensure that the closed-loop system peak magnitude (M_p) is less than or equal to 1.4 for each of the systems in Figure 6.42. Determine the range of α that satisfies this condition.

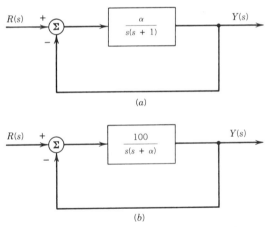

(a)

(b)

Figure 6.42 Closed-loop systems for Problem 6.4.

(§6.2) **6.5** Determine the three ranges of K that satisfy the following criteria for the system given in Figure 6.43.

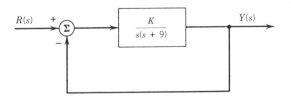

Figure 6.43 The closed-loop system for Problem 6.5.

    a. The closed-loop system bandwidth (BW) is less than 10 radians/second.

    b. The closed-loop system peak magnitude (M_p) is less than 1.2.

    c. The phase margin (PM) is greater than $50°$.

(§§3.4, **6.6** Consider the positional servomechanism of Figure 6.44 that has
4.2,   parameters given by $V = 20\pi$ volts, $n = 10$ turns, $R = 10$ Ω, and
6.2) $L = 0$. Let $K_T = 0.1$, $J_{eq} = 0.001$, and $B_{eq} = 0.008$ in SI units.
    Determine that range of positive K for which the following three
    design requirements are satisfied simultaneously:

    a. The percent overshoot of $\theta_c(t)$ due to a step input, that is, due
    to $\theta_r(s) = 1/s$, is not greater than 30%.

    b. The phase margin is between $40°$ and $50°$.

    c. The steady-state error due to a unit ramp input, that is, due to
    $\theta_r(s) = 1/s^2$, is not greater than 0.1 radian.

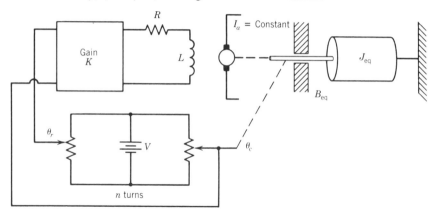

Figure 6.44 The positional servomechanism described in Problem 6.6.

(§6.2, **6.7** Approximate expressions are often useful in the design of a feed-
6.3)   back system. One such approximation between phase margin (PM)
    and the peak magnitude (M_p) of the closed-loop frequency response
    is

$$PM = \sin^{-1}\left(\frac{1}{M_p}\right)$$

For the system shown in Figure 6.45, demonstrate the accuracy of

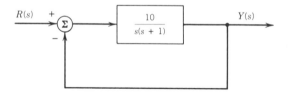

Figure 6.45 The system for Problem 6.7.

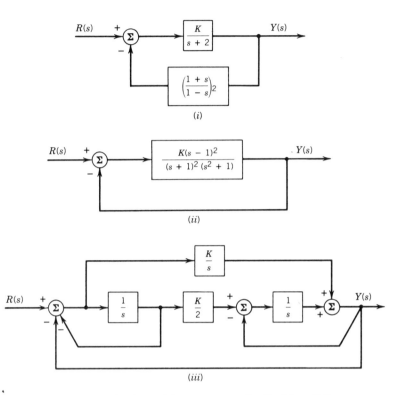

Figure 6.46 Closed-loop systems for Problem 6.8.

the above expression by comparing the result with that obtained from (6.12). In evaluating the above expression, first compute M_p from (6.9).

(§6.3) **6.8** Consider the closed-loop systems of Figure 6.46.

    a. Provide rough sketches only of Bode magnitude and phase plots for the open-loop transfer functions. Show asymptotes in all cases.

    b. Sketch polar plots of $GH(j\omega)$ in all cases.

(§6.3) **6.9** Sketch linear plots, Bode diagrams, log-log plots, and polar plots for each of the following open-loop transfer functions for $s = j\omega$.

    a. $GH(s) = \dfrac{10(s + 1)}{s + 2}$

    b. $GH(s) = \dfrac{100(s - 1)}{s^2 + 29s + 100}$

    c. $GH(s) = \dfrac{100(s + 10)}{s(s + 1)^3}$

(§6.4) **6.10** Use s-plane vectors to construct contour maps in the $F(s)$-plane for the contours shown in Figure 6.47 for each of the given $F(s)$

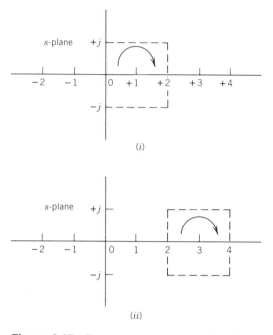

Figure 6.47 The s-plane contours for Problem 6.10.

functions. Determine the number of encirclements of the origin in the $F(s)$-plane in each case.

a. $F(s) = \dfrac{10(s - 1)}{(s - 3)^2}$

b. $F(s) = \dfrac{10(s - 1)^2}{s - 3}$

(§ 6.5) **6.11** For the system of Figure 6.48, determine the range of positive K for stability. Sketch the Nyquist plot (roughly) for $K = 2$ and $K = 10$, and use the Nyquist criterion to determine whether the closed-loop system is stable for each gain. A Bode plot (particularly the phase part) may be helpful as a preliminary step, but this is not necessary.

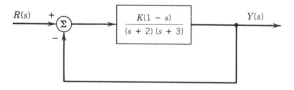

Figure 6.48 The closed-loop system for Problem 6.11.

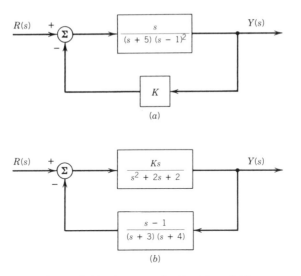

Figure 6.49 Systems for Problem 6.12.

(§6.5) **6.12** Construct the polar plots for the open-loop frequency responses of the systems given in Figure 6.49, and apply the Nyquist criterion to determine the number of closed-loop poles in the right-half plane. Draw a plot for positive K and another for negative K. Specify your ranges of K in a table for different stability conditions.

(§6.5) **6.13** Consider the feedback control system of Figure 6.50. Sketch the polar plot in the $GH(j\omega)$-plane, and use the Nyquist criterion to determine the number of closed-loop poles in the right-half s-plane as K varies from 0 to $+\infty$.

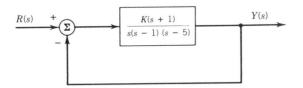

Figure 6.50 The feedback control system for Problem 6.13.

(§6.5) **6.14** Sketch Bode diagrams for the open-loop plant of the system in Figure 6.51, and then apply the Nyquist criterion for positive K

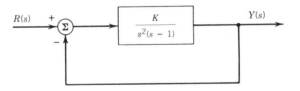

Figure 6.51 The system for Problem 6.14.

only to determine the number of closed-loop poles in the right half of the s-plane.

(§§6.5, **6.15** As K varies from 0 to $+\infty$ for the system of Figure 6.52,
5.3)

    a. Apply the Nyquist criterion to determine the number of closed-loop poles in the right-half s-plane.

    b. Sketch the root-locus of the closed-loop poles in the s-plane. Determine all $j\omega$-axis crossings, but do not specify exact breakaway points.

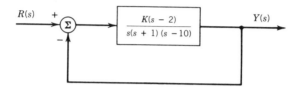

Figure 6.52 The system for Problem 6.15.

(§§6.5, **6.16** For the systems of Figure 6.53, apply the Nyquist criterion to
5.3,   determine the number of closed-loop poles in the right-half s-plane
5.4)   as K varies from $-\infty$ to $+\infty$. You are not to use the Routh table in this part of the problem. Sketch the root locus plots to verify your solutions.

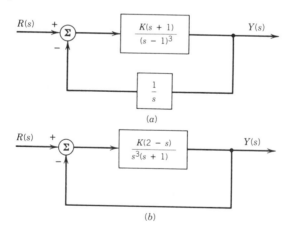

Figure 6.53 The systems for Problem 6.16.

(§§6.5, **6.17** Repeat the instructions of Problem 6.16 for the systems of Figure
5.3,   6.54.
5.4)

(§ 6.5) **6.18** Apply the Nyquist criterion to the closed-loop systems of Problem 5.22. Specify in a table the number of closed-loop poles in the right-half s-plane as K varies from $-\infty$ to $+\infty$. Show the polar plot in the $GH(j\omega)$-plane, and indicate clearly values for all real-axis crossings. Indicate in your table those ranges of K for stability.

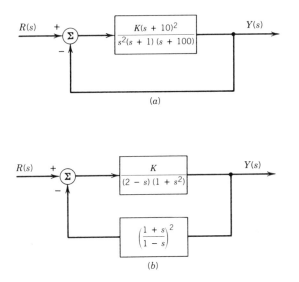

Figure 6.54 The systems for Problem 6.17.

(§6.5) **6.19** For the systems of Figure 6.55, use the Nyquist criterion to determine the number of closed-loop poles in the right-half of the
s-plane as K varies from −∞ to +∞. Tabulate your results in a
table showing the different stable and unstable ranges of K.

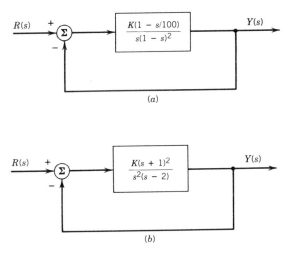

Figure 6.55 The systems for Problem 6.19.

(§§6.5, **6.20** Several methods may be used to investigate the stability of feed-
5.3) back systems. Using either the Nyquist criterion or the root locus
technique (not both), determine the stability of the system in Figure
6.56 as K varies from 0 to +∞. Specify the number of closed-loop
poles in the right-half s-plane as K varies.

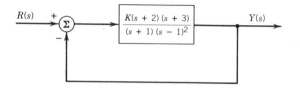

Figure 6.56 The feedback system for Problem 6.20.

(§6.6) **6.21** For the $G(j\omega)$-plane polar plot given in Figure 6.57, construct several constant-M circles that intersect the polar plot and sketch a linear closed-loop magnitude frequency plot of M versus ω. Use (6.9) to compute M_p for this second-order system, and compare the result with the maximum value of your curve obtained graphically.

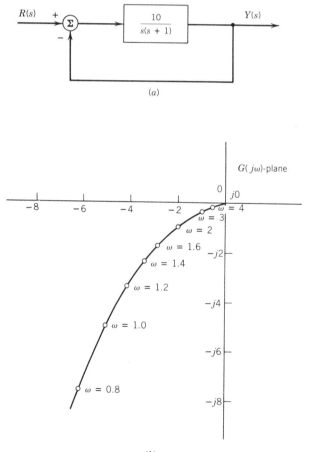

Figure 6.57 The feedback system and associated open-loop polar plot for Problem 6.21.

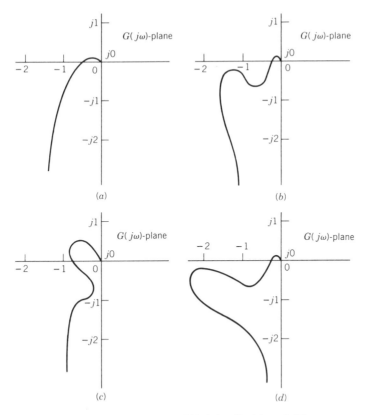

Figure 6.58 Polar plots of $G(j\omega)$ for Problem 6.22.

(§6.6) **6.22** For each of the polar plots of $G(j\omega)$ given in Figure 6.58, determine the gain margin, phase margin, and the maximum value of the closed-loop frequency magnitude curve.

(§6.6) **6.23** The closed-loop system bandwidth may be found graphically for those systems for which $M(\omega = 0) = 1$ by constructing the constant-M circle for $M = 0.707$ and noting the bandwidth as the frequency at the intersection of the polar plot. Determine and closed-loop bandwidth for the two cases given in Figure 6.59.

(§6.6) **6.24** Use Newton's method to compute the phase crossover (ω_{PC}) and gain crossover (ω_{GC}) frequencies for the following plants having negative unity feedback. Determine the gain margin and phase margin for each case.

  a. $G(s) = \dfrac{1}{s(s + 4)^2}$

  b. $G(s) = \dfrac{5}{s^3 + 10s^2 + s + 1}$

  c. $G(s) = \dfrac{1}{(s + 1)^3}$

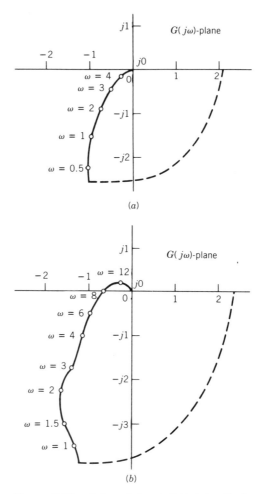

Figure 6.59 Polar plots of $G(j\omega)$ for Problem 6.23.

(§6.6) **6.25** Determine the closed-loop system peak magnitude (M_p) for the systems of Figure 6.60 by

    a. Using (6.64) and (6.60).
    b. Computing ζ and then using (6.9).

(§6.6) **6.26** Consider the system of Figure 6.61. Find M_p by using (6.63) to form an equation for ω_p, using Newton's method (if necessary) to solve for ω_p, and substituting that value into (6.60) to determine M_p.

(§6.7) **6.27** For the systems of Figure 6.62, sketch both root locus and Nyquist plots as K varies from $-\infty$ to $+\infty$. Determine the phase crossover frequency in each case by selecting an appropriate method from Table 6.7. Summarize in a table the results obtained by applying the Nyquist criterion to the given systems.

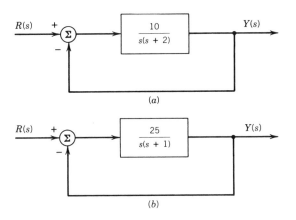

Figure 6.60 Closed-loop systems for Problem 6.25.

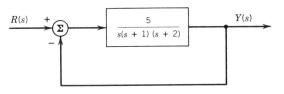

Figure 6.61 The system for Problem 6.26.

(§6.8) **6.28** Using constant-M and constant-N contours, plot curves of the closed-loop frequency response magnitude and phase for the negative unity-feedback systems with plant frequency response functions given by

  a. $\quad G(j\omega) = \dfrac{10}{j\omega(j\omega + 2)}$

  b. $\quad G(j\omega) = \dfrac{3}{j\omega(j\omega + 1)(j\omega + 2)}$

(§6.8) **6.29** Repeat the instructions of Problem 6.28 using the Nichols chart.

(§6.8) **6.30** Determine and plot $M(\omega)$ in decibels versus ω for the system of Problem 6.21 by using the Nichols chart. Determine and plot $\phi(\omega)$ versus ω for this system by using

  a. Constant-N circles.
  b. The Nichols chart.

(§6.8) **6.31** Determine and plot curves of $M(\omega)$ and $\phi(\omega)$ versus ω for the systems of Problem 6.28 by using

  a. Direct formulas in (6.81).
  b. Direct formulas resulting from (6.45).

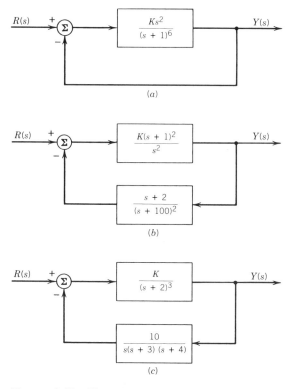

Figure 6.62 Closed-loop systems for Problem 6.27.

(§6.8) **6.32** Verify that the results in (6.84) are obtained by substituting (6.83) into (6.81) for all ω.

(§6.8) **6.33** Determine the partial derivatives in (6.85) for $\omega = 0, 1, 2$, and 3 for the closed-loop system having a plant frequency response function given by

$$G(j\omega) = \frac{10}{j\omega(j\omega + \alpha)}$$

and evaluate these derivatives for $\alpha_N = 1$. As in Figure 6.39, plot nominal, incremental, and exact magnitude and phase curves for $\Delta\alpha = \pm 0.4$.

PART III

CONTROL SYSTEM DESIGN

7

Classical Design by Series Compensation

The control engineer's possibilities for design are far more extensive than the comparatively routine methods of analysis. As described in Part II, analysis techniques are often preset step-by-step procedures used to determine system time responses (Chapter 4), stability (Chapter 5), and frequency responses (Chapter 6). In contrast to the analysis goal of obtaining the output response when given the system and its input, the goal of design is to determine one of the several possible systems yielding a desired output response for a given input. Analysis enables us to explain what can happen with an existing system under known circumstances; design requires that we create a new system that meets certain standards of performance. The control design task requires creativity and ingenuity in selecting both the structure and parameters of a feedback controller to be placed around the plant such that given time response or frequency response performance specifications are satisfied.

A mastery of analysis procedures prior to attempting control systems design is important because (1) it provides some experience on the output responses to be expected from certain classes of plants and controllers with given inputs, (2) it provides the background concepts, for example, root locus and frequency response techniques, necessary for solving design problems, and (3) it permits the control engineer to verify subsequently that the designed system model does meet the required performance specifications.

Controller design can be both challenging and rewarding. The challenge is selecting a controller structure and parameters that satisfy the design requirements and are relatively insensitive to plant parameter and input variations. A reward is observing the successful performance of the designed controller when applied to the actual plant in its operating environment. Further field testing may reveal a need for refining the plant model description. If so, a redesign of the controller may become necessary.

The material for design in this book is described in two chapters. Chapter 7 presents root locus and frequency response techniques for the classical design of series compensators to be inserted directly into the control loop. The resulting closed-loop system is of higher order than the original plant model. Chapter 8 describes state variable design using optimal control theory, which requires that more variables than only the plant output are accessible. Often, as in pole placement or state variable feedback (SVFB), all state variables are required for control. The closed-loop system resulting from state variable design is the same order as the original plant model.

7.1 DESIGN SPECIFICATIONS REVISITED

The time response and frequency response design specifications defined in Chapters 4 and 6 are examined in this section with respect to whether low or high values of the forward loop gain K are required. Recall that time response specifications are percent overshoot, rise time, delay time, settling time, and steady-state error. Frequency response specifications are closed-loop system bandwidth, system peak magnitude, phase margin, and gain margin.

7.1.1 Percent Overshoot and Steady-State Error

Depending on the order and structure of the closed-loop system, the adjustment of the forward loop gain K for an acceptable percent overshoot and steady-state error can become a design tradeoff situation. For example, consider a second-order Type 1 system that has negative unity feedback with open-loop poles at the origin and $-p_1$, that is,

$$G(s) = \frac{K}{s(s + p_1)} \qquad (7.1)$$

Figure 7.1 shows the root locus plot of the closed-loop poles in the s-plane as K varies from 0 to $+\infty$. The two loci start on the open-loop poles, move along the real axis toward each other, and form a breakaway point at $s_B = -p_1/2$ with a value of K equal to $p_1^2/4$. For larger values of K, the underdamped time response for a step input has a percent overshoot given by (4.25) in terms of the damping coefficient ζ. Lines of constant ζ in the s-plane are defined by

$$\theta = \cos^{-1}(\zeta) \qquad (7.2)$$

where θ is the angle shown in Figure 7.1. If the percent overshoot (%OS) is required to be less than or equal to some specified value %OS$_{\text{max}}$, then K must be sufficiently small ($\leq K_{\text{max}}$) to cause the closed-loop poles to lie within or on the boundaries of the wedge-shaped region in the s-plane.

For the Type 1 system having the $G(s)$ given in (7.1), we determine the steady-state error (e_{ss}) due to a unit ramp input as $1/K_v$, where K_v is the velocity error constant given by

$$K_v = \lim_{s \to 0} sG(s) = K/p_1 \qquad (7.3)$$

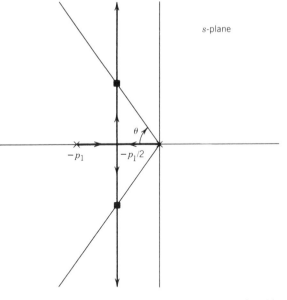

Figure 7.1 The root locus sketch for (7.1) with $K > 0$.

Therefore,

$$e_{ss} = 1/K_v = p_1/K \tag{7.4}$$

which requires a sufficiently large value of K to ensure that e_{ss} is not greater than some specified value $e_{ss_{max}}$. In brief, the percent overshoot design specification requires a low value of K and the steady-state error specification a high value of K.

EXAMPLE 7.1

For a system defined by the $G(s)$ in (7.1) with $p_1 = 2$ and negative unity feedback, determine the ranges of K satisfying each of the following design specifications.

$$\%OS \leq 15\%$$

$$e_{ss}\big|_{\substack{\text{unit} \\ \text{ramp}}} \leq 0.05 \text{ radians} \tag{7.5}$$

In satisfying the first requirement in (7.5), we use (4.32) to determine ζ_{min} corresponding to a 15% overshoot as 0.517 and then relate the range of K to ζ by identifying K as the square of the natural frequency ω_n^2 and p_1 as $2\zeta\omega_n$.

Therefore,

$$\zeta = \frac{p_1}{2\sqrt{K}} = \frac{1}{\sqrt{K}} \tag{7.6}$$

from which $K \leq K_{max} = 1/\zeta_{min}^2 = 3.74$.

Since $e_{ss} = 2/K$ from (7.4), we may satisfy the second requirement in (7.5) by setting $K \geq 2/0.05 = 40$. Obviously, the conflicting design requirements in (7.5) cannot be met simultaneously by adjusting only the gain K.

7.1.2 System Peak Magnitude, Phase Margin, and Gain Margin

Frequency response specifications for M_p, PM, and GM can often be satisfied by selecting a suitably low value of K. Figure 7.2 shows a typical $G(j\omega)$-plane plot for three values of K corresponding to the three equalities for the design specifications given by

$$M_p \leq M_{p\,max}$$

$$PM \geq PM_{min}$$

$$GM \geq GM_{min} \tag{7.7}$$

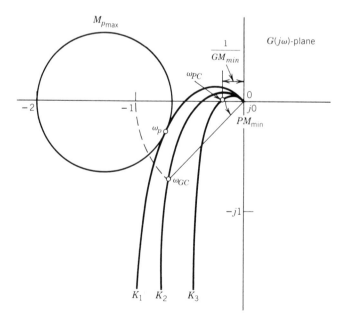

Figure 7.2 A typical polar plot of $G(j\omega)$ for three values of K corresponding to M_p, PM, and GM design specifications.

We may satisfy all three specifications simultaneously by selecting the most restrictive (lowest) value of K_{max}. For example, we would select $K = K_3$ for the case shown in Figure 7.2, since any value higher than K_3 violates the third requirement in (7.7). Incidentally, we note that any value of K higher than K_2 violates the second and third requirements in (7.7) and any value higher than K_1 violates all three requirements.

EXAMPLE 7.2

Determine the allowable range of K that simultaneously satisfies the three design specifications

$$M_p \leq 1.5$$

$$PM \geq 40°$$

$$GM \geq 4 \tag{7.8}$$

for the negative unity-feedback system having a plant transfer function $G(s)$ given by

$$G(s) = \frac{K}{s(s + 2)(s + 40)} \tag{7.9}$$

Figure 7.3 shows the $G(j\omega)$-plane plot for (7.9) with K set equal to 250, an approximate value selected by examining the phase margin specification. We first

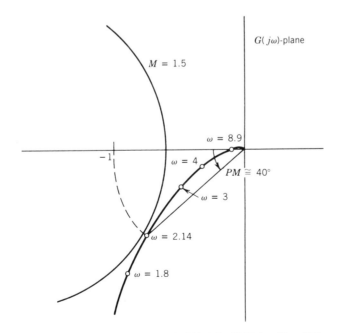

Figure 7.3 The polar plot of $G(j\omega)$ in (7.9) for $K = 250$.

determine the gain crossover frequency ω_{GC} that yields a phase margin of $40°$ for (7.9) from (6.2) as

$$PM = 40° = 180° - \left| \underline{/G(j\omega_{GC})} \right|$$

$$= 180° - \left| \left[-90° - \tan^{-1}\left(\frac{\omega_{GC}}{2} \right) - \tan^{-1}\left(\frac{\omega_{GC}}{40} \right) \right] \right| \qquad (7.10)$$

which reduces to

$$\tan^{-1}\left(\frac{\omega_{GC}}{2} \right) + \tan^{-1}\left(\frac{\omega_{GC}}{40} \right) = 50° \qquad (7.11)$$

Taking the tangent of both sides of (7.11) and using the two-angle formula yields

$$\frac{\dfrac{\omega_{GC}}{2} + \dfrac{\omega_{GC}}{40}}{1 - \left(\dfrac{\omega_{GC}}{2} \right)\left(\dfrac{\omega_{GC}}{40} \right)} = \tan 50° \qquad (7.12)$$

which can be solved to give $\omega_{GC} = 2.14$ radians/second. We now solve

$$\frac{K}{\omega_{GC}\sqrt{\omega_{GC}^2 + 2^2}\,\sqrt{\omega_{GC}^2 + 40^2}} = 1 \qquad (7.13)$$

to obtain 251.1 as the value of K for which $PM = 40°$. Therefore, the range of K satisfying the second requirement in (7.8) is $0 \le K \le 251.1$.

It appears from Figure 7.3 that the phase margin design specification is the most restrictive of the three given requirements. To examine this conjecture further, we calculate the values of M_p and GM corresponding to $K = 251.1$. Using (6.64), we solve for ω_p from

$$\left(K - 42\omega_p^2 \right) \frac{d}{d\omega_p} \left(K - 42\omega_p^2 \right) + \omega_p\left(80 - \omega_p^2 \right) \frac{d}{d\omega_p}\left[\omega_p\left(80 - \omega_p^2 \right) \right] = 0 \quad (7.14)$$

which simplifies to yield

$$3\omega_p^4 + 3208\omega_p^2 + (6400 - 84K) = 0 \qquad (7.15)$$

For $K = 251.1$, we obtain $\omega_p = 2.136$ radians/second and compute M_p from

$$M_p = \left| \frac{G(j\omega)}{1 + G(j\omega)} \right| \Bigg|_{\substack{K=251.1 \\ \omega = \omega_p = 2.136}}$$

$$= \frac{K}{\sqrt{\left(K - 42\omega_p^2 \right)^2 + \omega_p^2\left(80 - \omega_p^2 \right)^2}} \Bigg|_{\substack{K=251.1 \\ \omega_p = 2.136}} \qquad (7.16)$$

which gives $M_p = 1.46$. Therefore, the design requirement for M_p in (7.8) is barely satisfied for $K = 251.1$, indicating that only a slightly higher value of K would be allowed if $PM \ge 40°$ were not also a design requirement.

The gain margin specification $GM \geq 4$ in (7.8) is easily applied by using the Routh table to determine the maximum K for stability (K^*) as 3360. Therefore, for $K = 251.1$, we have a gain margin of $K^*/K = 3360/251.1 = 13.4$, which lies well within the specification. Moreover, it is easy to determine $0 \leq K \leq 840$ as the range of K that satisfies $GM \geq 4$ in (7.8). In summary, the range that satisfies all three design specifications simultaneously is $0 \leq K \leq 251.1$.

7.1.3 Rise, Delay, and Settling Times and System Bandwidth

Consider the time response $y(t)$ for a second-order underdamped system having a unit step input. We recall that $y(t)$ was given in (4.16) as

$$y(t) = 1 - \frac{e^{-\zeta\omega_n t}}{\sqrt{1 - \zeta^2}} \sin(\omega_d t + \theta) \tag{4.16}$$

Although exact normalized curves of rise time, delay time, and settling time were presented in Figure 4.6, we are now interested in examining related design specifications by forming conservative estimates based on only the upper and lower exponential envelope curves denoted by $y_e^+(t)$ and $y_e^-(t)$, respectively, and given by

$$y_e^+(t) = 1 + e^{-\zeta\omega_n t}$$

$$y_e^-(t) = 1 - e^{-\zeta\omega_n t} \tag{7.17}$$

For example, we may approximate the rise time t_r by the time required for the lower envelope $y_e^-(t)$ to move between values of 0.1 and 0.9, that is,

$$t_r \cong t\big|_{y_e^-(t)=0.9} - t\big|_{y_e^-(t)=0.1}$$

$$\cong \left[\frac{1}{-\zeta\omega_n} \ln(1 - 0.9)\right] - \left[\frac{1}{-\zeta\omega_n} \ln(1 - 0.1)\right]$$

$$\cong \frac{1}{\zeta\omega_n} \ln 9^1 \tag{7.18}$$

Similarly, approximate expressions for delay time t_d and settling time t_s are

$$t_d \cong t\big|_{y_e^-(t)=0.5} = \frac{1}{-\zeta\omega_n} \ln(1 - 0.5) = \frac{1}{\zeta\omega_n} \ln 2$$

$$t_s \cong t\big|_{\substack{y_e^-(t)=1-\delta_s \\ y_e^+(t)=1+\delta_s}} = \frac{1}{-\zeta\omega_n} \ln \delta_s = \frac{1}{\zeta\omega_n} \ln\left(\frac{1}{\delta_s}\right) \tag{7.19}$$

where $2\delta_s$ is the width of the settling time band, that is, $1 - \delta_s \leq y(t) \leq 1 + \delta_s$.

---

[1]As an alternate expression for rise time, Fortmann and Hitz [1] suggest the approximation $t_r \cong 2.5/\omega_n$.

These approximate expressions for t_r, t_d, and t_s are independent of the gain K for the second-order system described in (7.1) and Figure 7.1, since $\zeta\omega_n = p_1/2$. However, for the third-order system described in (7.9) the real parts of the dominant complex closed-loop pole pair are functions of K and, hence, these values of t_r, t_d, and t_s depend on K.

Suppose design specifications are given as

$$t_r \leq t_{r_{max}}$$

$$t_d \leq t_{d_{max}}$$

$$t_s \leq t_{s_{max}} \tag{7.20}$$

Substituting (7.18) and (7.19) into (7.20) and rearranging yields

$$-\zeta\omega_n \leq \frac{-\ln 9}{t_{r_{max}}}$$

$$-\zeta\omega_n \leq \frac{-\ln 2}{t_{d_{max}}}$$

$$-\zeta\omega_n \leq \frac{-\ln(1/\delta_s)}{t_{s_{max}}} \tag{7.21}$$

Figure 7.4 shows the acceptable (shaded) region in the s-plane corresponding to

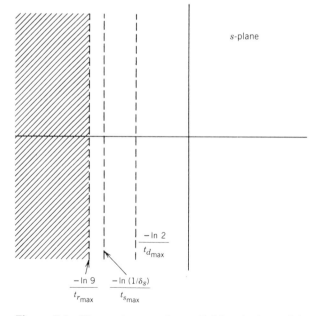

Figure 7.4 The s-plane region satisfying t_r, t_d, and t_s design specifications simultaneously.

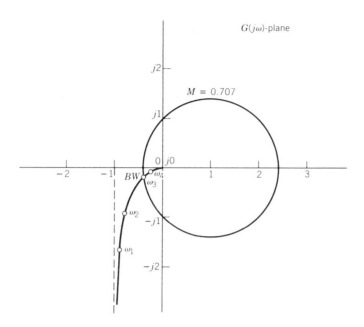

Figure 7.5 Using the constant-M circle for $M = 0.707$ to determine system bandwidth.

the most negative of the right-hand sides of the inequalities in (7.21). Using this most restrictive condition guarantees that all three design specifications are satisfied simultaneously.

A design requirement for the closed-loop system bandwidth BW may be expressed as

$$BW_{min} \le BW \le BW_{max} \qquad (7.22)$$

where the lower bound BW_{min} ensures that the system bandwidth is large enough to enable the system to respond to all command inputs of interest and the upper bound BW_{max} is selected to prohibit excessive amounts of high-frequency noise from interfering with proper system operations. Figure 7.5 shows the determination of system bandwidth for closed-loop systems having unity gain at $\omega = 0$ by using the constant-M circle for $M = 0.707$ in the $G(j\omega)$-plane.

EXAMPLE 7.3

For the second-order system of Example 7.1 having the $G(s)$ given in (7.1) with $p_1 = 2$, determine the ranges of K satisfying each of the design specifications listed in Table 7.1.

Of the nine requirements to be satisfied, ranges of K for the percent overshoot and steady-state error criteria were determined in Example 7.1. The three remaining time response specifications, that is, rise time, delay time, and settling time,

TABLE 7.1 Design Specifications and Ranges of K for Example 7.3

Design Specification	Design Formula $K_{max} = [p_1/2\zeta_{min}]^2$	Range of K
%OS ≤ 15%	$\zeta_{min} = \dfrac{\ln(100/\%OS_{max})}{\sqrt{\pi^2 + \ln^2(100/\%OS_{max})}}$	$K \leq 3.74$
$t_r \leq 3.0$ sec	$-p_1/2 = -\zeta\omega_n \leq -\ln 9/t_{r_{max}}$	$-p_1/2 = -1 \leq -0.73$
$t_d \leq 1.0$ sec	$-p_1/2 = -\zeta\omega_n \leq -\ln 2/t_{d_{max}}$	$-p_1/2 = -1 \leq -0.69$
$t_s \leq 4.0$ sec ($\delta_s = 5\%$)	$-p_1/2 = -\zeta\omega_n \leq -\ln(1/\delta_s)/t_{s_{max}}$	$-p_1/2 = -1 \leq -0.75$
$e_{ss}\big|_{\substack{unit\\ ramp}} \leq 0.05$ rad	$e_{ss} = 1/K_v = 1/(K/p_1) \leq 0.02$	$K \geq p_1/0.05 = 40$
$M_p \leq 1.5$	$\zeta_{min} = \dfrac{1}{\sqrt{2}}\sqrt{1 - \sqrt{1 - 1/M_p^2}}$	$K \leq 7.85$
$PM \geq 40°$	$\zeta_{min} = \dfrac{1}{2[(\alpha^2 + 1/2)^2 - 1/4]^{1/4}}$ where $\alpha = \tan(90° - PM_{min})$	$K \leq 7.42$
$GM \geq 4$	$K_{max} = K^*/GM_{min} = +\infty/4 = +\infty$	$K \leq +\infty$
$1.0 \leq BW \leq 5.0$ (rad/sec)	$K_{\substack{min\\(max)}} = BW_{\substack{min\\(max)}}\left[\sqrt{2BW^2_{\substack{min\\(max)}} + p_1^2} - BW_{\substack{min\\(max)}}\right]$	$1.45 \leq K \leq 11.74$

are independent of K but are satisfied for the given system, according to the design formulas in (7.21). Design formulas for closed-loop system peak magnitude, phase margin, and system bandwidth can be obtained by forming the algebraic inverses of the corresponding analysis equations in (6.9), (6.12), and (6.7), respectively. As described in Example 7.1, the design formula for ζ_{min} in terms of $\%OS_{max}$ was derived from (4.25) as (4.32). The gain margin is infinite for this example, since there are no $j\omega$-axis crossings for $K > 0$. It is not possible to satisfy all design requirements of Table 7.1 simultaneously by adjusting only the gain K. We emphasize again that the design formulas in Table 7.1 are valid only for underdamped second-order systems described by (7.1).

EXAMPLE 7.4

Results from Example 7.2 provided a range $0 \leq K \leq 251.1$ for satisfying the three frequency response specifications in (7.8) simultaneously. Determine the (possibly more restrictive) range of K that satisfies not only these three criteria but also the first four time response specifications listed in Table 7.1. What is the system bandwidth for the minimum and maximum values of K in this range?

Figure 7.6 shows the approximate s-plane region for which the time response specifications are satisfied. This approximation is based on using the dominant complex poles to yield an approximate second-order system. The intersection of the wedge-shaped region from Figure 7.1 with the most restrictive of the three semi-infinite regions for t_r, t_d, and t_s provides the allowable region for meeting all four requirements simultaneously. In this case we see from Table 7.1 that the settling time requirement, having $-\ln(1/\delta_s)/t_{s_{max}} = -0.75$, is the most restrictive.

The root locus for positive K is sketched in Figure 7.6. We next determine that range of K for which all three closed-loop poles are inside the shaded (acceptable) s-plane region. If we assume (and later verify) that the root locus emerges from the shaded region through the $\%OS$ boundary, then we can easily show from the angle requirement that $\theta_1 + \theta_2 + \theta_3 = 180°$ or $\theta_2 + \theta_3 = 180° - \theta_1 = \theta$. Taking the tangent of $\theta_2 + \theta_3$, we have

$$\tan(\theta_2 + \theta_3) = \frac{\tan\theta_2 + \tan\theta_3}{1 - \tan\theta_2 \tan\theta_3}$$

$$= \frac{\dfrac{\beta}{2 - \alpha} + \dfrac{\beta}{40 - \alpha}}{1 - \left(\dfrac{\beta}{2 - \alpha}\right)\left(\dfrac{\beta}{40 - \alpha}\right)} = \tan\theta \qquad (7.23)$$

where $\beta = \alpha \tan\theta = \alpha \tan[\cos^{-1}(\zeta_{min})]$, and ζ_{min} is determined as 0.517 from the $\%OS$ design formula in Table 7.1. Solving (7.23) yields $\alpha = 0.96$ and $\beta = 1.58$. Since $-\alpha = -0.96$ is less than -0.75 from the settling time design specification,

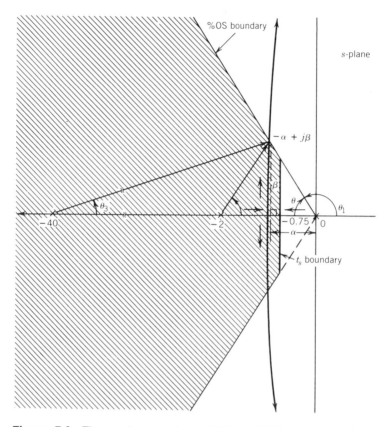

Figure 7.6 The s-plane region satisfying %OS, t_r, t_d, and t_s design specifications simultaneously.

we have shown that the assumed emergence of the root locus through the %OS boundary is indeed correct. Using s-plane vectors, we compute K_{max} as

$$K_{max} = \sqrt{\alpha^2 + \beta^2} \sqrt{(2 - \alpha)^2 + \beta^2} \sqrt{(40 - \alpha)^2 + \beta^2} \qquad (7.24)$$

which yields $K_{max} = 136.9$. Moreover, we determine K_{min} by s-plane vectors drawn to the point along the real axis where the shaded region is entered by the locus starting on the open-loop pole at the origin $(-0.75 + j0)$. Therefore,

$$K_{min} = (0.75)(2 - 0.75)(40 - 0.75) = 36.8 \qquad (7.25)$$

and the range of K that satisfies all seven criteria under consideration is $36.8 \leq K \leq 136.9$.

Figure 7.7 shows the constant-M circle for $M = 0.707$ in the $G(j\omega)$-plane with polar plots given for $K_{min} = 36.8$ and $K_{max} = 136.9$. We may use Newton's

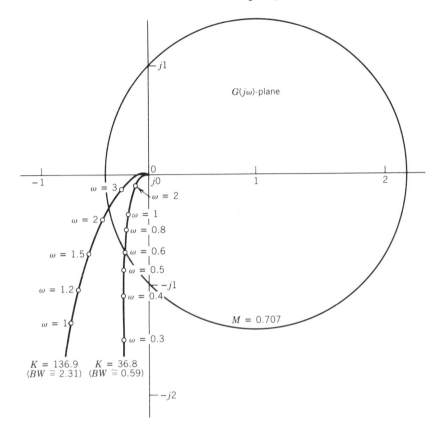

Figure 7.7   Using the $M = 0.707$ circle to determine system bandwidth for Example 7.4.

method to solve for accurate bandwidth values for each K from

$$\left| \frac{G(j\omega)}{1 + G(j\omega)} \right| \Bigg|_{\omega = BW} = \frac{K}{\sqrt{(K - 42\omega^2)^2 + \omega^2(80 - \omega^2)^2}} \Bigg|_{\omega = BW} = \frac{1}{\sqrt{2}} \quad (7.26)$$

The corresponding values of BW for $K_{min} = 36.8$ and $K_{max} = 136.9$ are approximately 0.59 radians/second and 2.31 radians/second, respectively.

7.1.4 Summary and Design Preview

The design formulas listed in Table 7.1 are valid for underdamped systems $(0 < \zeta < 1)$ or for higher-order systems having dominant closed-loop pole pairs that can be approximated by underdamped second-order systems, as demonstrated in Example 7.4. Moreover, the acceptable s-plane region of Figure 7.6 for

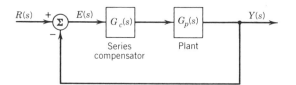

Figure 7.8 The configuration for series compensation.

closed-loop pole locations is also based on a second-order system approximation. In contrast, we should recognize that design in the $G(j\omega)$-plane is not restricted to second-order systems but is valid for general higher-order systems.

The purpose of this section has been to demonstrate conflicting requirements that can emerge when a single parameter is to be adjusted to satisfy several design specifications simultaneously. For example, low values of the forward loop gain K are often required to satisfy design specifications on percent overshoot, rise time, delay time, settling time, system (frequency response) peak magnitude, phase margin, and gain margin. The system bandwidth often requires a band of relatively low values of K. In contrast, a design specification on the steady-state error due to a unit ramp input requires large values of K. We conclude that the adjustment of K alone is not ordinarily satisfactory to meet all of these conflicting design requirements simultaneously.

7.2 THE SERIES COMPENSATION APPROACH

The traditional method of using series compensation in closed-loop systems to satisfy several design requirements simultaneously is presented in this section. Figure 7.8 shows the configuration for series compensation. The form of series compensator to be considered is

$$G_c(s) = \frac{1 + \tau_1 s}{1 + \tau_2 s} \tag{7.27}$$

which has unity gain for low frequencies. The design problem is first to select proper values of τ_1 and τ_2 and then to specify the elements of the physical device itself. We describe configurations of electrical networks having these desired properties in this section and present specific design procedures in the following sections.

7.2.1 Phase-Lead Networks

Consider the phase-lead network of Figure 7.9a. The current $I(s)$ resulting from the applied voltage $V_i(s)$ can be expressed as

$$I(s) = \frac{V_i(s)}{R_2 + Z_1(s)} \tag{7.28}$$

where $Z_1(s)$ is the complex impedance of the resistor-capacitor parallel combination, that is,

$$Z_1(s) = \frac{1}{\dfrac{1}{R_1} + sC} \tag{7.29}$$

By substituting (7.29) into (7.28), using $V_0(s) = R_2 I(s)$, and rearranging, we obtain the voltage transfer function $G_c(s)$ as

$$G_c(s) = \frac{V_0(s)}{V_i(s)} = \frac{s + \dfrac{1}{R_1 C}}{s + \dfrac{R_1 + R_2}{R_1 R_2 C}}$$

$$= \frac{s + z}{s + p} \tag{7.30}$$

In (7.30), we have defined z and p as

$$z = \frac{1}{R_1 C}$$

$$p = \frac{R_1 + R_2}{R_1 R_2 C} \tag{7.31}$$

Observe that the ratio $a = p/z$ can be determined as

$$a = p/z = (R_1 + R_2)/R_2 \tag{7.32}$$

Figure 7.9b shows the zero and pole locations of the phase-lead network at $-z$ and $-p$, respectively. Since $a > 1$, the relative positions are as indicated.

We may also express (7.30) in the standard form for use in constructing Bode diagrams with $s = j\omega$ as

$$G_c(j\omega) = \frac{V_0(j\omega)}{V_i(j\omega)} = \left(\frac{1}{a}\right) \frac{1 + j\omega/(1/aT)}{1 + j\omega/(1/T)} \tag{7.33}$$

where

$$T = \frac{1}{p} = \frac{R_1 R_2 C}{R_1 + R_2} \tag{7.34}$$

In other words, the zero is located at $-z = -1/aT$ and the pole at $-p = -1/T$. Figure 7.9c shows Bode magnitude and phase diagrams for $G_c(j\omega)$.

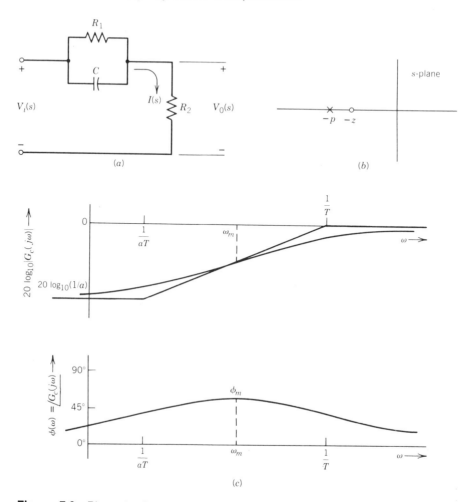

Figure 7.9 Phase-lead series compensation (a) network, (b) zero and pole s-plane locations, and (c) Bode magnitude and phase diagrams.

We need to solve for the maximum value of the phase curve as a function of a and T. Denoting the phase of $G_c(j\omega)$ as ϕ, we write

$$\phi = \angle G_c(j\omega) = \tan^{-1}(aT\omega) - \tan^{-1}(T\omega) \qquad (7.35)$$

Thus, setting the derivative of ϕ with respect to ω equal to zero, we have

$$\left.\frac{d\phi}{d\omega}\right|_{\omega=\omega_m} = \frac{aT}{1 + (aT\omega_m)^2} - \frac{T}{1 + (T\omega_m)^2} = 0 \qquad (7.36)$$

where ω_m is the frequency at which ϕ achieves its maximum. Solving (7.36) yields

$$\omega_m = \frac{1}{T\sqrt{a}} \qquad (7.37)$$

The maximum value of ϕ occurs at the geometric mean of the two corner frequencies $1/T$ and $1/aT$. This maximum phase $\phi(\omega_m) = \phi_m$ is

$$\phi_m = \tan^{-1}\left(aT\frac{1}{T\sqrt{a}}\right) - \tan^{-1}\left(T\frac{1}{T\sqrt{a}}\right)$$

$$= \tan^{-1}(\sqrt{a}) - \tan^{-1}(1/\sqrt{a}) \qquad (7.38)$$

Taking the tangent of ϕ_m in (7.38) and using the two-angle tangent formula, we have

$$\tan\phi_m = \frac{\sqrt{a} - 1/\sqrt{a}}{1 + \sqrt{a}(1/\sqrt{a})}$$

$$= \frac{1}{2}(\sqrt{a} - 1/\sqrt{a}) = \frac{a - 1}{2\sqrt{a}} \qquad (7.39)$$

In Figure 7.9c, as the ratio a between the two corner frequencies approaches unity, the two corners coincide, the compensator pole and zero cancel, and the amount of phase lead is reduced to zero. As a becomes very large, the corners separate by a large amount and ϕ_m approaches $90°$. As a rule of thumb, we normally try to limit a to the range $1 < a < 15$, since values greater than approximately 15 tend to permit unwanted noise disturbances to enter and influence the system dynamics adversely. The precise limiting value of a allowed depends on the particular application.

Using the identity $\sin\phi_m = \tan\phi_m / \sqrt{(1 + \tan^2\phi_m)}$, we may form

$$\sin\phi_m = \frac{a - 1}{a + 1} \qquad (7.40)$$

Solving (7.40) for a gives the design equation

$$a = \frac{1 + \sin\phi_m}{1 - \sin\phi_m} \qquad (7.41)$$

Equation (7.41), rather than (7.39) or (7.40), will be useful in the design procedure, since we will want to determine the value of a for which a given additional phase lead (ϕ_m) is obtained. The remaining problem of deciding where to place the two corners of the compensator will be treated in subsequent sections.

7.2.2 Phase-Lag Networks

The phase-lag network of Figure 7.10a has the voltage transfer function given by

$$G_c(s) = \frac{V_0(s)}{V_i(s)} = \frac{R_2 + \dfrac{1}{sC}}{R_1 + R_2 + \dfrac{1}{sC}} \qquad (7.42)$$

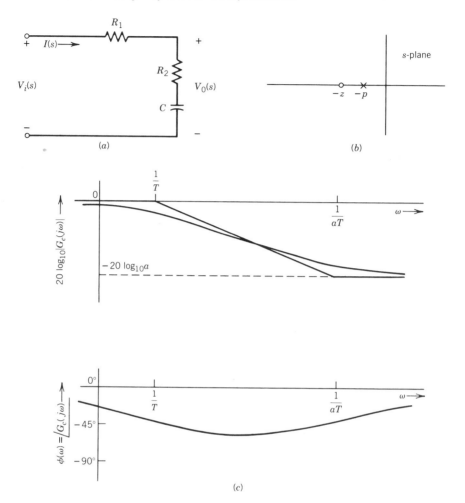

Figure 7.10 Phase-lag series compensation (*a*) network, (*b*) zero and pole s-plane locations, and (*c*) Bode magnitude and phase diagrams.

If we define z, p, a, and T as

$$z = 1/R_2C$$

$$p = 1/(R_1 + R_2)C$$

$$a = R_2/(R_1 + R_2)$$

$$T = (R_1 + R_2)C \tag{7.43}$$

then (7.42) can be rearranged in standard forms for root locus and Bode diagrams (with $s = j\omega$) as

$$G_c(s) = a\left(\frac{s + z}{s + p}\right)$$

$$G_c(j\omega) = \frac{1 + j\omega/(1/aT)}{1 + j\omega/(1/T)} \tag{7.44}$$

The s-plane zero and pole locations are shown in Figure 7.10b and Bode diagrams in Figure 7.10c.

7.3 PHASE-LEAD DESIGN BY ROOT LOCUS

We describe in this section how to design a phase-lead series compensator by using the root locus technique [1–6]. Initially, we convert the design requirements into an allowable s-plane region and then verify that the simple adjustment of the gain K will not permit us to satisfy the design requirements. We next specify the desired dominant closed-loop pole locations and attempt to place the lead network zero and pole such that the root locus passes through these desired points. Selecting the location of the network zero automatically fixes the design result, since the lead network pole must subsequently be placed to yield a net angle of 180° for the open-loop compensated plant $G_c(s)G_p(s)$ at the desired location of the closed-loop pole. The gain K of the compensated plant is calculated from s-plane vectors, and the velocity error constant K_v, for example, is then determined. Suppose we are attempting to guarantee that the steady-state error for a unity-feedback Type 1 system resulting from a unit ramp input does not exceed some specified value. We can compute the steady-state error as the reciprocal of K_v. If the resulting e_{ss} from the compensator pole and zero placement is too large, we establish new desired dominant closed-loop poles and repeat the above design procedure.

Figure 7.11 shows the steps of the recommended design algorithm. In Step 4, the compensator zero is placed somewhere to the left of the second open-loop pole. If this zero were placed precisely on this pole, we would have cancellation compensation. Dorf [2] recommends placing this zero immediately below the desired closed-loop pole location. While this choice makes the resulting computations slightly easier, we suggest a zero location somewhere to the left of this point to improve the assumption that the desired quadratic poles tend to dominate the closed-loop response. We show the limits of this approximation in the following examples.

EXAMPLE 7.5

Consider the positional servomechanism with the field-controlled dc motor shown in Figure 7.12. We showed in Section 2.1 that the plant transfer function can be

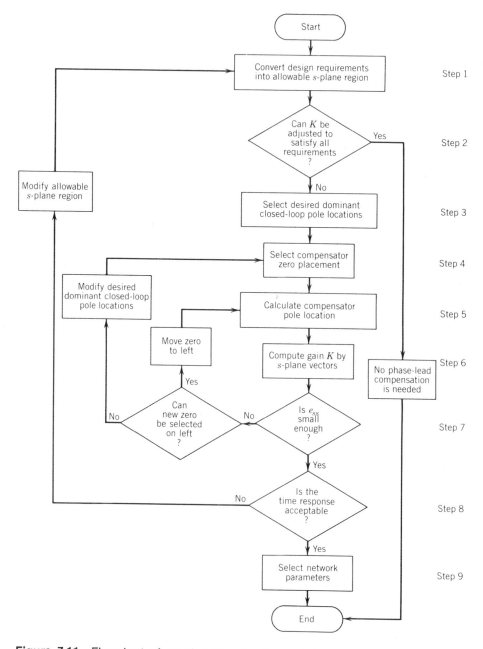

Figure 7.11 Flowchart of an algorithm for the root locus design of phase-lead compensation networks.

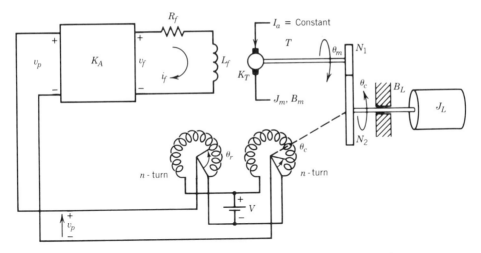

Figure 7.12 Positional servomechanism with field-controlled dc motor.

expressed in the form

$$G_p(s) = \frac{K_v}{s(1 + \tau_m s)(1 + \tau_e s)} \tag{7.45}$$

where K_v is the Bode gain corresponding to the velocity error constant for this Type 1 system. The mechanical time constant τ_m is usually several times larger than the electrical time constant τ_e, which places the corner frequency at $1/\tau_e$ at a higher frequency than the one at $1/\tau_m$. The parameters in (7.45) are given by $\tau_m = J/B$ and $\tau_e = L_f/R_f$. The gain K_v is equivalent to $K_p K_A K_T (N_1/N_2) B R_f$, where these parameters can be identified from the transfer function of Figure 2.12d.

Suppose we consider the case where $\tau_m = 0.5$ and $\tau_e = 0$. We have assumed that the inductance of the motor field circuit (L_f) is negligible, resulting in a second-order plant described by

$$G_p(s) = \frac{K_v}{s(1 + 0.5s)} \tag{7.46}$$

which is equivalent to

$$G_p(s) = \frac{K}{s(s + 2)} \tag{7.47}$$

where $K = K_v/\tau_m = K_v/0.5 = 2K_v$.

It is required to design a phase-lead network for a negative unity-feedback system having the plant transfer function in (7.46) or (7.47) subject to the design

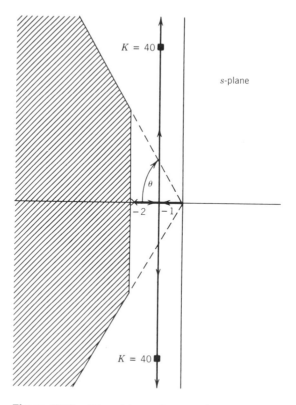

Figure 7.13 Allowable *s*-plane region to satisfy %OS ≤ 15° and t_r ≤ 1.0 second.

specifications given by

$$\%OS \leq 15\%$$

$$t_r \leq 1.0 \text{ second}$$

$$e_{ss}\big|_{\substack{\text{unit} \\ \text{ramp}}} \leq 0.05 \text{ radians} \tag{7.48}$$

Figure 7.13 shows the allowable region in the *s*-plane (Step 1) to satisfy the %OS and t_r design requirements in (7.48). We have superimposed the wedge-shaped region shown in Figure 7.1 for a %OS ≤ 15% with the *s*-plane region lying on or to the left of the vertical line $\sigma = -\ln 9/t_{r_{max}} = -(\ln 9)/1 = -2.2$, which was identified for the general case in Figure 7.4. Since K_v must be equal to or greater than 20 to satisfy the steady-state error requirement ($K_v \geq 1/e_{ss_{max}} = 1/0.05 = 20$), we require $K \geq 40$. Figure 7.13 shows that the closed-loop pole locations for the uncompensated system with $K = 40$ lie outside the *s*-plane region for which %OS ≤ 15% and $t_r < 1.0$ second (Step 2). Therefore, it is evident that compensation is required to meet all three design specifications simultaneously.

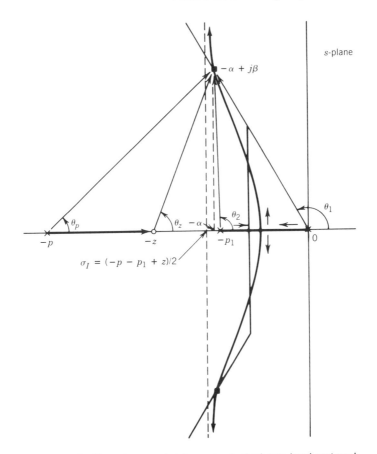

Figure 7.14 Root locus plot for a typical phase-lead network pole / zero placement.

In Step 3, we arbitrarily select the closed-loop pole locations for the phase-lead compensated system at $-\alpha \pm j\beta$. We choose $\alpha = 3$ arbitrarily and calculate β as $\alpha \tan \theta$, where $\theta = \cos^{-1}(\zeta_{min})$ from (7.2), and ζ_{min} is determined from the %OS design formula in Table 7.1. Alternatively, we can show that β equals $\alpha\sqrt{1 - \zeta_{min}^2}/\zeta_{min}$. Both methods yield $\beta = 4.97$. The points $-3 \pm j4.97$ lie on the boundary of the allowable s-plane region determined in Step 1.

In Step 4, we place the zero somewhere to the left of the second pole. A typical phase-lead network zero placement is shown in Figure 7.14. Once the zero location has been arbitrarily specified, we can compute the angle θ_p from

$$\sum \underline{/\text{Zeros of } G_c G_p(s)} - \sum \underline{/\text{Poles of } G_c G_p(s)} = (2k + 1)180°$$

$$\theta_z - \left[\theta_p + \theta_1 + \theta_2\right] = (2k + 1)180° \qquad (7.49)$$

where k is zero or any positive or negative integer. Values of θ_z, θ_1, and θ_2 can be determined as

$$\theta_z = \tan^{-1}\left(\frac{\beta}{z - \alpha}\right)$$

$$\theta_1 = 180° - \tan^{-1}\left(\frac{\beta}{\alpha}\right)$$

$$\theta_2 = 180° - \tan^{-1}\left(\frac{\beta}{\alpha - p_1}\right) \qquad (7.50)$$

where the desired dominant closed-loop poles are at $-\alpha \pm j\beta$ with $\alpha = 3$ and $\beta = 4.97$, and the second open-loop pole is at $-p_1$ with $p_1 = 2$. By using (7.50) in (7.49) with $k = -1$, we find that

$$\theta_p = \theta_z - \theta_1 - \theta_2 - (2k + 1)180°$$

$$= \tan^{-1}\left(\frac{\beta}{z - \alpha}\right) - 121.1° - 101.4° + 180°$$

$$= \tan^{-1}\left(\frac{4.97}{z - 3}\right) - 42.5° \qquad (7.51)$$

The compensator pole location at $s = -p$ may then be determined in Step 5 by using

$$p = \frac{\beta}{\tan\theta_p} + \alpha$$

$$= \frac{4.97}{\tan\theta_p} + 3 \qquad (7.52)$$

Table 7.2 shows the results obtained by using the phase-lead algorithm of Figure 7.11. The first try places the zero at $-p_1 = -2$ to cancel the second

TABLE 7.2 Pole / Zero Placement Attempts for Example 7.5

Dominant Poles	z	θ_p	p	a	K	K_v	e_{ss}
$-3 \pm j4.97$	2.0	58.9°	6.0	3.0	33.7	5.6	0.178
	3.0	47.5°	7.6	2.5	39.9	7.9	0.126
	6.0	16.4°	19.9	3.3	89.3	13.5	0.074
	7.977	2.4°	119.6	15.0	488.0	16.3	0.061
$-4 \pm j6.62$	6.5	21.4°	20.9	3.2	137.3	21.3	0.047
	9.42	2.8°	141.2	15.0	858.9	28.6	0.035

open-loop pole, the second places the zero at $-\alpha = -3$ as suggested by Dorf in [2], and the third puts the zero arbitrarily at -6. We may use (7.51) and (7.52) to show that the zero can be placed no farther to the left than -7.977 without violating the rule-of-thumb restriction that the ratio $a = p/z$ between pole and zero locations must not exceed 15. At $z = 7.977$, (7.51) and (7.52) yield $p = 119.6$, which gives

$$G_cG_p(s) = \frac{K(s + 7.977)}{s(s + 2)(s + 119.6)} \tag{7.53}$$

Evaluating K at $s = -3 \pm j4.97$ by s-plane vectors (Step 6) gives $K = 488.0$. The value of K_v is $Kz/p_1 p = K/2a = 488.0/30.0 = 16.3$, since $G_cG_p(s)$ may be expressed in standard Bode form as

$$G_cG_p(s) = \frac{\dfrac{488.0(7.977)}{2(119.6)}(1 + s/7.977)}{s(1 + s/2)(1 + s/119.6)}$$

$$= \frac{16.3(1 + s/7.977)}{s(1 + s/2)(1 + s/119.6)} \tag{7.54}$$

where $a = p/z = 119.6/7.977 = 15.0$. Therefore, the steady-state error for a unit ramp input is $1/K_v = 1/16.3 = 0.061$ radians, which is too large (Step 7) to meet the third design specification in (7.48).

Since e_{ss} is not small enough in Step 7 of the algorithm (Figure 7.11), we enter the feedback loop in the flowchart, which modifies the desired dominant closed-loop pole locations. We next place the desired dominant closed-loop poles at $-\alpha \pm j\beta = -4 \pm j6.62$, which is again on the boundary of the acceptable %OS and t_r region in the s-plane. The resulting equations corresponding to (7.51) and (7.52) for θ_p and p are

$$\theta_p = \tan^{-1}\left(\frac{6.62}{z - 4}\right) - 47.9°$$

$$p = \frac{6.62}{\tan \theta_p} + 4 \tag{7.55}$$

Selecting z arbitrarily as 6.5 yields $p = 20.9$, $a = 3.2$, and $K = 137.3$. For this choice, $G_cG_p(s)$ becomes

$$G_cG_p(s) = \frac{137.3(s + 6.5)}{s(s + 2)(s + 20.9)}$$

$$= \frac{21.3(1 + s/6.5)}{s(1 + s/2)(1 + s/20.9)} \tag{7.56}$$

which yields $e_{ss} = 1/K_v = 1/21.3 = 0.047$ radians, and the steady-state error design specification in (7.48) is satisfied. An additional calculation in Table 7.2 for

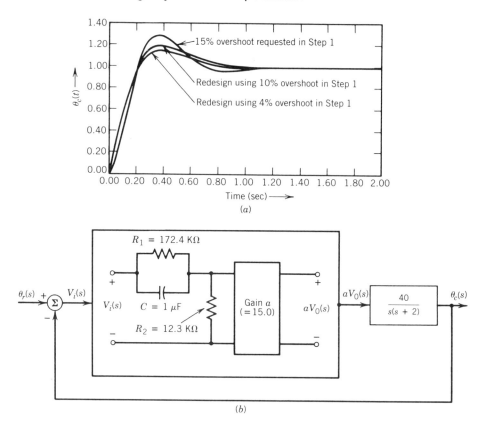

Figure 7.15   Phase-lead design results for Example 7.5.

the modified desired closed-loop pole locations shows that the zero can be placed no farther to the left than -9.42 for $a \leq 15$.

In Step 8, we examine the time response of the phase-lead compensated system to determine whether it does indeed meet the required performance specifications. We observe that the actual %OS may be somewhat greater than 15% because of the dominant closed-loop pole approximation in Step 3 and the presence of a zero in the compensator-plant transfer function. Computer simulation results in Figure 7.15a show that the design in (7.56) yields a 28% overshoot. Redesigns using 10% and 4% in Step 1 provide overshoots of 19% and 15%, respectively. Thus, a substantial safety margin, that is, the use of 4% (with $\alpha = 5.0$) rather than 15% in Step 1, was needed in this example to obtain a design that meets the given specifications. The resulting $G_cG_p(s)$ becomes

$$G_cG_p(s) = \frac{600(s + 5.8)}{s(s + 2)(s + 86.8)}$$

$$= \frac{20.0(1 + s/5.8)}{s(1 + s/2)(1 + s/86.8)} \tag{7.57}$$

for which $a = p/z = 15.0$.

The remaining step is to determine parameters for the phase-lead compensation network (Step 9) in terms of z and p. Since these are three network parameters (R_1, R_2, and C) and only two design parameters (z and p), the resulting solution is not unique. Often the capacitance C is selected arbitrarily for convenience in physical realization. If we select C as $1\,\mu F$ for this example, then (7.31) and (7.32) may be used for the choice in (7.57) to yield $R_1 = 172.4\,K\Omega$ and $R_2 = 12.3\,K\Omega$. We must also adjust the value of K to be 600, which includes the gain factor $a = 15.0$. Figure 7.15b shows the resulting phase-lead network and the additional gain ($a = 15.0$) needed to ensure $K_v \geq 20$. In a positional servomechanism, such as in Figure 7.12, this network is usually inserted at the amplifier input and the amplifier gain K_A is increased by a factor a.

EXAMPLE 7.6

Consider again the system of Figure 7.12 but with $\tau_m = 0.5$ and $\tau_e = 0.025$, which yields a third-order plant described by

$$G_p(s) = \frac{K_v}{s(1 + 0.5s)(1 + 0.025s)} \tag{7.58}$$

or equivalently

$$G_p(s) = \frac{K}{s(s + 2)(s + 40)} \tag{7.59}$$

where $K = K_v / \tau_m \tau_e = 80K_v$. The problem is to design a phase-lead network to satisfy the design specifications in (7.48).

Figure 7.16a shows a root locus sketch for the uncompensated system with the allowable s-plane region for %OS and t_r requirements (Step 1). We require $K \geq 1600$ to meet the steady-state error requirement, since K_v must be greater than or equal to 20, as in Example 7.5. Obviously, the %OS and t_r requirements cannot be satisfied by adjusting only the gain K (Step 2), and therefore a compensation network is needed. Observing the similarity with Example 7.5, we select the desired dominant closed-loop pole locations for the compensated system as $-4 \pm j6.62$ (Step 3) and select the compensator zero at $-z = -6.5$ (Step 4). Figure 7.16b shows a sketch of the resulting root locus plot. As in (7.51), we determine θ_p from

$$\theta_p = \theta_z - \theta_1 - \theta_2 - \theta_3 - (2k + 1)180°$$

$$= \tan^{-1}\left(\frac{6.62}{z - 4}\right)\Big|_{z=6.5} - 121.1° - 106.8° - 10.4° + 180°$$

$$= \tan^{-1}\left(\frac{6.62}{6.5 - 4}\right) - 58.4° = 11.0° \tag{7.60}$$

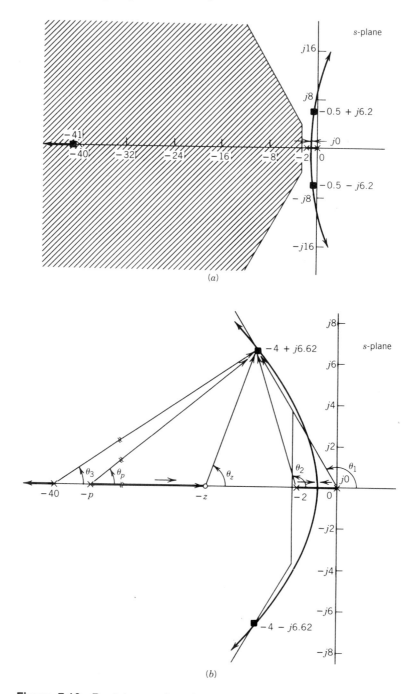

(a)

(b)

Figure 7.16 Root locus plots for the (a) uncompensated and (b) phase-lead compensated systems of Example 7.6.

Therefore, in Step 5 the compensator pole at $-p$ is determined from

$$p = \frac{6.62}{\tan \theta_p} + 4 = 38.2 \qquad (7.61)$$

Using s-plane vectors in Step 6, we determine K for the phase-lead compensated system as $K = 9636.1$. Moreover, we compute K_v as $K/(80a) = 9636.1/[80(38.3)/6.5] = 20.5$ and e_{ss} in Step 7 as $1/K_v = 0.049$ radians, which meets the design specification in (7.48). In Step 8, we find from computer simulations that the actual %OS is 30% for this design attempt. Therefore, we must impose more stringent design requirements and return to Step 1 to begin the redesign procedure. The succeeding steps are identical to those examined here and in Example 7.5 and will not be repeated. Once an acceptable design is obtained, R_1 and R_2 can be calculated in Step 9 and the phase-lead network inserted for series compensation.

7.4 PHASE-LEAD COMPENSATION USING BODE DIAGRAMS

Design by using Bode diagrams allows us to satisfy frequency response specifications directly without converting them to desired closed-loop pole locations in the s-plane [1–6]. In fact, the designer's choice between root locus and Bode design procedures can usually be based on which design specifications are given. Primarily, we examine phase margin or peak magnitude requirements in this section and show some correspondence between the resulting poles and zeros of phase-lead networks designed here and those obtained in Section 7.3.

Consider the polar plot in the $G(j\omega)$-plane of Figure 7.17. The plot for the uncompensated system lies too near the $-1 + j0$ point as measured by either

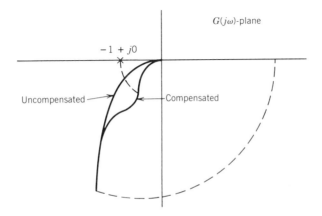

Figure 7.17 A typical polar plot for uncompensated and phase-lead compensated systems.

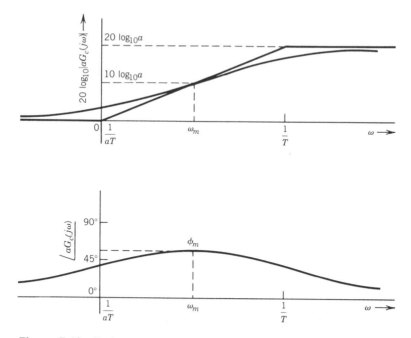

Figure 7.18 Bode diagrams for the phase-lead series compensation network with gain a included.

phase margin (PM) or peak magnitude (M_p) specifications. Placing a phase-lead series compensation network in the forward path can often give acceptable values for these design specifications. We need to select the pole and zero of the phase-lead network such that the maximum phase lead occurs at the new gain crossover frequency, that is, at the frequency where the phase margin is measured for the compensated system.

We set the gain K for the uncompensated system to give the desired steady-state error, for example, for a unit ramp input. To maintain the low-frequency gain when the series compensation network is added, we increase the amplifier gain by a factor of a, which is the ratio between the network corner frequencies. The resulting Bode diagrams are shown in Figure 7.18.

Figure 7.19 shows a flowchart of the phase-lead design procedure using Bode diagrams. If we are designing a phase-lead network for a Type 1 system subject to steady-state error and phase margin specifications, then in Step 1 we compute

$$K_{v_{min}} = 1/e_{ss_{max}}\Big|_{\substack{\text{unit} \\ \text{ramp}}} \tag{7.62}$$

We adjust the forward path gain K to yield this value of K_v and then attempt to satisfy the given phase margin design requirement. In Step 2, we determine the gain crossover frequency for the uncompensated system and the associated phase

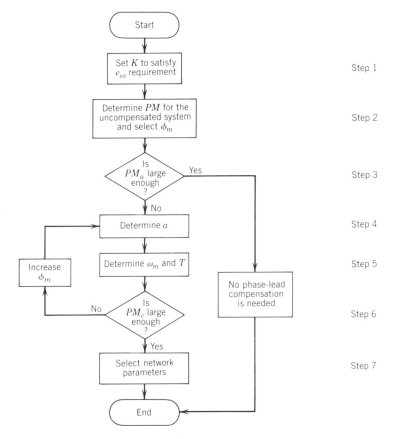

Figure 7.19 Flowchart of a Bode phase-lead design procedure.

margin. This permits us to decide just how much additional phase lead we need to meet the design specification. If the design specification is met with no additional phase being needed, then Step 3 terminates the design algorithm. If $\phi_m > 0$, we proceed to design the phase-lead network. In practice we add approximately 5° to 10° (or even more in some cases) to this preliminary value of ϕ_m. This additional amount is necessary because the phase curve usually is decreasing somewhat at the gain crossover frequency for the uncompensated system. We will find that the new gain crossover frequency lies to the right of the gain crossover frequency for the uncompensated system, as shown in Figure 7.20. Thus, the phase margin at the new gain crossover frequency would be smaller than previously indicated unless we include additional phase to offset this decrease in phase. A gentle negative slope usually requires only 5° to 10° additional phase. On the other hand, a sharply decreasing phase curve in the vicinity of the gain crossover frequency for the uncompensated system requires a much greater addition of phase for ϕ_m. In fact, it may not be possible to add enough phase for systems with extremely large phase slopes in this vicinity. In such cases, the phase-lag design of the following two sections is more appropriate.

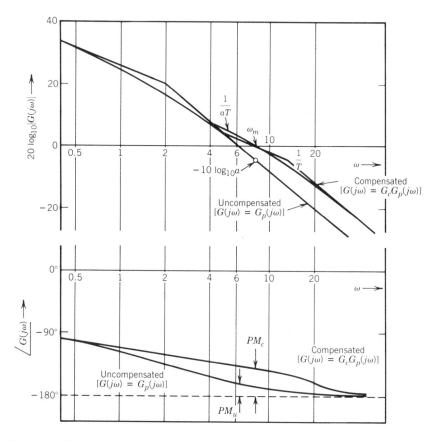

Figure 7.20 Application of the design procedure in Figure 7.19 for phase-lead design.

Step 4 solves for the value of a from (7.41) as

$$a = \frac{1 + \sin \phi_m}{1 - \sin \phi_m} \qquad (7.41)$$

Next, since the maximum phase being added occurs at ω_m, we wish to place ω_m at the new gain crossover frequency. Figure 7.18 shows that the Bode magnitude curve has a value of $10 \log_{10} a$ at ω_m. Thus, in Step 5 we locate that frequency on the Bode magnitude curve for the uncompensated system in Figure 7.20 where the magnitude is $-10 \log_{10} a$. By placing ω_m at this frequency, we will have used the phase-lead network most effectively, that is, added the maximum phase at that frequency where it contributes most in satisfying the phase margin design requirement.

Once ω_m is determined, we solve for T from (7.37) as

$$T = \frac{1}{\omega_m \sqrt{a}} \qquad (7.63)$$

We calculate the new phase margin for the compensated system and compare it with the design requirement in Step 6. If a larger PM_c is needed, we increase the ϕ_m appropriately and repeat Steps 4 and 5 until the specification is met. Finally, we determine the network parameters R_1, R_2, and C in terms of the design parameters a and T (Step 7) by using (7.32) and (7.34).

EXAMPLE 7.7

It is required to design a phase-lead compensation network for the system in (7.46) of Example 7.5 such that the phase margin is greater than or equal to 40° and the steady-state error due to a unit ramp input is less than or equal to 0.05 radians.

We construct the Bode magnitude and phase diagrams, as shown in Figure 7.20, for $K_v = 20$, since $K_v = 1/e_{ss_{max}} = 1/0.05 = 20$. The value of K in (7.47) is $2K_v = 40$. Since this system is a second-order case, we may identify K as ω_n^2, compute ζ as $-p_1/2\omega_n = 2/(2\sqrt{40}) = 0.158$, and then use the expression in (6.12) to obtain PM for the uncompensated system as $PM_u = 18.0°$. Equation (6.10) gives the gain crossover frequency as 6.2 radians/second. Because of the gentle negative slope at the gain crossover frequency, we add 5° as a safety margin to the difference in the required 40° and the 18.0° just obtained. Resulting values of ϕ_m and a, obtained by using (7.41), are given by

$$\phi_m = 40° - 18.0° + 5° = 27°$$

$$a = \frac{1 + \sin 27°}{1 - \sin 27°} = 2.7 \qquad (7.64)$$

We now locate the new gain crossover frequency at that point where the Bode magnitude curve has a value of $-10\log_{10} a = -10\log_{10}(2.7) = -4.3$ decibels. We set $20\log_{10}|G_p(j\omega)| = -4.3$ and obtain

$$|G_p(j\omega)| = \frac{40}{\omega\sqrt{\omega^2 + 4}} = 10^{-4.3/20} = 0.613 \qquad (7.65)$$

Squaring (7.65) and rearranging, we have

$$\omega^4 + 4\omega^2 - \left(\frac{40}{0.613}\right)^2 = 0 \qquad (7.66)$$

which has the solution

$$\omega = \sqrt{-2 + \sqrt{(2)^2 + \left(\frac{40}{0.613}\right)^2}}$$

$$= 8.0 \text{ radians/second} \qquad (7.67)$$

Thus, we choose $\omega_m = 8.0$ radians/second and compute T from (7.63) as

$$T = \frac{1}{\omega_m \sqrt{a}} = \frac{1}{8.0\sqrt{2.7}} = 0.076 \text{ second} \qquad (7.68)$$

from which

$$\frac{1}{T} = \frac{1}{0.076} = 13.1 \text{ radians/second}$$

$$\frac{1}{aT} = \frac{1.31}{2.7} = 4.9 \text{ radians/second} \qquad (7.69)$$

Therefore, the compensated system has the combined compensator-plant transfer function given by

$$G_cG_p(s) = \frac{20(1 + s/4.9)}{s(1 + s/2)(1 + s/13.1)} \qquad (7.70)$$

Finally, we compute the phase of $G_cG_p(j\omega)$ at $\omega = \omega_m = 8.0$ radians/second to obtain

$$\angle G_cG_p(j8.0) = -138.9° \qquad (7.71)$$

from which the phase margin for the compensated system is

$$PM_c = 180° - 138.9° = 41.1° \qquad (7.72)$$

Thus, we have shown that the design specification of $PM \geq 40°$ is satisfied by the compensated system.

If we arbitrarily select the capacitance C as $1\,\mu F$, then from (7.32) and (7.34), we have

$$R_1 = \frac{aT}{C} = \frac{(4.9)^{-1}}{10^{-6}} = 204.1 \text{ K}\Omega$$

$$R_2 = \frac{R_1}{a - 1} = 120.1 \text{ K}\Omega \qquad (7.73)$$

Recall once again that the amplifier gain of 40 must be increased by a factor of $a = 2.7$ to yield the required transfer function in (7.70).

EXAMPLE 7.8

Consider again the negative unity-feedback system of Example 7.6 that has a plant transfer function given by

$$G_p(s) = \frac{K}{s(s + 2)(s + 40)} \qquad (7.59)$$

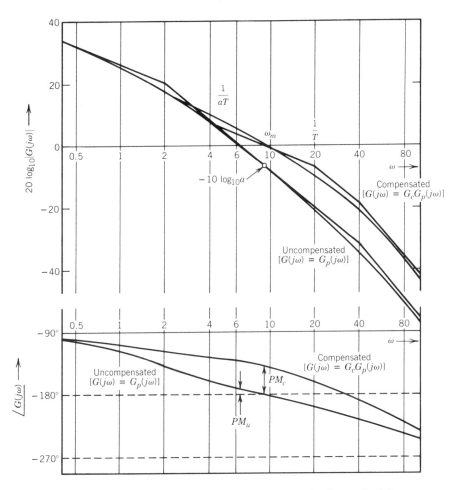

Figure 7.21 Bode magnitude and phase curves for Example 7.8.

A phase-lead compensation network is to be designed to meet the same criteria specified in Example 7.7, that is, $PM \geq 40°$ and $e_{ss}|_{\substack{\text{unit} \\ \text{ramp}}} \leq 0.05$ radians.

We determine $K_v = 20$ as before to meet the steady-state error requirement and note that $K = 80K_v = 1600$. Using this value of gain, we construct the Bode magnitude and phase diagrams in Figure 7.21. The gain crossover frequency cannot be found by using simple closed-form expressions in this third-order case. Instead, we must solve

$$\frac{1600}{\omega\sqrt{\omega^2 + (2)^2}\sqrt{\omega^2 + (40)^2}}\Bigg|_{\omega = \omega_{GC}} = 1 \qquad (7.74)$$

Simplifying (7.74) yields

$$f(\omega_{GC}^2) = \omega_{GC}^6 + 1604\omega_{GC}^4 + 6400\omega_{GC}^2 - (1600)^2 = 0 \qquad (7.75)$$

which may be solved numerically in only a few iterations by using Newton's method. The resulting value of ω_{GC} is 6.1 radians/second, which gives $PM_u = 9.4°$. Therefore, since $PM_u < 40°$, a compensation network is needed to meet this design specification.

As a first attempt, we select a 5° safety margin to add to the difference in 40° and 9.4° as

$$\phi_m = 40° - 9.4° + 5° = 35.6° \qquad (7.76)$$

In Step 4 of Figure 7.19, the value of a from (7.41) is 3.8, and $-10\log_{10} a = -5.8$. We next solve for ω_m in Step 5 from

$$|G(j\omega_m)| = \frac{1600}{\omega_m\sqrt{\omega_m^2 + (2)^2}\sqrt{\omega_m^2 + (40)^2}} = 10^{-5.8/20} = 0.513 \qquad (7.77)$$

which can be simplified and solved by Newton's method to yield $\omega_m = 8.6$ radians/second. From (7.63), we determine $1/T = 16.8$ radians/second, from which $1/aT = 4.4$ radians/second. Thus, this first design attempt yields the combined compensator-plant transfer function

$$G_cG_p(s) = \frac{20(1 + s/4.4)}{s(1 + s/2)(1 + s/40)(1 + s/16.8)} \qquad (7.78)$$

where the amplifier gain in (7.78) has been increased by a factor of $a = 3.8$. Setting $s = j\omega$ in (7.78) and calculating the phase of $G_cG_p(j\omega)$ for $\omega = \omega_m = 8.6$ radians/second enables us to determine $PM_c = 36.8°$, which is too small to meet the design specification as determined in Step 6 of the algorithm.

We next return to Step 4 and arbitrarily add a safety margin of 10°, instead of 5° as in (7.76), to give $\phi_m = 40.6°$. The new value of a is 4.7, $-10\log_{10} a = -6.7$, and $\omega_m = 9.1$ radians/second. We proceed to determine $1/T = 19.8$ radians/second and $1/aT = 4.2$ radians/second. If the forward loop gain K is increased by a factor of $a = 4.7$, we may express the resulting $G_cG_p(s)$ as

$$G_cG_p(s) = \frac{20(1 + s/4.2)}{s(1 + s/2)(1 + s/40)(1 + s/19.8)} \qquad (7.79)$$

which yields $PM_c = 40.2°$ as desired. Curves for the design in (7.79) are shown as the compensated case in Figure 7.21.

Compensation network parameters for (7.79) are determine from (7.32) and (7.34) as

$$R_1 = \frac{aT}{C} = \frac{(4.2)^{-1}}{10^{-6}} = 238\,\mathrm{K\Omega}$$

$$R_2 = \frac{R_1}{a - 1} = 64\,\mathrm{K\Omega} \qquad (7.80)$$

where C has been selected arbitrarily as $1\,\mu\mathrm{F}$.

This example on phase-lead compensation for a third-order system has demonstrated the difficulty in solving for the gain crossover frequencies for both uncompensated and compensated cases. Although formulas in terms of ζ and ω_n were available for the second-order system of Example 7.7, no such simple closed-form expressions could be used for higher-order systems. Newton's method enabled us to solve iteratively for the unknown frequencies.

This example has also shown that a safety margin of only 5° may not be large enough and one or more additional trials may be needed. While a 10° safety margin was barely acceptable in the present example, much larger safety margins might be required in some applications. A limiting value of ϕ_m is reached when the ratio $a = p/z$ becomes 15.

In brief, two limitations of series compensation using phase-lead networks are (1) the requirement of increasing the forward loop gain by a factor of a and (2) the restriction that $a = p/z$ must not be greater than 15. In the first case, boosting the amplifier gain as required can sometimes result in a saturation condition for which linear analysis and design procedures are invalid. In the second case, plants having sharply decreasing phase curves in the vicinity of the gain crossover frequency tend to require larger safety margins. However, it is the total ϕ_m, rather than only that part due to a safety margin, which can result in an excessive value of a. Therefore, for some systems we see that it may not be possible to satisfy the design specifications by using phase-lead networks.

7.5 PHASE-LAG COMPENSATION BY ROOT LOCUS

The phase-lag network of Figure 7.10a can be used for series compensation to satisfy the same design specifications of many cases for which the phase-lead network was used in Sections 7.3 and 7.4 [1–6]. There are selected cases for which one network is better suited for series compensation than the other. At the end of Section 7.4, we indicated systems where the phase-lead network is not applicable. Moreover, we can find systems where the phase-lag network is not applicable, as we will see in the following section.

The motivation for phase-lag compensation by a root locus procedure can be described by first determining the forward loop gain K_u for the uncompensated system. This gain is calculated by using s-plane vectors drawn to the dominant closed-loop poles that are selected to satisfy a %OS design specification. A typical plant transfer function is given by

$$G_p(s) = \frac{K_u(s + z_1)\dots}{s(s + p_1)(s + p_2)\dots} \tag{7.81}$$

Phase-lag places a pole-zero pair near the origin on the negative real axis at $s = -z$ and at $s = -p = -az$, where $0 < a < 1$. Using (7.44), we express the combined plant-compensator transfer function as

$$G_cG_p(s) = \frac{Ka(s + z)(s + z_1)\dots}{s(s + p)(s + p_1)(s + p_2)\dots} \tag{7.82}$$

We than calculate the value of Ka as approximately K_u from s-plane vectors drawn to the dominant closed-loop poles. Therefore, we may select a as approximately K_u divided by the desired K—that is, by that value of K needed to satisfy the steady-state error design requirement. However, we might expect the calculated Ka to be even larger than K_u, since the s-plane vector from the compensator pole is slightly longer than the one from the compensator zero. An offsetting effect results from the root locus for the compensated system passing slightly to the right of the intended dominant pole locations because of the addition of the compensator pole-zero pair. For this reason, we often determine a by

$$a = \frac{K_u}{K_{\text{desired}}}\left(\frac{1}{1 + SM}\right) \tag{7.83}$$

where SM is a safety margin of 10% or so, which decreases a to account for the new locations of the dominant closed-loop poles of the compensated system.

Figure 7.22 shows a flowchart of the step-by-step procedure for phase-lag compensation by the root locus method. In Step 1, we convert the %OS design specification into an allowable s-plane region. We proceed in Step 2 to construct the root locus for the uncompensated system and then to calculate K_u at dominant closed-loop poles selected on the boundary of the allowable s-plane region. We use K_u in Step 3 to determine whether the steady-state error design specification is satisfied without requiring series compensation. If it is not satisfied, we calculate a from (7.83) with $SM = 10\% = 0.10$. In Step 5, we arbitrarily locate the compensator zero on the real axis at a point that is 5% to 10% of the distance from the origin to the second open-loop pole and place the compensator pole at $s = -az$, where $0 < a < 1$. We next determine new dominant closed-loop pole locations at the boundary of the allowable s-plane region by constructing the root locus for the compensated system. The value of Ka (and hence K) is then obtained from s-plane vectors drawn to these new closed-loop poles. In Step 7, we calculate e_{ss} and test to see whether it meets the design specification. If it does not, we decrease a [corresponding to a further increase in the safety margin SM in (7.83)] and return to Step 5. If e_{ss} is small enough in Step 7, we next obtain the time response curve by computer simulation and determine whether all design requirements have been met. If the %OS is excessive, we decrease z, that is, move the zero to the right, decrease a, and return to Step 5 to calculate the new pole location. Once the design requirements are satisfied in Step 8, we complete the phase-lag design with the calculation of network parameters in Step 9. Rearranging (7.43) gives

$$R_2 = \frac{1}{zC}$$

$$R_1 = \frac{1}{pC} - R_2 \tag{7.84}$$

where the value of C is selected arbitrarily as before.

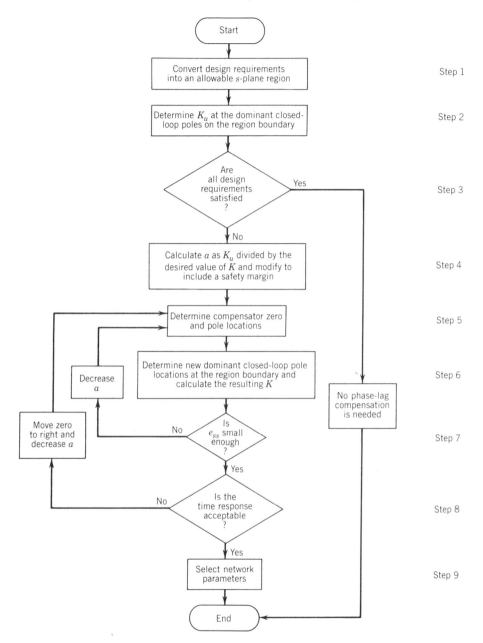

Figure 7.22 Flowchart of an algorithm for the root locus design of phase-lag compensation networks.

EXAMPLE 7.9

Design a phase-lag series compensator for the negative unity-feedback system with plant transfer function given by

$$G_p(s) = \frac{K}{s(s + 2)} \tag{7.47}$$

to satisfy the design specifications

$$\%OS \leq 15\%$$

$$e_{ss}|_{\substack{\text{unit} \\ \text{ramp}}} \leq 0.05 \text{ radians} \tag{7.85}$$

This system was considered earlier in Example 7.5 with an additional design constraint on rise time. Since time responses for phase-lag compensated systems are generally much slower than those for phase-lead compensated systems, design requirements on rise time are not included.

Figure 7.23a shows the root locus plot for the uncompensated system described by (7.47) and the allowable s-plane region for closed-loop poles. This region differs from the one in Figure 7.13 only because the rise time design requirement has been omitted here. We refer to Step 2 (Figure 7.22) and select the dominant closed-loop poles at $-\alpha \pm j\beta$. We choose $\alpha = 1$ arbitrarily and compute β as $\alpha \tan[\cos^{-1}(\zeta_{\min})]$ or as $\alpha\sqrt{1 - \zeta_{\min}^2}/\zeta_{\min}$ to obtain $\beta = 1.656$. The points $-1 \pm j1.656$ are located on the boundary of the allowable region. We next use s-plane vectors to compute K_u for the uncompensated system as

$$K_u = \left[\sqrt{1^2 + (1.656)^2}\right]^2 = 3.74 \tag{7.86}$$

which is the value shown in Table 7.1 for a 15% overshoot requirement and $p_1 = 2$. Therefore, since $K_v = K/p_1 = K/2 = 1.87$, we determine e_{ss} for the uncompensated system as $1/K_v = 1/1.87 = 0.53$ radians and note that this value of K_u is much too small (Step 3).

The calculation of a in Step 4 is achieved from (7.83) by first dividing K_u in (7.86) by that value of K which yields $e_{ss} = 0.05$ radians, that is, by $K = 2K_{v_{\min}} = 2(1/0.05) = 40$. Arbitrarily choosing 10% for SM in (7.83) yields

$$a = \frac{3.74}{40}\left(\frac{1}{1 + 0.10}\right) = 0.085 \tag{7.87}$$

In Step 5, we arbitrarily place the compensator zero at -0.2, which is 10% of the distance from the origin to the second open-loop pole (at $s = -p_1 = -2$). Therefore, the compensator pole is at $-az = -0.017$, and the combined com-

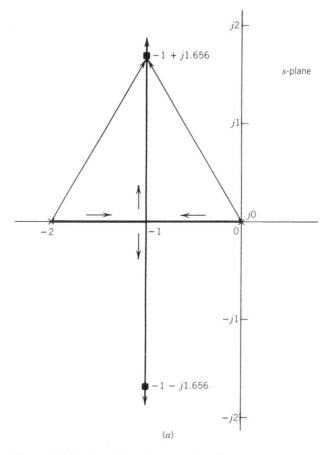

$-1 + j1.656$

s-plane

$-1 - j1.656$

(a)

Figure 7.23 (*a*) *S*-plane vectors for the uncompensated and (*b*) phase-lag compensated systems of Example 7.9.

pensator-plant transfer function becomes

$$G_cG_p(s) = \frac{Ka(s + 0.2)}{s(s + 2)(s + 0.017)} \tag{7.88}$$

Figure 7.23*b* shows the *s*-plane vectors for computing the value of *Ka* for the compensated system. This calculation utilizes the new dominant closed-loop pole locations at $-0.90 \pm j1.49$, which are determined from (7.88) as the intersection of the %OS allowable *s*-plane region boundary with the root locus of the compensated system. In other words, $s = 0.90 \pm j1.49$ satisfies the two equations given by

$$\beta/\alpha = \tan\left[\cos^{-1}(\zeta_{min})\right] = 1.656$$

$$\underline{/G_cG_p(s)}\,\Big|_{s=-\alpha+j\beta} = (2k + 1)180° \tag{7.89}$$

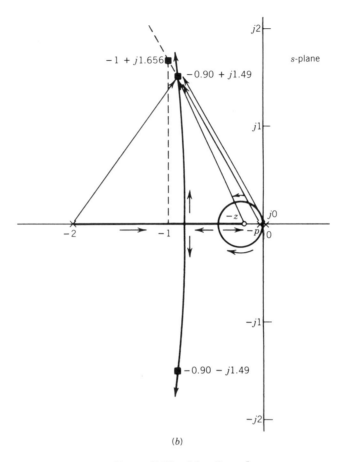

(b)

Figure 7.23   (*Continued*)

The value of 0.90 was obtained numerically by moving along the line having a negative slope at an angle $\theta = \cos^{-1}(\zeta_{min})$ with the negative real axis in the s-plane. Using $\alpha = 0.90$, we easily find $\beta = 1.656(0.90) = 1.49$. Using the s-plane vectors in Figure 7.23b drawn to $s = -0.90 + j1.49$ yields $Ka = 3.393$, from which we may express (7.88) as

$$G_cG_p(s) = \frac{3.393(s + 0.2)}{s(s + 2)(s + 0.017)}$$

$$= \frac{20.0(1 + s/0.2)}{s(1 + s/2)(1 + s/0.017)} \qquad (7.90)$$

which gives $K_v = 20.0$ and $e_{ss} = 0.050$ radians. Thus, the steady-state error design specification has been satisfied.

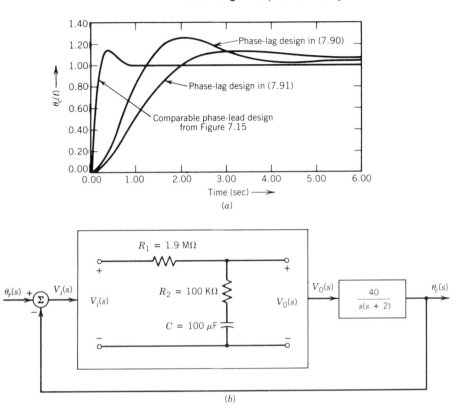

Figure 7.24 Phase-lag design results for Example 7.9.

Computer simulation results are shown in Figure 7.24*a* for the design indicated in (7.90) and for $G_cG_p(s)$ given by

$$G_cG_p(s) = \frac{2.00(s + 0.1)}{s(s + 2)(s + 0.005)}$$

$$= \frac{20.0(1 + s/0.1)}{s(1 + s/2)(1 + s/0.005)} \tag{7.91}$$

Figure 7.24*a* indicates that the %OS is 30% for the design in (7.90) and 14% for the design in (7.91). The time curve for phase-lead design from Figure 7.15 is also shown to emphasize that phase-lag designs yield significantly slower time responses than phase-lead designs.

Selecting C equal to 100 μF, we use (7.84) to obtain $R_1 = 1.9$ MΩ and $R_2 = 100$ KΩ in Step 9 of the phase-lag design algorithm. These element values for (7.91) are shown for the network in Figure 7.24*b*.

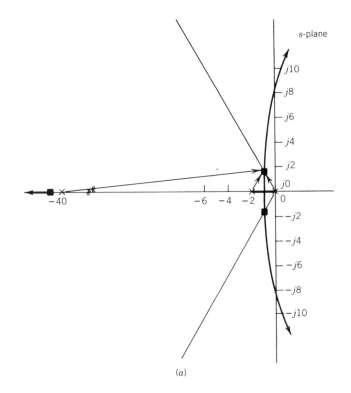

Figure 7.25 Root locus plots for (*a*) the uncompensated and (*b*) phase-lag compensated systems of Example 7.10.

EXAMPLE 7.10

For the third-order system of Examples 7.6 and 7.8, design a phase-lag compensation network to satisfy the specifications in (7.85) of Example 7.9, that is, %OS $\leq$ 15% and e_{ss} (due to a unit ramp) ≤ 0.05 radians. For convenience, we recall here that the plant transfer function is

$$G_p(s) = \frac{K}{s(s + 2)(s + 40)} \tag{7.59}$$

Figure 7.25*a* shows the root locus plot for (7.59). We determine the dominant closed-loop poles at the intersection of the root locus with the boundary of the allowable *s*-plane region for %OS $\leq$ 15% as $-0.96 \pm j1.59$. At these closed-loop poles, we next use *s*-plane vectors to calculate $K_u = 137.8$, from which $K_v = 1.723$ and $e_{ss} = 0.58$ radians. Therefore, e_{ss} is much too large in Step 3 of Figure 7.27. We determine the desired value of K as $2(40)/0.05 = 1600$, from which a,

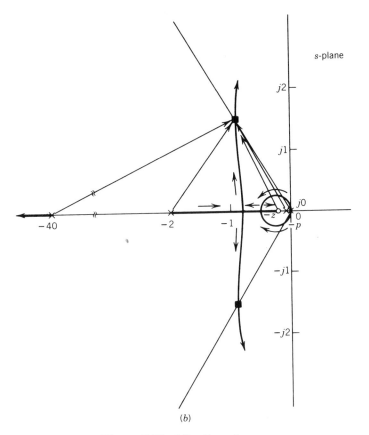

(b)

Figure 7.25   (*Continued*)

according to (7.83), may be selected as

$$a = \frac{137.8}{1600}\left(\frac{1}{1 + 0.10}\right) = 0.078 \qquad (7.92)$$

where *SM* has been chosen arbitrarily as 10% = 0.10. As in Example 7.9, we arbitrarily select $z = 0.2$, and p may be computed to be $p = az = 0.078(0.2) = 0.0156$. As before, the new dominant closed-loop poles (for the compensated system) must satisfy (7.89) for $G_c G_p(s)$ given by

$$G_c G_p(s) = \frac{Ka(s + 0.2)}{s(s + 2)(s + 40)(s + 0.0156)} \qquad (7.93)$$

which yields $s = -0.85 \pm j1.41$. Figure 7.25*b* shows the *s*-plane vectors used to determine *Ka* as 120.3, from which $K = 120.3/a = 120.3/0.078 = 1542.4$, $K_v = K/80 = 1542.4/80 = 19.3$, and $e_{ss} = 0.052$ radians. The test in Step 7 of Figure 7.22 fails, and we must decrease *a* further and return to Step 5.

We next select 15% for the *SM* in (7.83) to yield

$$a = \frac{137.8}{1600}\left(\frac{1}{1 + 0.15}\right) = 0.075 \qquad (7.94)$$

which gives $p = az = 0.015$ for z selected arbitrarily as 0.2. We again determine the dominant closed-loop poles for the compensated system as $-0.85 \pm j1.41$. We proceed to use s-plane vectors to determine $Ka = 123.6$, which gives

$$G_cG_p(s) = \frac{123.6(s + 0.2)}{s(s + 2)(s + 40)(s + 0.015)}$$

$$= \frac{20.6(1 + s/0.2)}{s(1 + s/2)(1 + s/40)(1 + s/0.015)} \qquad (7.95)$$

Therefore, $e_{ss} = 0.049$ radians, which satisfies the design requirement and passes the test in Step 7.

In Step 8, computer simulation results indicated that the %OS for the design in (7.95) is 27%. A redesign yielding an overshoot of 15% has the transfer function given by

$$G_cG_p(s) = \frac{80.0(s + 0.1)}{s(s + 2)(s + 40)(s + 0.005)}$$

$$= \frac{20.0(1 + s/0.1)}{s(1 + s/2)(1 + s/40)(1 + s/0.005)} \qquad (7.96)$$

Again selecting $C = 100$ μF, we determine in Step 9 values for R_1 and R_2 as 1.9 MΩ and 100 KΩ, respectively, which completes the required design of the phase-lag network.

7.6 PHASE-LAG COMPENSATION BY BODE PROCEDURES

The Bode design of phase-lag series compensation networks is perhaps the simplest of the four procedures described in Sections 7.3 through 7.6 [1–6]. As in Section 7.4, typical design specifications for which Bode procedures are most useful include restrictions on the phase margin and on the steady-state error resulting from a unit ramp input. We begin, as before, by constructing Bode magnitude and phase diagrams for the uncompensated system when the forward loop gain K is selected to barely satisfy the steady-state error requirement. If the phase margin for the uncompensated system is too small, the phase curve is, by

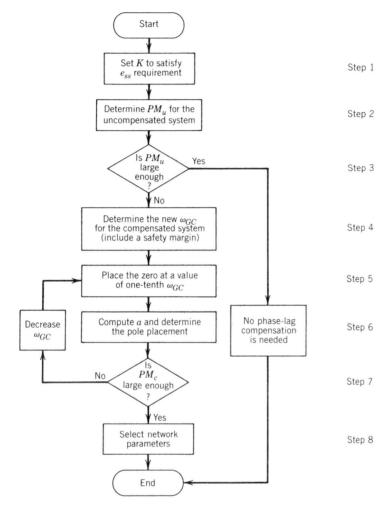

Figure 7.26 Flowchart of the Bode phase-lag design procedure.

definition, too low at that frequency (ω_{GC}) where the Bode magnitude curve crosses the 0 db level. There are two basic series compensation approaches for correcting this situation: phase-lead or phase-lag compensation. We may use phase-lead compensation to add sufficient phase without altering the magnitude curve greatly. We described the Bode design of phase-lead networks in Section 7.4. On the other hand, we may use phase-lag compensation to attenuate the magnitude curve, resulting in a somewhat lower gain crossover frequency at which the phase curve yields the proper amount of phase to meet the phase margin requirement. We describe the details of this second approach in this section.

Figure 7.26 provides a flowchart of the Bode phase-lag design procedure, and Figure 7.27 shows a typical case. The first three design steps are identical to those for the phase-lead design algorithm in Figure 7.19. If the phase margin requirement is not satisfied in Step 3, we proceed in Step 4 to determine that frequency

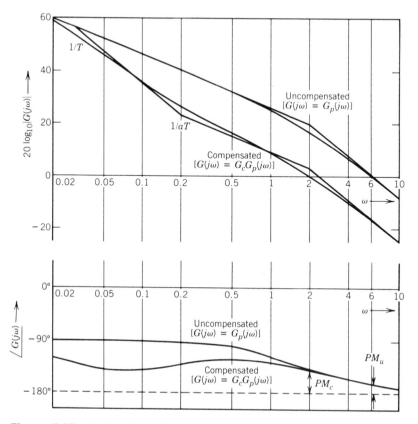

Figure 7.27 Application of the design procedure in Figure 7.26 for phase-lag network design.

at which the phase is large enough to meet the design specification, provided the new gain crossover frequency (ω_{GC}) occurs at that point. An additional amount of phase, perhaps 5° or so, should be included as a safety margin, since the phase-lag network that achieves the desired attenuation also lowers the phase curve slightly at ω_{GC}. In Step 5, we arbitrarily place the compensator zero at a frequency of one-tenth the desired new gain crossover frequency ω_{GC}. In Step 6, we determine a to provide the necessary attenuation as the reciprocal of $|G_p(j\omega)|$, evaluated at the new gain crossover frequency for the compensated system (ω_{GC}).[2] The compensator pole (at $1/T$) is located by multiplying the zero location $(1/aT)$ by a. If PM_c is not large enough in Step 7, we decrease ω_{GC} (increase the safety margin) and return to Step 5. However, if PM_c is adequate, we proceed to determine the phase-lag network parameters by selecting C

---

[2]Alternatively, if A is the necessary attenuation in decibels, we set $-20\log_{10} a = A$ and determine a as $10^{-A/20}$

arbitrarily and then calculating R_1 and R_2 from (7.43) as

$$R_2 = \frac{aT}{C}$$

$$R_1 = \left(\frac{1-a}{a}\right) R_2 \tag{7.97}$$

EXAMPLE 7.11

Design a phase-lag compensation network to yield $PM \geq 40°$ and e_{ss} due to a unit ramp input ≤ 0.05 radians for the negative unity-feedback system with plant transfer function given by

$$G_p(s) = \frac{K}{s(s+2)} \tag{7.47}$$

This system and design requirements were considered earlier in Example 7.7 for phase-lead compensation, where Steps 1, 2, and 3 yielded $K = 40$ and $PM_u = 18°$, which gives $e_{ss} = 0.05$ radians but is not large enough to satisfy the design specification on the phase margin. Figure 7.27 shows Bode diagrams for this case. If we include a safety margin of 5° in Step 4, we seek to determine the ω for which

$$\angle G_p(j\omega) = -180° + (40° + 5°) = -135° \tag{7.98}$$

Using (7.47) in (7.98) allows us to solve easily for ω as 2 radians/second, which we select as the new gain crossover frequency for the compensated system (ω_{GC}). Therefore, in Step 5, we arbitrarily place the zero $(1/aT)$ at one-tenth ω_{GC}, that is, $1/aT = 0.2$. In Step 6, we determine a as

$$a = \frac{1}{|G_p(j\omega_{GC})|} = \frac{1}{\left.\frac{40}{\omega_{GC}\sqrt{\omega_{GC}^2 + 2^2}}\right|_{\omega_{GC}=2}} = \frac{\sqrt{2}}{10} = 0.141 \tag{7.99}$$

We next calculate $1/T$ as $(1/aT)a = 0.0283$, from which

$$G_cG_p(s) = \frac{20(1 + s/0.2)}{s(1 + s/2)(1 + s/0.0283)} \tag{7.100}$$

In Step 7, we calculate $\angle G_cG_p(j2) = -139.9°$, yielding a PM_c of 40.1°, which satisfies the design specification. In Step 8, we arbitrarily select $C = 100\ \mu F$ and use (7.97) to give

$$R_2 = \frac{aT}{C} = \frac{1/0.2}{100 \cdot 10^{-6}} = 50 \text{ K}\Omega$$

$$R_1 = \left(\frac{1-a}{a}\right) R_2 = \left(\frac{1-0.141}{0.141}\right) 50 \text{ K}\Omega = 304 \text{ K}\Omega \qquad (7.101)$$

which completes the phase-lag network design.

EXAMPLE 7.12

Design a phase-lag compensation network to satisfy $PM \geq 40°$ and $e_{ss}|_{\substack{\text{unit} \\ \text{ramp}}} \leq 0.05$ radians for the negative unity-feedback system having

$$G_p(s) = \frac{K}{s(s+2)(s+40)} \qquad (7.59)$$

This problem was considered for phase-lead compensation in Example 7.8, when it was found the $e_{ss} = 0.05$ radians requires $K = 1600$, and the value of phase margin for the uncompensated system is 9.4°, which is too small (Step 3). Again including a safety margin of 5° in Step 4, we use (7.59) to calculate ω_{GC} from

$$\angle G_p(j\omega_{GC}) = -90° - \tan^{-1}(\omega_{GC}/2) - \tan^{-1}(\omega_{GC}/40)$$

$$= -180° - (40° + 5°) = -135° \qquad (7.102)$$

which yields $\omega_{GC} = 1.83$ radians/second. We select $1/aT = 0.183$ and calculate

$$a = \frac{1}{\dfrac{1600}{\omega_{GC}\sqrt{\omega_{GC}^2 + 2^2}\sqrt{\omega_{GC}^2 + 40^2}}}\Bigg|_{\omega_{GC}=1.83} = 0.124 \qquad (7.103)$$

from which $1/T = (1/aT)a = 0.0226$. Therefore,

$$G_c G_p(s) = \frac{20(1 + s/0.183)}{s(1 + s/2)(1 + s/40)(1 + s/0.0226)} \qquad (7.104)$$

from which $\angle G_c G_p(1.83) = -140.0°$ and $PM_c = 40.0°$. We complete the design by setting $C = 100 \ \mu\text{F}$ and computing values of R_1 and R_2 as

$$R_2 = \frac{aT}{C} = \frac{1/0.183}{100 \cdot 10^{-6}} = 55 \text{ K}\Omega$$

$$R_1 = \left(\frac{1-a}{a}\right) R_2 = \left(\frac{1 - 0.124}{0.124}\right) 55 \text{ K}\Omega = 388 \text{ K}\Omega \qquad (7.105)$$

7.7 DESIGN COMPARISONS AND EXTENSIONS

In this section we make phase-lead and phase-lag compensation design comparisons by first describing the usefulness and limitations of each approach and then identifying selected cases for which only one of the design procedures is applicable. We introduce the lag-lead compensation network as an extension of the concepts in Sections 7.3 through 7.6 and suggest the use of automated digital computer design routines, perhaps involving computer graphics, for this compensation design.

7.7.1 Phase-Lead and Phase-Lag Design Comparisons

For both phase-lead and phase-lag design, we select between root locus and Bode procedures according to the design specifications to be satisfied. As indicated in the opening remarks of Section 7.4, root locus procedures are more directly applicable when given time response performance specifications such as %OS, rise time, delay time, or settling time. We can use the root locus approach to place dominant closed-loop poles within s-plane regions in which these design specifications are only approximately satisfied. We then select compensator zeros and poles to yield a sufficiently large error constant, for example, K_v for Type 1 systems, to satisfy steady-state error requirements. Computer simulations of time responses are needed to enable us to properly evaluate these designs and determine whether a redesign is required. On the other hand, we find that Bode procedures are more directly applicable when frequency response performance specifications such as phase margin, gain margin, system peak magnitude, or system bandwidth are given. Initially, we set the forward-loop gain as the value needed to meet the steady-state error requirement and proceed to either increase the phase (phase-lead) or decrease the magnitude (phase-lag) in the vicinity of the gain crossover frequency (for *PM* specification) for the compensated system.[3]

---

[3] We follow these same procedures when a gain margin requirement is specified, except conditions must be satisfied at the phase crossover frequency for the compensated system. See Problem 7.18.

TABLE 7.3 Phase-Lead and Phase-Lag Design Characteristics

Phase-lead design:
- Provides series compensation by the placement of dominant s-plane closed-loop poles to satisfy time response specifications and by additional phase to satisfy frequency response specifications.
- Yields faster rise, delay, and settling times.
- Results in an increased system bandwidth.
- Requires additional forward-loop gain.
- Not applicable to those systems requiring more than approximately 61° of additional phase at the new gain crossover frequency.

Phase-lag design:
- Provides series compensation by increasing the error constant for given dominant s-plane closed-loop poles to satisfy time response specifications and by magnitude attenuation to satisfy frequency response specifications.
- Yields slower transient responses.
- Results in a decreased system bandwidth.
- Not applicable to those systems having no frequency at which the phase exists to yield the desired phase margin.

When both time response and frequency response specifications are given for a particular design, such as %OS, *PM*, and steady-state error (e_{ss}) requirements, we select what we consider to be the most stringent subset of these specifications for which either root locus or Bode procedures apply, perform the required design, and then verify that the remaining specifications are also satisfied. Iterations may be necessary to meet all specifications simultaneously.

Table 7.3 summarizes the characteristics of phase-lead and phase-lag compensation. As indicated earlier, a phase-lead compensated system has faster rise, delay, and settling times and a corresponding larger system bandwidth. How large the system bandwidth should be depends on the particular application. If it is too large, high-frequency noise can become a problem; if too small, the speed of the time response can become too slow. The rule of thumb that limits a in phase-lead design to a value of approximately 15 is intended to prohibit excessive system bandwidths. This value of $a = 15$ limits the allowable amount of phase that can be added (ϕ_m) to approximately 61.° Problem 7.19 shows a case where phase-lead design is not possible because the phase is decreasing rapidly at ω_{GC} for the uncompensated system, indicating the need for such a large safety margin that $\phi_m > 61°$ would be necessary. In brief, the limitations on phase-lead design are the requirement for additional forward-loop gain (by a factor of a) and the restriction that $\phi_m \leq 61°$ to prohibit an excessive system bandwidth.

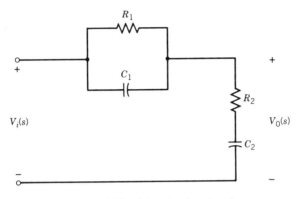

Figure 7.28 A lag-lead network.

A phase-lag compensated system attenuates the frequency response magnitude in the vicinity of the gain crossover frequency, resulting in a reduced system bandwidth and, consequently, a slower time response—that is, higher values of rise, delay, and settling times. Phase-lag design is applicable only for those cases in which there exists some frequency having the desired phase. Problem 7.20 provides a situation where phase-lag compensation cannot be used.

7.7.2 Lag-Lead Compensation

As a useful extension of the procedures developed thus far in the chapter, we may combine phase-lead and phase-lag design concepts to yield a lag-lead network that retains many of the desirable features of the separate networks. In lag-lead design, we both attenuate the magnitude curve (from lag design) and provide additional phase (from lead design) at the gain crossover frequency. The advantages of lag-lead compensation are (1) the system bandwidth is increased (yielding a faster transient response) beyond that which would be achieved with only phase-lag compensation and (2) the requirement of additional forward-loop gain inherent in phase-lead compensation can be avoided.

Figure 7.28 shows a lag-lead network for which the voltage transfer function can be expressed as[4]

$$G_c(s) = \frac{V_0(s)}{V_i(s)} = \frac{(1 + a_1 T_1 s)(1 + a_2 T_2 s)}{(1 + T_1 s)(1 + T_2 s)} \qquad (7.106)$$

---

[4] The lag-lead network is sometimes designated as a lead-lag network because the lead part (R_1 and C_1) is encountered at the input terminals. We prefer the terminology *lag-lead* because the lag effect occurs at lower frequencies.

where

$$a_1 T_1 = R_2 C_2$$

$$a_2 T_2 = R_1 C_1$$

$$a_1 a_2 = 1$$

$$T_1 + T_2 = R_1 C_1 + R_2 C_2 + R_1 C_2 \tag{7.107}$$

The zero and pole resulting from the lag network (R_2 and C_2) occur at $1/a_1 T_1$ and $1/T_1$, respectively, where $0 < a_1 < 1$. Furthermore, the zero and pole resulting from the lead network (R_1 and C_1) occur at $1/a_2 T_2$ and $1/T_2$, respectively, where $a_2 > 1$. In brief, we require $1/T_1 < 1/a_1 T_1 < 1/a_2 T_2 < 1/T_2$.

Suppose the design specifications include a phase margin requirement, a limit on the steady-state error due to a unit ramp input (Type 1 system), and a closed-loop system bandwidth that is the same for compensated and uncompensated systems. We may use the flowchart described in Figure 7.29 for lag-lead

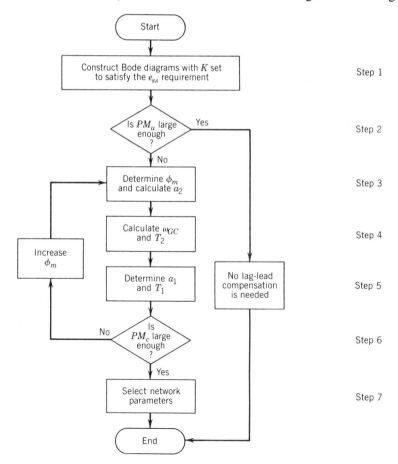

Figure 7.29 Flowchart of an algorithm for lag-lead network design.

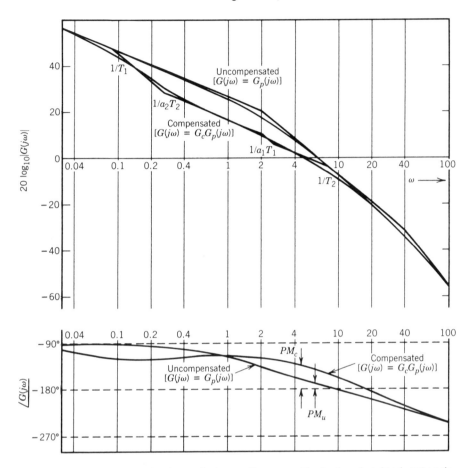

Figure 7.30 Bode magnitude and phase diagrams illustrating lag-lead network design.

design. In Step 1, we set K to satisfy the steady-state error requirement and plot the Bode magnitude and phase curves, as in Figure 7.30. We determine PM_u in Step 2 and compare it to the phase margin specification. If PM_u is large enough, we do not need to add a compensation network, and thus we terminate the procedure at this point. If PM_u is too small, we determine the amount of additional phase needed from

$$\phi_m = PM_{\min} - PM_u \qquad (7.108)$$

where $PM_{\min}$ is the minimum allowable phase margin according to the design specification. Since the new gain crossover frequency for the compensated system occurs at a slightly lower value than the corresponding ω_{GC} for the uncompensated system (presumably with greater phase at that new frequency), we usually can apply (7.108) without an additional safety margin. As in phase-lead design, we next calculate a_2 in Step 3 from

$$a_2 = \frac{1 + \sin \phi_m}{1 - \sin \phi_m} \tag{7.109}$$

In Step 4, we locate the new ω_{GC} by using

$$|G_p(j\omega_{GC})| = \sqrt{a_2} \tag{7.110}$$

As an approximation, we may express

$$1/T_2 = \omega_{GC}\sqrt{a_2}$$

$$1/aT_2 = (1/T_2)a_2 \tag{7.111}$$

Moreover, we arbitrarily select the zero corresponding to the lag part of the network in Step 5 at

$$1/a_1 T_1 = 0.1(1/a_2 T_2) \tag{7.112}$$

from which

$$1/T_1 = (1/a_1 T_1)a_1 = (1/a_1 T_1)/a_2 \tag{7.113}$$

since $a_1 = 1/a_2$. In Step 6, we determine whether PM_c is large enough. If not, we increase ϕ_m and recalculate a_2 from (7.109) in Step 3. If so, we proceed to select lag-lead network parameters in Step 7 by choosing C_1 arbitrarily and then using

$$R_1 = \frac{a_2 T_2}{C_1}$$

$$C_2 = \frac{T_1(1 - a_1) + T_2(1 - a_2)}{R_1}$$

$$R_2 = \frac{a_1 T_1}{C_2} \tag{7.114}$$

which are determined from (7.108).

EXAMPLE 7.13

Design a lag-lead compensator network to satisfy

$$PM \geq 40°$$

$$e_{ss}|_{\substack{\text{unit} \\ \text{ramp}}} \leq 0.05 \text{ radians} \tag{7.115}$$

and the requirement that the attenuation for large ω should be approximately the same for both compensated and uncompensated systems. The plant transfer function for this negative unity-feedback system is given by

$$G_p(s) = \frac{K}{s(s + 2)(s + 40)} \qquad (7.59)$$

Phase-lead and phase-lag designs were considered in Examples 7.8 and 7.12 for this system and the first two design requirements.

The typical Bode magnitude and phase curves used for illustration in Figure 7.30 apply to the current design problem. In Steps 1 and 2, we calculate $K_v = 20$ (or $K = 1600$), construct the Bode diagrams just described, and determine $PM_u = 9.4°$, as in Examples 7.8 and 7.12. In Step 3 we calculate $a_2 = 3.07$ from (7.109) and in Step 4 determine $\omega_{GC} = 4.6$ radians/second from (7.110), that is,

$$\frac{1600}{\omega_{GC}\sqrt{\omega_{GC}^2 + 2^2}\sqrt{\omega_{GC}^2 + 40^2}} = \sqrt{3.07} \qquad (7.116)$$

Therefore, from (7.111), we have

$$1/T_2 = \omega_{GC}\sqrt{a_2} = 4.6\sqrt{3.07} = 8.0$$

$$1/a_2 T_2 = (1/T_2)/a_2 = 8.0/3.07 = 2.6 \qquad (7.117)$$

In Step 5, we use (7.112) and (7.113) to calculate

$$1/a_1 T_1 = 0.1(1/a_2 T_2) = 0.1(2.6) = 0.26$$

$$1/T_1 = (1/a_1 T_1)/a_2 = 0.26/3.07 = 0.085 \qquad (7.118)$$

We may now express $G_c G_p(s)$ as

$$G_c G_p(s) = \frac{20(1 + s/0.26)(1 + s/2.6)}{s(1 + s/2)(1 + s/40)(1 + s/0.085)(1 + s/8.0)} \qquad (7.119)$$

From (7.119), we determine $PM_c = 44.5°$ (at $\omega_{GC} = 4.6$ radians/second), which satisfies the phase margin design requirement.

Selecting $C_1 = 1\ \mu\text{F}$ arbitrarily, we use (7.114) in Step 7 to yield

$$R_1 = \frac{(2.6)^{-1}}{10^{-6}} = 385\ \text{K}\Omega$$

$$C_2 = \frac{(0.085)^{-1} - (0.26)^{-1} + (8.0)^{-1} - (2.6)^{-1}}{385\ \text{K}} = 42.1\ \mu\text{F}$$

$$R_2 = \frac{(0.26)^{-1}}{42.1 \cdot 10^{-6}} = 91.5\ \text{K}\Omega \qquad (7.120)$$

which completes the lag-lead network design.

7.8 DISCUSSION AND SUMMARY

An important feature in the practical design of phase-lead, phase-lag, and lag-lead compensation networks is the sensitivity of the system performance to each network parameter. For example, if design specifications on %OS or PM are given, it is useful to calculate their sensitivities with respect to R_1, R_2, and C for phase-lead and phase-lag designs and with respect to R_1, R_2, C_1, and C_2 for lag-lead design. Having determined these sensitivities about nominal network parameters, the control system designer can then determine the deleterious effects on system performance resulting from the (hopefully slight) mismatch when using off-the-shelf components. A careful sensitivity analysis during the design stage can often prevent the need for a system redesign after extensive field testing has begun.

The series compensation procedures described in this chapter for single-input, single-output systems are directly applicable to decoupled multivariable control systems [7]. Such systems have the same number of inputs and outputs with decoupling accomplished by multivariable feedback to allow the first output to be influenced by only the first input, the second output by the second input, and so forth. Therefore, series compensation can be introduced independently in each separate channel of the decoupled multivariable control system. Additional analysis and design procedures for multivariable systems are described by Rosenbrock [8] and MacFarlane [9].

We can select from a wide range of computational aids to automate the steps of the design algorithms presented in this chapter. We can easily place the algorithm steps on programmable calculators, which have an advantage of portability for classroom testing purposes. We might also choose to form control system design software packages on digital computers ranging from personal computers to large-scale centralized computers. Finally, we could employ interactive computer graphics facilities not only to perform the system design itself, but also to simulate the system's time and frequency responses and verify that all sensitivity considerations have been included.

In summary, we have introduced the series compensation approach in this chapter for designing feedback control systems that satisfy given time and frequency response performance specifications. We presented algorithms for phase-lead and phase-lag design based on both root locus and Bode procedures. We illustrated each design procedure with two examples of a positional servomechanism (second-order and third-order systems). We described a lag-lead design procedure that combined the phase-lead and phase-lag networks to yield a passive network having the advantages of each. Finally, we discussed the importance of sensitivity analysis for the designed control system, the extension of design results to uncoupled multivariable control systems, and the variety of computational aids that can be employed for design purposes. In the following chapter, we compare this transfer-function feedback design based on dynamic series compensators with state variable feedback design based on optimal control theory.

REFERENCES

1. T. E. Fortmann and K. L. Hitz. *An Introduction to Linear Control Systems.* New York: Marcel Dekker, 1977.
2. R. C. Dorf. *Modern Control Systems,* 3rd ed. Chapter 10. Reading, Mass.: Addison-Wesley, 1980.
3. B. C. Kuo. *Automatic Control Systems,* 4th ed. Chapters 8 and 10. Englewood Cliffs, N.J.: Prentice-Hall, 1982.
4. J. J. D'Azzo and C. H. Houpis. *Linear Control System Analysis and Design,* 2nd ed. Chapters 10 and 11. New York: McGraw-Hill, 1981.
5. A. P. Sage. *Linear Systems Control.* Chapters 5 and 6. Champaign, Ill.: Matrix Publishers, 1978.
6. J. L. Melsa and D. G. Schultz. *Linear Control Systems.* Chapter 10. New York: McGraw-Hill, 1969.
7. D. H. Owens. *Multivariable and Optimal Systems.* London, England: Academic Press, 1981.
8. H. H. Rosenbrock. *State-Space and Multivariable Theory.* New York: Wiley, 1970.
9. A. G. J. MacFarlane, editor. *Frequency-Response Methods in Control Systems.* New York: IEEE Press, 1979.

PROBLEMS

(§ 7.1)   **7.1** For the system in Figure 7.31, determine the ranges of K that satisfy each of the design specifications given by

$$\%OS \leq 20\%$$

$$e_{ss}\big|_{\substack{\text{unit} \\ \text{ramp}}} \leq 0.02 \text{ radians}$$

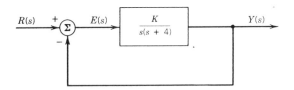

Figure 7.31   The system for Problem 7.1.

(§ 7.1)   **7.2** What are conservative estimates of t_r, t_d, and t_s for the second-order system in Figure 7.32 when $K = 7.0$? Determine the ranges of K satisfying each of the other design specifications listed in Table 7.1.

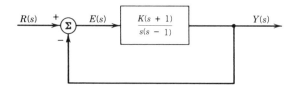

Figure 7.32 The system for Problem 7.2.

(§7.1)   **7.3** For the negative unity-feedback system having

$$G(s) = \frac{K}{s(s + 1)(s + 40)}$$

determine the range of K that simultaneously satisfies the three design specifications given by

$$M_p \leq 1.6$$

$$PM \geq 4.5°$$

$$GM \geq 5$$

(§7.1)   **7.4** Determine the region in the s-plane for which the following specifications are satisfied simultaneously.

$$\%OS \leq 20\%$$

$$t_r \leq 4.0 \text{ seconds}$$

$$t_d \leq 2.0 \text{ seconds}$$

$$t_s \leq 6.0 \text{ seconds}$$

For the system described in Problem 7.3, what range of K (if any) is needed to satisfy each of the specifications?

(§7.2)   **7.5** Suppose we have designed a phase-lead network with

$$G_c(s) = \frac{V_0(s)}{V_i(s)} = \frac{s + 10}{s + 20}$$

What values of R_1 and R_2 should be selected if $C = 1 \ \mu F$? Repeat for $C = 10 \ \mu F$.

(§7.2) **7.6** Repeat Problem 7.5 for a phase-lag network with

$$G_c(s) = \frac{V_0(s)}{V_i(s)} = \frac{1 + s/20}{1 + s/10}$$

(§7.3) **7.7** Design a phase-lead network by the root locus procedure for a negative unity-feedback system having a plant transfer function given by

$$G_p(s) = \frac{K}{s(s + 4)}$$

The design specifications are

$$\%OS \leq 20\%$$

$$t_r \leq 3.0 \text{ seconds}$$

$$e_{ss}|_{\substack{\text{unit} \\ \text{ramp}}} \leq 0.04 \text{ radians}$$

Specify values of R_1 and R_2, if $C = 1$ μF. Use computer simulations to verify that the design specifications are satisfied.

(§7.3) **7.8** Determine the smallest value possible for e_{ss} in Problem 7.7 if the dominant closed-loop poles are to have a negative real part of -6. The ratio $a = p/z$ must not be greater than 15.

(§7.3) **7.9** Repeat Problem 7.7 if the plant transfer function is given by

$$G_p(s) = \frac{K}{s(s + 1)(s + 40)}$$

(§7.4) **7.10** Design a phase-lead compensation network if

$$G_p(s) = \frac{K}{s(s + 4)}$$

and it is required that $PM \geq 45°$ and e_{ss} due to a unit ramp is 0.04 radians or less. Determine network parameters if $C = 1$ μF.

(§7.4) **7.11** Design a phase-lead compensation network if

$$G_p(s) = \frac{K}{s(s + 4)(s + 80)}$$

and the design specifications are $PM \geq 45°$ and e_{ss} due to a unit ramp input is 0.04 radians or less. Determine network parameters if $C = 1$ μF.

(§7.5) **7.12** Design a phase-lag network for the system of Problem 7.10 if the plant and steady-state error requirement are unchanged but $PM \geq 45°$ is replaced by $\%OS \leq 12\%$. Determine network parameters if $C = 100 \ \mu F$. Use computer simulations to verify your design.

(§7.5) **7.13** Design a phase-lag network for the system of Problem 7.11 if the plant and e_{ss} requirement are unchanged but $PM \geq 45°$ is replaced by $\%OS \leq 12\%$. Determine network parameters if $C = 100 \ \mu F$. Verify your design by using computer simulations.

(§7.6) **7.14** Design a phase-lag network for the system of Problem 7.10. Determine network parameters if $C = 100 \ \mu F$.

(§7.6) **7.15** Design a phase-lag network for the system of Problem 7.11. Determine network parameters if $C = 100 \ \mu F$.

(§7.3, **7.16** Let the design specifications for a negative unity-feedback system
7.5) having

$$G_p(s) = \frac{K}{s(s+3)}$$

be given by

$$\%OS \leq 20\%$$

$$e_{ss}\big|_{\substack{\text{unit} \\ \text{ramp}}} \leq 0.06 \text{ radians}$$

Design both a phase-lead network ($C = 1 \ \mu F$) and a phase-lag network ($C = 100 \ \mu F$) to meet these design specifications. Show time response curves by using computer simulations.

(§7.7) **7.17** Design a series compensation network for the negative unity-feedback system having

$$G_p(s) = \frac{K}{s(s+1)}$$

subject to

$$\%OS \leq 30\%$$

$$PM \geq 45°$$

$$e_{ss}\big|_{\substack{\text{unit} \\ \text{ramp}}} \leq 0.05 \text{ radians}$$

(§7.7) **7.18** Design a series compensation network for the negative unity-feed-back system having

$$G_p(s) = \frac{K}{s(s + 2)(s + 40)}$$

subject to

$$GM \geq 5$$

$$e_{ss}|_{\substack{\text{unit} \\ \text{ramp}}} \leq 0.05 \text{ radians}$$

(§7.7) **7.19** Design a series compensation network for the negative unity-feed-back system having

$$G_p(s) = \frac{K}{s(s^2 + 20s + 10^6)}$$

subject to

$$PM \geq 50°$$

$$e_{ss}|_{\substack{\text{unit} \\ \text{ramp}}} \leq 0.05 \text{ radians}$$

(§7.7) **7.20** Design a series compensation network for the negative unity-feed-back system having

$$G_p(s) = \frac{K}{s^2(s + 1)}$$

subject to

$$\%OS \leq 20\%$$

$$e_{ss}|_{\substack{\text{unit} \\ \text{parabola}}} \leq 0.05 \text{ radians/second}$$

(§7.7) **7.21** Design a lag-lead network if

$$G_p(s) = \frac{K}{s(s + 4)}$$

and it is required that $PM \geq 45°$, e_{ss} due to a unit ramp input is 0.04 radians or less, and the lag-lead compensated system is to have approximately the same attenuation for large ω as the uncompensated system. Specify network parameters.

8

State Variable Design Using Optimal Control Theory

Experience with alternative design concepts provides the control engineer flexibility in approaching the design task. This knowledge can be useful, for example, when deciding between a percent overshoot constraint and the integral of squared error as a criterion for aircraft autopilot design. Moreover, decisions can be made more easily on whether a dynamic or static controller is appropriate for a given application. Finally, the design approach may be chosen to simplify algorithmic steps for large-scale system design or to utilize automated design routines. In brief, familiarity with alternative designs permits choices regarding (1) the aspects of system behavior to be emphasized, (2) controller structure and characteristics, and (3) operational steps leading to an acceptable controller.

The series compensation techniques of Chapter 7 are based on classical s-plane (root locus) or frequency response (Bode) methods that utilize only the plant output for feedback with a dynamic controller. In this final chapter on design, we employ modern state-space procedures that require the availability of all state variables to form linear static controllers. Rather than using percent overshoot or settling time as performance specifications, we select performance measures formed as the integrals of weighted functions of the state variables and plant input. After minimizing these performance measures, we can compare the resulting closed-loop systems with those obtained earlier from classical design procedures.

8.1 REFORMULATION OF THE DESIGN PROBLEM

The relative quality of a system design can be determined by a performance measure that provides a *figure of merit* as a single number. Consider the general performance measure given by

$$J(u) = \int_{t_0}^{t_f} g(\mathbf{x}, u, t)\, dt \tag{1}$$

where $g(\mathbf{x}, u, t)$ is a functional of the state $\mathbf{x}$, the plant control u, and time t. Also

458

TABLE 8.1   Typical Performance Measures

Criterion	Mathematical Expression
Integral squared error (ISE)	$J(u) = \int_{t_0}^{t_f} e^2 \, dt$
Integral absolute error (IAE)	$J(u) = \int_{t_0}^{t_f} |e| \, dt$
Integral time squared error (ITSE)	$J(u) = \int_{t_0}^{t_f} (t - t_0) e^2 \, dt$
Integral time absolute error (ITAE)	$J(u) = \int_{t_0}^{t_f} (t - t_0) |e| \, dt$
Minimum energy	$J(u) = \int_{t_0}^{t_f} u^2 \, dt$
Minimum fuel	$J(u) = \int_{t_0}^{t_f} |u| \, dt$
Minimum time	$J(u) = \int_{t_0}^{t_f} dt$
Quadratic in state and control	$J(u) = \int_{t_0}^{t_f} (\mathbf{x}^T Q \mathbf{x} + \gamma u^2) \, dt$

referred to as a performance index, cost functional, or penalty function, the performance measure indicates how well the system performs between an initial time t_0 and a final time t_f. Our design goal is to minimize the performance measure [1–4].

The selection of a performance measure is based on which aspects of system behavior in a particular application are deemed most important to the control system designer. If the output time response is required to approach its final value rapidly, then errors in the state variables should be weighted heavily in the performance measure. On the other hand, if a major design concern is to conserve energy or fuel in a spacecraft application, the plant input should be a prominent part of the performance measure.

We identify several typical performance measures in Table 8.1. The first two criteria are based on minimizing integrals of the squared error or absolute value of the error. Time weights are incorporated into the next two criteria to emphasize a reduction of errors occurring later in the time response. The next three performance measures focus on minimizing energy, fuel, or time. The last entry in Table 8.1 is a quadratic performance measure composed of the integral of a weighted combination of the squares of state variables and the plant input. Both the selection of the type of performance measure and the choice of weights are often decisions to be made by the control system designer.

8.1.1 Parameter Optimization

When the controller structure or configuration is specified, we may adjust controller parameters to minimize the given performance measure. We refer to this design procedure as *parameter optimization*. For example, we prove later in this chapter that the optimal feedback control law for a linear plant with a quadratic performance measure is itself linear. At this point, we arbitrarily assume a linear controller configuration and apply parameter optimization.

The mathematical operation involved in solving this linear quadratic control problem by parameter optimization evaluates the quadratic performance measure given by

$$J(u) = \int_{t_0}^{t_f} (\mathbf{x}^T Q \mathbf{x} + \gamma u^2) \, dt \tag{8.2}$$

under the assumption of a linear controller

$$u = \mathbf{k}^T \mathbf{x} \tag{8.3}$$

Substituting (8.3) into (8.2) yields

$$J(u) = \int_{t_0}^{t_f} \mathbf{x}^T [Q + \gamma \mathbf{k} \mathbf{k}^T] \mathbf{x} \, dt \tag{8.4}$$

We then select the controller parameters in the vector $\mathbf{k}$ to minimize $J(u)$ in (8.4). In general, we refer to the resulting feedback control law as a specific optimal control law because the controller configuration has been specified. Moreover, we say that the design is suboptimal if one or more of the controller parameters in (8.3) have been fixed and the optimization is performed by adjusting the remaining free controller parameters. While the specific optimal control law in (8.3) may be either optimal or suboptimal for the linear quadratic control problem, depending on whether all parameters in $\mathbf{k}$ are free for adjustment, this linear controller can only be suboptimal for those problems in which the optimal controller configuration is nonlinear.

EXAMPLE 8.1

Consider the linear system of Figure 8.1 having a plant transfer function given by

$$\frac{Y(s)}{U(s)} = \frac{1}{s(s+1)} \tag{8.5}$$

The error e is the difference in the input $r(t)$ and the output y, which is fed back negatively as shown. We now add tachometer feedback to this system to form a linear controller $u(\mathbf{x})$ as

$$u = e - K_t \dot{y}$$

$$= r(t) - x_1 - K_t x_2 \tag{8.6}$$

where we have selected phase variables $x_1 = y$ and $x_2 = \dot{y}$. Starting with $x_1(0) = 0$

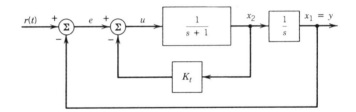

Figure 8.1 The linear system for parameter optimization in Example 8.1.

and $x_2(0) = 0$, we apply a unit step input $[r(t) = 1]$, which drives $x_1(t)$ toward $+1$ as $t \to \infty$. As a performance measure, we select

$$J(u) = \int_0^\infty (e^2 + \gamma u^2)\, dt$$

$$= \int_0^\infty \left[(1 - x_1)^2 + \gamma(1 - x_1 - K_t x_2)^2 \right] dt \qquad (8.7)$$

From either (8.5) or Figure 8.1, we observe that $\omega_n = 1$ and $\zeta = 1/2$ for the system without tachometer feedback. When tachometer feedback is inserted, we might obtain either an underdamped, critically damped, or overdamped system time response. An underdamped response has the form

$$y(t) = x_1(t) = 1 - \frac{e^{-\zeta_t t}}{\sqrt{1 - \zeta_t^2}} \sin(\omega_d t + \theta) \qquad (8.8)$$

where ζ_t is the system damping coefficient and ω_d and θ satisfy

$$\omega_d = \sqrt{1 - \zeta_t^2}$$

$$\theta = \cos^{-1}(\zeta_t) \qquad (8.9)$$

We form the derivative of $y(t)$ in (8.8) as

$$\dot{y}(t) = x_2(t) = \frac{e^{-\zeta_t t}}{\sqrt{1 - \zeta_t^2}} \sin \omega_d t \qquad (8.10)$$

Using (8.8) and (8.10) in (8.7) gives

$$J(K_t) = \int_0^\infty \left\{ \left[\frac{e^{-\zeta_t t}}{\sqrt{1 - \zeta_t^2}} \sin(\omega_d t + \theta) \right]^2 \right.$$

$$\left. + \gamma \left[\frac{e^{-\zeta_t t}}{\sqrt{1 - \zeta_t^2}} \sin(\omega_d t + \theta) - K_t \frac{e^{-\zeta_t t}}{\sqrt{1 - \zeta_t^2}} \sin \omega_d t \right]^2 \right\} dt \quad (8.11)$$

The performance measure J in (8.11) is to be minimized by adjusting the parameter K_t in Figure 8.1. However, according to (4.46), the relationship between K_t and ζ_t is $2\zeta_t\omega_n = 2\zeta\omega_n + K_t\omega_n^2$. For $\omega_n = 1$ and $\zeta = 1/2$, we have

$$\zeta_t = \tfrac{1}{2}(K_t + 1) \tag{8.12}$$

Consequently, we have an underdamped response $(0 < \zeta_t < 1)$ for $-1 < K_t < 1$. We note that the system response is unstable for $K_t < -1$ and overdamped for $K_t > 1$. Substituting (8.9) and (8.12) into (8.11) yields an integral for J as a function of K_t and γ. Expanding the integrand by using trigonometric identities, we can evaluate the resulting expressions with the aid of integral tables to obtain

$$J(K_t) = \frac{1 + 2\gamma + (K_t + 1)^2}{2(K_t + 1)} \tag{8.13}$$

Curves of J versus K_t are shown in Figure 8.2 for $\gamma = 0$, 0.5, 1, and 1.5 with minimum points occurring for $K_t = 0$, 0.41, 0.73, and 1, respectively. For $\gamma > 1.5$, minima of J occur for values of K_t greater than unity and, hence, corresponding optimal system time responses for those performance measures are overdamped.

The results in Figure 8.2 are generally suboptimal since only one state variable (x_2) is fed back through an adjustable gain determined by parameter optimization. We observe that it may be possible to obtain even lower values of $J(u)$ in (8.7) by feeding back both x_1 and x_2 through adjustable gains k_1 and k_2.

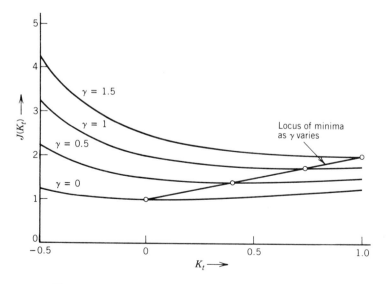

Figure 8.2 Curves of J versus K_t for Example 8.1.

Although these gains may also be found by parameter optimization, we prefer the direct procedure of the following sections for their calculation.

8.2 INTRODUCTION TO OPTIMAL CONTROL

We presented the concept of a performance measure in the last section as a means of assessing the relative quality of a system design. Once a particular controller configuration is assumed, we may use parameter optimization to determine the best feedback solution (lowest value of J) under that assumption. A basic problem is to determine which of all possible configurations yields the lowest value of the performance measure J. Even if we do not choose to build that optimal controller structure for our particular application, we would nevertheless like to know how the controller we are using compares with this optimal solution.

8.2.1 Problem Formulation

In deriving necessary conditions for the general optimization problem, we consider the performance measure [1–4] defined by

$$J(u) = \int_0^\infty g(\mathbf{x}, u, t)\, dt \tag{8.14}$$

Let the plant be described by

$$\dot{\mathbf{x}} = f(\mathbf{x}, u, t) \tag{8.15}$$

with known boundary conditions on the state $\mathbf{x}(t)$ at $t = 0$, that is, $\mathbf{x}(0)$. We assume that the class of plant inputs is unconstrained and, consequently, any value of u is admissible.

The optimization problem is to minimize the performance measure $J(u)$ in (8.14) subject to the dynamic plant equations in (8.15) and the associated boundary conditions.

8.2.2 Euler-Lagrange Equations

As a preliminary step in the solution of this unconstrained optimal control problem, consider the performance measure given by

$$J(y) = \int_{t_0}^{t_f} g(y, \dot{y}, t)\, dt \tag{8.16}$$

where $y(t)$ is a continuous scalar function of t on the interval $[t_0, t_f]$ with

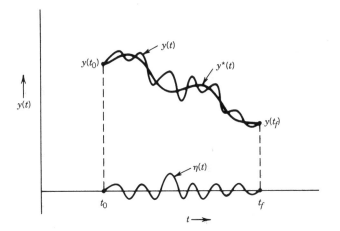

Figure 8.3 Illustrating the time variation $\eta(t)$ about an optimal solution $y*(t)$.

piecewise continuous derivatives. Let $y*(t)$ be an optimal solution and $\eta(t)$ an arbitrary variation about this optimal solution, that is,

$$y(t) = y*(t) + \varepsilon\eta(t) \tag{8.17}$$

where ε is a small, positive scalar constant. We require that $y*(t)$ must satisfy the given boundary conditions on $y(t)$, that is, $y*(t_0) = y(t_0)$ and $y*(t_f) = y(t_f)$. Therefore, $\eta(t)$ must satisfy $\eta(t_0) = \eta(t_f) = 0$. We indicate a typical time response of $y(t)$, $y*(t)$, and $\eta(t)$ in Figure 8.3.

To obtain a necessary condition for the optimality of $y*(t)$, we first substitute (8.17) into (8.16) to give

$$J(y) = \int_{t_0}^{t_f} g(y* + \varepsilon\eta, \dot{y}* + \varepsilon\dot{\eta}, t) \, dt \tag{8.18}$$

We then determine the derivative of $J(y)$ with respect to ε as

$$\frac{dJ}{d\varepsilon} = \int_{t_0}^{t_f} \left[\frac{\partial g}{\partial y} \eta + \frac{\partial g}{\partial \dot{y}} \dot{\eta} \right] dt \tag{8.19}$$

Integrating (8.19) by using the parts formula of elementary integral calculus, we obtain

$$\frac{dJ}{d\varepsilon} = \int_{t_0}^{t_f} \left[\frac{\partial g}{\partial y} \eta - \frac{d}{dt} \left(\frac{\partial g}{\partial \dot{y}} \right) \eta \right] dt$$

$$= \frac{\partial g}{\partial \dot{y}} \eta \Big|_{t_0}^{t_f} + \int_{t_0}^{t_f} \left[\frac{\partial g}{\partial y} - \frac{d}{dt} \left(\frac{\partial g}{\partial \dot{y}} \right) \right] \eta(t) \, dt \tag{8.20}$$

where the first term is zero because $\eta(t_0) = \eta(t_f) = 0$, leaving only the integral term. As a necessary condition to minimize $J(y)$ in (8.16), we set the derivative in (8.20) equal to zero and simultaneously set ε equal to zero, thereby forcing $y(t)$ to be the optimal solution $y*(t)$. Since this condition must hold for any arbitrary variation $\eta(t)$, we obtain

$$\left[\frac{\partial g}{\partial y} - \frac{d}{dt}\left(\frac{\partial g}{\partial \dot{y}} \right) \right]\Bigg|_{y=y*} = 0 \tag{8.21}$$

which is referred to as the Euler-Lagrange equation for the case of a scalar variable y in (8.16). For the vector case, (8.21) becomes

$$\left[\frac{\partial g}{\partial \mathbf{y}} - \frac{d}{dt}\left(\frac{\partial g}{\partial \dot{\mathbf{y}}} \right) \right]\Bigg|_{\mathbf{y}=\mathbf{y}*} = \mathbf{0} \tag{8.22}$$

We are now ready to solve the optimal control problem formulated in (8.14) and (8.15). We begin by defining a function L as

$$L(\mathbf{x}, \boldsymbol{\lambda}, u, t) = g(\mathbf{x}, u, t) + \boldsymbol{\lambda}^T[\mathbf{f}(\mathbf{x}, u, t) - \dot{\mathbf{x}}] \tag{8.23}$$

where $\boldsymbol{\lambda}$ represents an n-vector of Lagrange multipliers which serve as differential constraints associated with the n plant equations in (8.15). The n-vector $\boldsymbol{\lambda}$ is a function of time t, since the state $\mathbf{x}$ is itself a function of t.

Consider the performance measure given by

$$J(u) = \int_{t_0}^{t_f} L(\mathbf{x}, \boldsymbol{\lambda}, u, t) \, dt \tag{8.24}$$

It is well known that the minimization of (8.14) with the additional constraints in (8.15) is equivalent to the minimization of (8.24) with no additional constraints [1–4]. Using (8.23), we write the Euler-Lagrange equations for (8.22) as

$$\frac{d}{dt}\left(\frac{\partial L}{\partial \dot{\mathbf{x}}} \right) = \frac{\partial L}{\partial \mathbf{x}}$$

$$\frac{d}{dt}\left(\frac{\partial L}{\partial \dot{\boldsymbol{\lambda}}} \right) = \frac{\partial L}{\partial \boldsymbol{\lambda}}$$

$$\frac{d}{dt}\left(\frac{\partial L}{\partial \dot{u}} \right) = \frac{\partial L}{\partial u} \tag{8.25}$$

where variations about optimal values have been included for the state $\mathbf{x}$, Lagrange multipliers $\boldsymbol{\lambda}$, and plant input u. Evaluating the equations in (8.25) at these optimal values (denoted by the superscript *) yields

$$-\dot{\boldsymbol{\lambda}}^* = \left.\frac{\partial g}{\partial \mathbf{x}}\right|_* + \left(\frac{\partial \mathbf{f}}{\partial \mathbf{x}}\right)^T\bigg|_* \boldsymbol{\lambda}^*$$

$$\dot{\mathbf{x}}^* = \mathbf{f}(\mathbf{x}^*, u^*, t)$$

$$0 = \left.\frac{\partial g}{\partial u}\right|_* + \left(\frac{\partial \mathbf{f}}{\partial u}\right)^T\bigg|_* \boldsymbol{\lambda}^* \tag{8.26}$$

The third equation in (8.26) is an algebraic equation that relates u^* to $\mathbf{x}^*$ and $\boldsymbol{\lambda}^*$. The first two equations are each of order n and, consequently, require a total of $2n$ boundary conditions. Usually, some of the known boundary conditions are at t_0 and some at t_f. For this reason, we refer to the problem of determining time solutions for the Euler-Lagrange equations as a two-point boundary value problem.

EXAMPLE 8.2

Consider the first-order linear plant described by

$$\dot{x} = -3x + 4u \tag{8.27}$$

with boundary conditions given as $x(0) = x_0$ and $x(\infty) = 0$. Let the performance measure to be minimized be

$$J(u) = \int_0^\infty (x^2 + u^2)\, dt \tag{8.28}$$

The problem is to determine the optimal control u that will transfer the state x from x_0 (at $t = 0$) to the origin (at $t = \infty$) while minimizing $J(u)$.

We first form the functional L in (8.23) as

$$L(x, \lambda, u, t) = x^2 + u^2 + \lambda(-3x + 4u - \dot{x}) \tag{8.29}$$

Using (8.26), we obtain

$$-\dot{\lambda}^* = 2x^* - 3\lambda^*$$

$$\dot{x}^* = -3x^* + 4u^*$$

$$0 = 2u^* + 4\lambda^* \tag{8.30}$$

We solve the third equation in (8.30) for u^* and then rewrite the first two equations as

$$\dot{x}^* = -3x^* - 8\lambda^*$$

$$\dot{\lambda}^* = -2x^* + 3\lambda^* \tag{8.31}$$

Using any of the state transition matrix methods of Section 4.4, we determine the unforced solution of the equations in (8.31) as

$$\begin{pmatrix} x^*(t) \\ \lambda^*(t) \end{pmatrix} = \begin{pmatrix} \frac{4}{5}e^{-5t} + \frac{1}{5}e^{5t} & \frac{4}{5}e^{-5t} - \frac{4}{5}e^{5t} \\ \frac{1}{5}e^{-5t} - \frac{1}{5}e^{5t} & \frac{1}{5}e^{-5t} + \frac{4}{5}e^{5t} \end{pmatrix} \begin{pmatrix} x^*(0) \\ \lambda^*(0) \end{pmatrix} \tag{8.32}$$

We are given $x^*(0) = x_0$ and $x^*(\infty) = 0$ as boundary conditions. The first of these is satisfied automatically in (8.32). Moreover, we may use $x^*(\infty) = 0$ to solve for $\lambda^*(0)$ by first writing the component equation for $x^*(t)$, that is

$$x^*(t) = \left(\frac{4}{5}e^{-5t} + \frac{1}{5}e^{5t}\right)x^*(0) + \left(\frac{4}{5}e^{-5t} + \frac{4}{5}e^{5t}\right)\lambda^*(0) \tag{8.33}$$

As $t \to \infty$, we neglect terms involving e^{-5t} in (8.33) and set $x^*(t) = 0$ to obtain

$$0 = \frac{1}{5}e^{5t}x^*(0) - \frac{4}{5}e^{5t}\lambda^*(0)$$

$$= \left[\frac{1}{5}x^*(0) - \frac{4}{5}\lambda^*(0)\right]e^{5t} \tag{8.34}$$

We set the bracketed term in (8.34) equal to zero and solve for $\lambda^*(0)$ in terms of $x^*(0)$ to yield

$$\lambda^*(0) = \frac{1}{4}x^*(0) = \frac{1}{4}x_0 \tag{8.35}$$

Therefore, the time solutions for $x^*(t)$, $\lambda^*(t)$, and $u^*(t)$ are

$$x^*(t) = x_0 e^{-5t}$$

$$\lambda^*(t) = \lambda^*(0)e^{-5t} = \frac{1}{4}x_0 e^{-5t}$$

$$u^*(t) = -2\lambda^*(t) = -\frac{1}{2}x_0 e^{-5t} \tag{8.36}$$

The solution to the Euler-Lagrange equations in (8.30) provides the optimal open-loop control function $u^*(t)$ in (8.36). We show this open-loop result in Figure 8.4a.

Seldom are we able to determine the optimal closed-loop control configuration from the calculus of variations approach. Yet for this first-order system, we see from (8.36) that

$$u^*(x) = kx^* = -\frac{1}{2}x^* \tag{8.37}$$

Figure 8.4b shows this linear feedback structure.

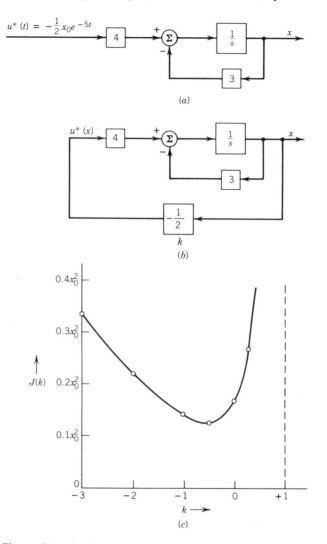

$u^*(t) = -\frac{1}{2}x_0 e^{-5t}$

(a)

$u^*(x)$

(b)

$J(k)$

$0.4x_0^2$

$0.3x_0^2$

$0.2x_0^2$

$0.1x_0^2$

-3 -2 -1 0 $+1$

$k \longrightarrow$

(c)

Figure 8.4 Optimal (*a*) open-loop and (*b*) closed-loop solutions with (*c*) a curve of $J(k)$ versus k for Example 8.2.

Suppose we had initially selected a linear feedback configuration $u = kx$ with the intent of using the parameter optimization procedure of the last section to obtain the optimal value of k. Evaluating $J(u)$ in (8.28) for $u = kx$ gives

$$J(u) = \int_0^\infty \left[x_0^2 e^{2(-3+4k)t} + k^2 x_0^2 e^{2(-3+4k)t} \right] dt$$

$$= \frac{x_0^2(1 + k^2)}{2(3 - 4k)} \tag{8.38}$$

Figure 8.4*c* shows a curve of $J(k)$ versus k. We next calculate dJ/dk from (8.38) and set this derivative equal to zero to obtain k equal to $-\frac{1}{2}$ or $+2$. We must choose $k = -\frac{1}{2}$ to yield a stable system. The corresponding value of J in (8.38) is $x_0^2/8$. In summary, we have used the Euler-Lagrange equations to determine the optimal open-loop control function $u^*(t)$ for a first-order linear system with a quadratic performance measure. Moreover, we have determined that the optimal feedback control structure is a constant gain $(k = -\frac{1}{2})$ and have verified this optimal value of k by parameter optimization for an assumed linear feedback configuration.

8.2.3 The Linear Quadratic Control Problem

Consider the linear, time-varying plant described by

$$\dot{\mathbf{x}} = A(t)\mathbf{x} + \mathbf{b}(t)u \qquad (8.39)$$

with initial conditions $\mathbf{x}(t_0) = \mathbf{x}_0$. We use the quadratic performance measure already given in (8.2), that is,

$$J(u) = \int_{t_0}^{t_f} (\mathbf{x}^T Q \mathbf{x} + \gamma u^2)\, dt \qquad (8.2)$$

where Q is a positive semidefinite matrix, γ is a positive constant, and both t_0 and t_f are specified. The problem is to determine the optimal open-loop control function $u^*(t)$ and the optimal closed-loop control law $u^*(\mathbf{x})$ for the linear quadratic control problem.

The optimal open-loop control function $u^*(t)$ can be obtained by solving the Euler-Lagrange equations. Identifying g and $\mathbf{f}$ in (8.2) and (8.39), we may use the third equation in (8.26) to yield $2\gamma u^* + \boldsymbol{\lambda}^T \mathbf{b}(t) = 0$, from which

$$u^* = -\frac{1}{2\gamma}\boldsymbol{\lambda}^T \mathbf{b}(t) \qquad (8.40)$$

Performing the indicated operations in the second and first equations in (8.26) and substituting (8.40) for u^* yields

$$\dot{\mathbf{x}}^* = A(t)\mathbf{x}^* - \frac{1}{2\gamma}\mathbf{b}(t)\mathbf{b}^T(t)\boldsymbol{\lambda}^*$$

$$\dot{\boldsymbol{\lambda}}^* = -2Q\mathbf{x}^* - A^T(t)\boldsymbol{\lambda}^* \qquad (8.41)$$

which are the required Euler-Lagrange equations.

This formulation would be complete if $2n$ boundary conditions were known. We have provided $\mathbf{x}(t_0) = \mathbf{x}_0$ as n of those boundary conditions. If t_f is specified and $\mathbf{x}(t_f)$ is free, we state without proof that transversality conditions require

$\lambda(t_f) = \mathbf{0}$ for optimality. If $t_f = \infty$, this condition corresponds to $\mathbf{x}(\infty) = \mathbf{0}$, as in the scalar case considered in Example 8.2. In solving (8.41) using these boundary conditions, we obtain $\mathbf{x}^*(t)$ and $\lambda^*(t)$, from which $u^*(t)$ can be obtained from (8.40).

To determine the optimal closed-loop control law $u^*(\mathbf{x})$ for the linear quadratic control problem, let us assume that there exists a symmetric, time-varying, positive definite matrix $P(t)$ satisfying

$$\lambda = 2P(t)\mathbf{x} \tag{8.42}$$

for all $\mathbf{x}$. Using (8.42) in (8.41) yields

$$\dot{P} + Q + A^T(t)P + PA(t) - \frac{1}{\gamma}Pb(t)b^T(t)P = 0 \tag{8.43}$$

which is referred to as the matrix Riccati equation [1–4]. There are $n(n + 1)/2$ coupled component differential equations in (8.43), with an identical number of unknown variables in the matrix P. The boundary condition for (8.43) is $P(t_f) = 0$. Substituting (8.42) into (8.40) gives the optimal closed-loop control law $u^*(\mathbf{x})$ as

$$u^*(\mathbf{x}) = -\frac{1}{2\gamma}(2P\mathbf{x})^T b(t) = -\frac{1}{\gamma}b^T(t)P\mathbf{x} \tag{8.44}$$

For the special case where the linear plant is time-invariant and the process is of infinite duration, that is, $t_f = \infty$, the n by n matrix P is constant. Since $\dot{P}$ is zero for this case, (8.43) becomes the nonlinear algebraic matrix equation given by

$$Q + A^T P + PA - \frac{1}{\gamma}Pbb^T P = 0 \tag{8.45}$$

EXAMPLE 8.3

Let the plant be described by

$$\dot{x}_1 = x_2$$

$$\dot{x}_2 = -2x_1 - 3x_2 + u \tag{8.46}$$

with $x_1(0) = 1$ and $x_2(0) = 0$. Suppose we select a quadratic performance measure (with $t_f = \infty$) as

$$J(u) = \frac{1}{2}\int_0^\infty \left(5x_1^2 + 5x_2^2 + u^2\right) dt \tag{8.47}$$

We wish to determine the optimal open-loop control function $u^*(t)$.

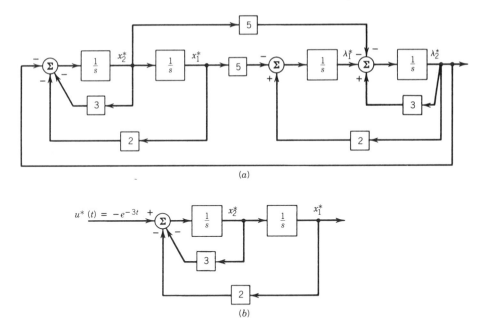

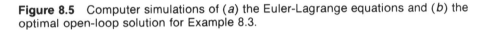

Figure 8.5 Computer simulations of (a) the Euler-Lagrange equations and (b) the optimal open-loop solution for Example 8.3.

From (8.41), we have

$$\dot{x}_1^* = x_2^*$$

$$\dot{x}_2^* = -2x_1^* - 3x_2^* - \lambda_2^*$$

$$\dot{\lambda}_1^* = -5x_1^* + 2\lambda_2^*$$

$$\dot{\lambda}_2^* = -5x_2^* - \lambda_1^* + 3\lambda_2^* \qquad (8.48)$$

with boundary conditions $x_1^*(0) = 1$, $x_2^*(0) = 0$, $\lambda_1^*(t_f) = 0$, and $\lambda_2^*(t_f) = 0$.

Figure 8.5 shows a computer simulation diagram for the fourth-order system of equations in (8.48). Using any of the methods of Section 4.4, preferably the impulsed-integrator procedure for determining the resolvent matrix, we may obtain the state transition matrix $\Phi(t)$ to yield the form of the time solution for (8.48) as

$$\begin{pmatrix} x_1^*(t) \\ x_2^*(t) \\ \lambda_1^*(t) \\ \lambda_2^*(t) \end{pmatrix} = \begin{pmatrix} \phi_{11} & \phi_{12} & \phi_{13} & \phi_{14} \\ \phi_{21} & \phi_{22} & \phi_{23} & \phi_{24} \\ \phi_{31} & \phi_{32} & \phi_{33} & \phi_{34} \\ \phi_{41} & \phi_{42} & \phi_{43} & \phi_{44} \end{pmatrix} \begin{pmatrix} 1 \\ 0 \\ \lambda_1^*(0) \\ \lambda_2^*(0) \end{pmatrix} \qquad (8.49)$$

where we have inserted $x_1^*(0) = 1$ and $x_2^*(0) = 0$ in (8.49). We determine the characteristic equation for the closed-loop system in Figure 8.5a as

$$s^4 - 10s^2 + 9 = 0 \tag{8.50}$$

which yields roots of -1, $+1$, -3, and $+3$. Therefore, each element ϕ_{ij} of the state transition matrix in (8.49) has the form

$$\phi_{ij}(t) = K_1 e^{-t} + K_2 e^{+t} + K_3 e^{-3t} + K_4 e^{3t} \tag{8.51}$$

We express the component equations for $\lambda_1^*(t)$ and $\lambda_2^*(t)$ in (8.49) and then use the remaining boundary conditions $\lambda_1^*(\infty) = \lambda_2^*(\infty) = 0$ to obtain $\lambda_1^*(0) = 6$ and $\lambda_2^*(0) = 1$. The resulting time expressions for x_1^*, x_2^*, λ_1^*, and λ_2^* are

$$x_1^*(t) = 1.5e^{-t} - 0.5e^{-3t}$$

$$x_2^*(t) = -1.5e^{-t} + 1.5e^{-3t}$$

$$\lambda_1^*(t) = 7.5e^{-t} - 1.5e^{-3t}$$

$$\lambda_2^*(t) = e^{-3t} \tag{8.52}$$

Using (8.40), we obtain the optimal open-loop control $u^*(t)$ as

$$u^*(t) = -\lambda_2^*(t) = -e^{-3t} \tag{8.53}$$

which is shown in Figure 8.5b.

In solving for the optimal closed-loop control law using the matrix Riccati equation, we first identify

$$A = \begin{pmatrix} 0 & 1 \\ -2 & -3 \end{pmatrix} \qquad b = \begin{pmatrix} 0 \\ 1 \end{pmatrix}$$

$$Q = \begin{pmatrix} 2.5 & 0 \\ 0 & 2.5 \end{pmatrix} \qquad \gamma = 0.5 \tag{8.54}$$

Using (8.54) in (8.45) gives

$$\begin{pmatrix} 2.5 & 0 \\ 0 & 2.5 \end{pmatrix} + \begin{pmatrix} 0 & -2 \\ 1 & -3 \end{pmatrix}\begin{pmatrix} p_{11} & p_{12} \\ p_{12} & p_{22} \end{pmatrix} + \begin{pmatrix} p_{11} & p_{12} \\ p_{12} & p_{22} \end{pmatrix}\begin{pmatrix} 0 & 1 \\ -2 & -3 \end{pmatrix}$$

$$-\frac{1}{0.5}\begin{pmatrix} p_{11} & p_{12} \\ p_{12} & p_{22} \end{pmatrix}\begin{pmatrix} 0 \\ 1 \end{pmatrix}(0 \ \ 1)\begin{pmatrix} p_{11} & p_{12} \\ p_{12} & p_{22} \end{pmatrix} = \begin{pmatrix} 0 & 0 \\ 0 & 0 \end{pmatrix} \tag{8.55}$$

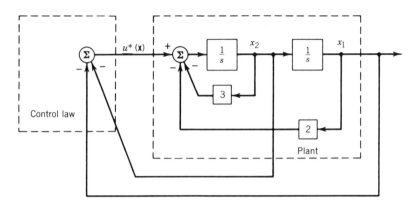

Figure 8.6 The linear plant and optimal feedback control law for Example 8.3.

Simplifying (8.55) gives

$$\begin{pmatrix} 2.5 - 4p_{12} - 2p_{12}^2 & p_{11} - 3p_{12} - 2p_{22} - 2p_{12}p_{22} \\ p_{11} - 3p_{12} - 2p_{22} - 2p_{12}p_{22} & 2.5 + 2p_{12} - 6p_{22} - 2p_{22}^2 \end{pmatrix} = \begin{pmatrix} 0 & 0 \\ 0 & 0 \end{pmatrix}$$

$$(8.56)$$

from which

$$P = \begin{pmatrix} 3 & 0.5 \\ 0.5 & 0.5 \end{pmatrix} \tag{8.57}$$

We observe that P is positive definite. From (8.44) and (8.57), the optimal closed-loop control law is given by

$$u^*(\mathbf{x}) = -\frac{1}{0.5}(0 \quad 1)\begin{pmatrix} 3 & 0.5 \\ 0.5 & 0.5 \end{pmatrix}\begin{pmatrix} x_1 \\ x_2 \end{pmatrix}$$

$$= -x_1 - x_2 \tag{8.58}$$

Figure 8.6 shows a computer simulation diagram for the linear system in (8.46) with its optimal feedback control law in (8.58). A comparison may be made with the simulation diagram in Figure 8.5b, which provides an open-loop control function $u^*(t)$ that is valid only for the given initial conditions $x_1(0) = 1$ and $x_2(0) = 0$. On the other hand, the closed-loop control law $u^*(\mathbf{x})$ in Figure 8.6 is optimal for all initial conditions for the state vector $\mathbf{x}$. Moreover, as demonstrated in Section 3.4, noise disturbance effects are minimized with a feedback configuration.

EXAMPLE 8.4

We return to the design problem of Example 8.1, where we used parameter optimization to adjust a tachometer gain K_t for suboptimal control. We now want to determine the optimal feedback control law $u^*(\mathbf{x})$ obtained by feeding back, through appropriate gains, both state variables for this second-order plant. We can apply (8.45) to solve for the matrix P and then obtain $u^*(\mathbf{x})$ from (8.44).

Using the direct phase variable form shown in Figure 8.1, we express the state variable equations as

$$\dot{x}_1 = x_2$$

$$\dot{x}_2 = -x_2 + u \tag{8.59}$$

We again use the performance measure defined earlier in Example 8.1 as

$$J(u) = \int_0^\infty (e^2 + \gamma u^2)\, dt \tag{8.7}$$

where the error e is the difference between the unit step input $[r(t) = 1]$ and x_1, that is, $e = 1 - x_1$. Since $J(u)$ is in terms of e and not $\mathbf{x}$, we must have differential equations describing e. Letting $e_1 = e$ and $e_2 = \dot{e}$, we now define a new set of state variables as

$$\dot{e}_1 = e_2$$

$$\dot{e}_2 = -e_2 - u \tag{8.60}$$

We obtain the second equation in (8.60) by calculating $\dot{e}_1 = -\dot{x}_1$ from $e_1 = 1 - x_1$ and then identifying e_2 as $-x_2$ and $\dot{e}_2$ as $-\dot{x}_2$, which from (8.59) is equal to $-(-x_2 + u)$ or $-e_2 - u$.

Using (8.60) and (8.7) with $e = e_1$, we identify

$$A = \begin{pmatrix} 0 & 1 \\ 0 & -1 \end{pmatrix} \quad \mathbf{b} = \begin{pmatrix} 0 \\ -1 \end{pmatrix} \quad Q = \begin{pmatrix} 1 & 0 \\ 0 & 0 \end{pmatrix} \tag{8.61}$$

Using (8.61) in (8.45) gives

$$\begin{pmatrix} 1 & 0 \\ 0 & 0 \end{pmatrix} + \begin{pmatrix} 0 & 0 \\ 1 & -1 \end{pmatrix}\begin{pmatrix} p_{11} & p_{12} \\ p_{12} & p_{22} \end{pmatrix} + \begin{pmatrix} p_{11} & p_{12} \\ p_{12} & p_{22} \end{pmatrix}\begin{pmatrix} 0 & 1 \\ 0 & -1 \end{pmatrix}$$

$$-\frac{1}{\gamma}\begin{pmatrix} p_{11} & p_{12} \\ p_{12} & p_{22} \end{pmatrix}\begin{pmatrix} 0 \\ -1 \end{pmatrix}(0 \quad -1)\begin{pmatrix} p_{11} & p_{12} \\ p_{12} & p_{22} \end{pmatrix} = \begin{pmatrix} 0 & 0 \\ 0 & 0 \end{pmatrix} \tag{8.62}$$

Simplifying (8.62) yields

$$
\begin{pmatrix}
1 - \dfrac{1}{\gamma}p_{12}^2 & p_{11} - p_{12} - \dfrac{1}{\gamma}p_{12}p_{22} \\[3mm]
p_{11} - p_{12} - \dfrac{1}{\gamma}p_{12}p_{22} & 2p_{12} - 2p_{22} - \dfrac{1}{\gamma}p_{22}^2
\end{pmatrix}
= \begin{pmatrix} 0 & 0 \\ 0 & 0 \end{pmatrix}
\qquad (8.63)
$$

from which we obtain

$$
P = \begin{pmatrix}
\sqrt{\gamma + 2\sqrt{\gamma}} & \sqrt{\gamma} \\[2mm]
\sqrt{\gamma} & -\gamma + \sqrt{\gamma^2 + 2\sqrt{\gamma}}
\end{pmatrix}
\qquad (8.64)
$$

It is again easy to show that the matrix P is positive definite as required. From (8.64) and (8.44), we determine $u^*(\mathbf{e})$ as

$$
u^*(\mathbf{e}) = -\frac{1}{\gamma}\mathbf{b}^T P \mathbf{e}
$$

$$
= \frac{1}{\sqrt{\gamma}}e_1 + \left(-1 + \sqrt{1 + 2/\sqrt{\gamma}}\,\right)e_2
\qquad (8.65)
$$

Figure 8.7 shows the optimal closed-loop controller configuration as a function of the performance measure parameter γ.

We may compare the results from Example 8.1 for the suboptimal controller configuration of Figure 8.1 with corresponding results for the optimal feedback control law $u^*(\mathbf{e})$ in (8.65). We determine the minimum value of $J(u)$, denoted as J^*, as $\sqrt{\gamma + 2\sqrt{\gamma}}$. Curves of J^* versus γ are shown in Figure 8.8 for optimal and suboptimal feedback structures. The optimal feedback control law has both e and $\dot{e}$ fed back through gains identified in (8.65), while the suboptimal controller utilized for feedback only $\dot{e}$ through an adjustable tachometer gain K_t. We observe that J^* for the optimal controller is lower for all values of γ except $\gamma = 1$, for which the two controllers are identical.

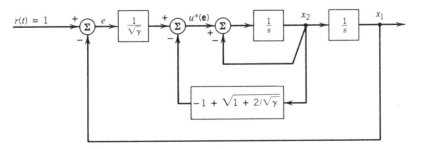

Figure 8.7   The optimal feedback control law for Example 8.4.

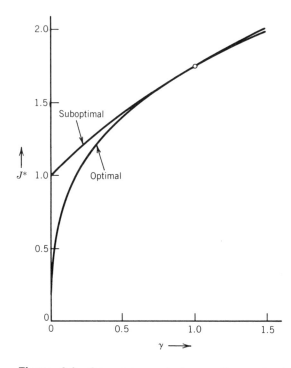

Figure 8.8 Comparisons between the optimal
solution of Example 8.4 and the suboptimal solution of Example 8.1.

8.3 DESIGN COMPARISONS

Kalman [5] considered the inverse optimal control problem of determining when
a linear closed-loop system is optimal. The matrix A and vector $\mathbf{b}$ completely
describe the linear plant, and the matrix P and scalar γ define the optimal
feedback control law in (8.44). If these descriptors of the linear plant and
feedback control law are specified, we may solve (8.45) for the weighting matrix Q
to determine the performance measure for which this closed-loop system is
optimal. These remarks also apply to the time-varying case, except that the matrix
Q, which may also be time varying, may be determined from (8.43). Therefore, we
conclude that all linear systems are optimal according to some quadratic performance measure to be determined from either (8.43) or (8.45).

We use the foregoing concept of an inverse optimal control problem in this
section to aid in comparing results obtained by the methods of optimal control,
state variable feedback (SVFB) control, and series compensation. We present two

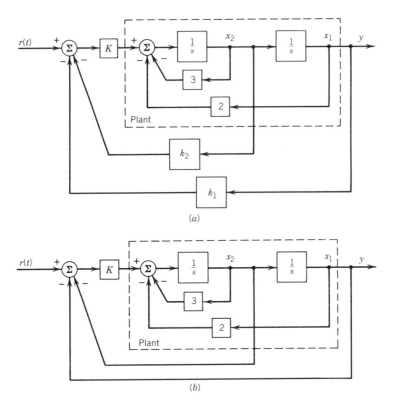

Figure 8.9 Development of linear SVFB control for Example 8.5.

examples: the first compares optimal control and SVFB control (pole placement), and the second compares optimal control with series compensation.

EXAMPLE 8.5

Let the transfer function of a linear plant be given by

$$G(s) = \frac{1}{(s + 1)(s + 2)} = \frac{1}{s^2 + 3s + 2} \tag{8.66}$$

We are to design a linear SVFB control law as shown in Figure 8.9a using the direct phase variable form of state variables such that the resulting closed-loop poles are placed at -1 and -3. Moreover, we are to determine a quadratic performance measure for which optimization theory yields this same feedback control law.

Using the procedures developed in Section 3.6, we obtain a closed-loop transfer function for Figure 8.9a as

$$\frac{Y(s)}{R(s)} = \frac{K}{s^2 + (3 + Kk_2)s + (2 + Kk_1)} \tag{8.67}$$

Requiring pole placements at -1 and -3 yields a desired closed-loop transfer function as

$$\frac{Y(s)}{R(s)} = \frac{1}{(s + 1)(s + 3)} = \frac{1}{s^2 + 4s + 3} \tag{8.68}$$

Equating (8.67) and (8.68) gives $K = 1$ and

$$3 + Kk_2 = 4$$

$$2 + Kk_1 = 3 \tag{8.69}$$

from which $k_1 = k_2 = 1$. This linear SVFB control law is shown in Figure 8.9b. For the corresponding optimal control law, we first identify for the plant

$$A = \begin{pmatrix} 0 & 1 \\ -2 & -3 \end{pmatrix}$$

$$\mathbf{b} = \begin{pmatrix} 0 \\ 1 \end{pmatrix} \tag{8.70}$$

Since we want to obtain the same control law as for the SVFB control case above, we require

$$u^*(\mathbf{x}) = -x_1 - x_2 = \begin{pmatrix} -1 \\ -1 \end{pmatrix}^T \begin{pmatrix} x_1 \\ x_2 \end{pmatrix} \tag{8.71}$$

Equating (8.71) and (8.44), we obtain

$$\begin{pmatrix} -1 \\ -1 \end{pmatrix} = -\frac{1}{\gamma} \begin{pmatrix} p_{11} & p_{12} \\ p_{12} & p_{22} \end{pmatrix} \begin{pmatrix} 0 \\ 1 \end{pmatrix} \tag{8.72}$$

which yields $p_{12} = p_{22} = \gamma$. Using (45), we may write

$$Q + \begin{pmatrix} 0 & -2 \\ 1 & -3 \end{pmatrix} \begin{pmatrix} p_{11} & \gamma \\ \gamma & \gamma \end{pmatrix} + \begin{pmatrix} p_{11} & \gamma \\ \gamma & \gamma \end{pmatrix} \begin{pmatrix} 0 & 1 \\ -2 & -3 \end{pmatrix}$$

$$-\frac{1}{\gamma} \begin{pmatrix} \gamma \\ \gamma \end{pmatrix} (\gamma \quad \gamma) = 0 \tag{8.73}$$

Solving (8.73) for Q gives

$$Q = \begin{pmatrix} 5\gamma & 6\gamma - p_{11} \\ 6\gamma - p_{11} & 5\gamma \end{pmatrix} \tag{8.74}$$

We may select $p_{11} = 6\gamma$ in (8.74) without loss of generality. We then obtain

$$Q = \begin{pmatrix} 5\gamma & 0 \\ 0 & 5\gamma \end{pmatrix} \tag{8.75}$$

Therefore, we have an optimal feedback control law that is identical to the linear SVFB control obtained in Figure 8.9b when we select a quadratic performance measure in (8.2) given by

$$J(u) = \gamma \int_0^\infty \left(5x_1^2 + 5x_2^2 + u^2 \right) dt \tag{8.76}$$

We see that optimal control with a selected quadratic performance measure yields the same results as pole placement for linear time-invariant plants. Moreover, optimal control provides a generalized format for the control of time-varying or nonlinear plants.

EXAMPLE 8.6

Consider the linear plant with phase-lead series compensation shown in Figure 8.10a. In Example 7.5 of Section 7.3, we determined values of the parameters a and T to satisfy design specifications of $\%OS \leq 15\%$ and $t_r \leq 1.0$ second for a step input and $e_{ss} \leq 0.05$ radians for a unit ramp input. We pinpoint in this example the general difficulty that prevents us from using optimal control [in particular (8.45)] to yield the same feedback control law as phase-lead compensation.

We let a and T remain unspecified here and attempt to determine $J(u)$ in terms of these parameters. As shown in Figure 8.10b, we define state variables as

$$\dot{x}_1 = x_2$$
$$\dot{x}_2 = -2x_2 + Ku$$
$$\dot{x}_3 = -\frac{1}{T}x_3 + e \tag{8.77}$$

Letting $e_1 = e = 1 - x_1$, $e_2 = \dot{e}$, and $e_3 = \ddot{e}$, we may express an alternate form of state variables as

$$\dot{e}_1 = e_2$$
$$\dot{e}_2 = -2e_2 - Ku$$
$$\dot{e}_3 = -e_1 - \frac{1}{T}e_3 \tag{8.78}$$

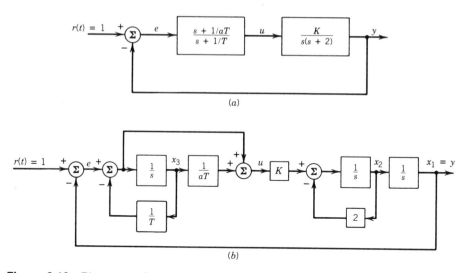

(a)

(b)

Figure 8.10 Phase variable selection for the phase-lead series compensated system of Example 8.6.

Therefore, we identify A and $\mathbf{b}$ as

$$A = \begin{pmatrix} 0 & 1 & 0 \\ 0 & -2 & 0 \\ -1 & 0 & -1/T \end{pmatrix} \qquad \mathbf{b} = \begin{pmatrix} 0 \\ -K \\ 0 \end{pmatrix} \tag{8.79}$$

From Figure 8.10b, we have

$$u = \dot{x}_3 + (1/aT)x_3$$

$$= e - [(1 - 1/a)/T]x_3$$

$$= e_1 + [(1 - 1/a)/T]e_3 \tag{8.80}$$

Using (8.80) and (8.44), we obtain

$$\begin{pmatrix} 1 \\ 0 \\ (1 - 1/a)/T \end{pmatrix} = -\frac{1}{\gamma}\begin{pmatrix} p_{11} & p_{12} & p_{13} \\ p_{12} & p_{22} & p_{23} \\ p_{13} & p_{23} & p_{33} \end{pmatrix}\begin{pmatrix} 0 \\ -K \\ 0 \end{pmatrix} \tag{8.81}$$

Solving (8.81) with $\gamma = 1$ yields $p_{12} = 1/K$, $p_{22} = 0$, and $p_{23} = (1 - 1/a)/TK$. Since $p_{22} = 0$, the matrix P will not be positive definite as required by optimal control theory. Therefore, we conclude for this example that it is not possible to determine a quadratic performance measure $J(u)$ in (8.2) that yields the same optimal control law as the one specified by phase-lead series compensation.

Let us arbitrarily choose the feedback control law to be

$$
u^*(\mathbf{e}) = \begin{pmatrix} 1 \\ \rho \\ (1 - 1/a)/T \end{pmatrix}^T \begin{pmatrix} e_1 \\ e_2 \\ e_3 \end{pmatrix} \tag{8.82}
$$

where ρ can be selected such that the matrix P is positive definite and the resulting matrix Q from (8.45) is positive semidefinite. Using (8.82) in (8.45) and solving for Q yields

$$
Q = \begin{pmatrix} 1 + 2p_{13} & \rho + 1/K + \xi/K - p_{11} & \xi + p_{33}p_{13}/T \\ \rho + (1 + \xi)/K - p_{11} & \rho^2 + \rho - 2/K & \xi\rho + \xi(1 + T)/K - p_{13} \\ \xi + p_{33} + p_{13}/T & \xi\rho + \xi/K + \xi/TK - p_{13} & \xi^2 + 2p_{33}/T \end{pmatrix}
$$

$$\tag{8.83}$$

where we have let $\xi = (1 - 1/a)/TK$. When only the output y (i.e., x_1) is available for feedback, we can use observer theory [6] to form approximations for other state variables. In this example, only x_2 was inaccessible. For higher-order plants, we may need to form approximations for several additional state variables before the linear quadratic control problem can be formulated and the resulting linear feedback law obtained. Finally, we should recognize that the phase-lead series compensation design yields a specific optimal control law since the feedback configuration has been fixed. The design is also suboptimal with respect to the $J(u)$ resulting from (8.83).

8.3.1 Design Procedures in Retrospect

We summarize in Figure 8.11 the various control system design procedures described in this and previous chapters. We identify proportional control, PID control, and SVFB control as design techniques, primarily from Chapter 3, that can be performed by both transfer function and state variable methods. Series compensation in Chapter 7—that is, phase-lead, phase-lag, and lag-lead compensation—depends on the transfer function procedures of root locus or Bode design. On the other hand, the optimization methods described in the earlier sections of Chapter 8 are based on state variable design. Multivariable control system design can be accomplished by series compensation for noninteracting control after decoupling or by optimal control theory to determine the m-vector feedback control law $\mathbf{u}^*(\mathbf{x})$ as described in the following section.

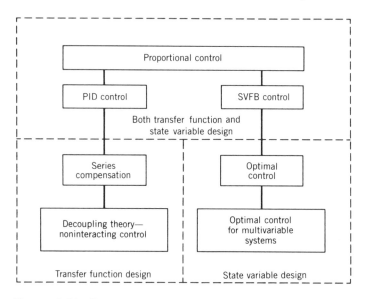

Figure 8.11 Relationships between control system design procedures.

8.4 MULTIVARIABLE OPTIMAL CONTROL

Consider the multivariable plant described by

$$\dot{\mathbf{x}} = A\mathbf{x} + B\mathbf{u} \qquad (8.84)$$

where $\mathbf{u}$ is an m-dimensional control vector and B is an associated n by m matrix. Let the performance measure to be minimized be given by

$$J(\mathbf{u}) = \int_0^\infty \left(\mathbf{x}^T Q \mathbf{x} + \mathbf{u}^T R \mathbf{u}\right) dt \qquad (8.85)$$

where R is an m by m symmetric, positive definite matrix. Thus, we see that (8.84) and (8.85) define the linear quadratic multivariable control problem [1–4].

Euler-Lagrange equations for the multivariable case in (8.84) and (8.85) yield (8.26) with the scalar u replaced by a vector $\mathbf{u}$ to give

$$\dot{\mathbf{x}}^* = A\mathbf{x}^* - \left(\tfrac{1}{2}\right) BR^{-1}B^T\boldsymbol{\lambda}^*$$

$$\dot{\boldsymbol{\lambda}}^* = -2Q\mathbf{x}^* - A^T\boldsymbol{\lambda}^*$$

$$\mathbf{u}^* = -\left(\tfrac{1}{2}\right) R^{-1}B^T\boldsymbol{\lambda}^* \qquad (8.86)$$

with $\mathbf{x}(0) = \mathbf{x}_0$ and $\boldsymbol{\lambda}(\infty) = \mathbf{0}$. The solution of (8.86) yields the optimal open-loop control function $\mathbf{u}^*(t)$ and associated $\mathbf{x}^*(t)$ and $\boldsymbol{\lambda}^*(t)$.

The Riccati equation for the multivariable case is

$$Q + A^T P + PA - PBR^{-1}B^T P = 0 \qquad (8.87)$$

where P is a constant matrix because A and B are not functions of time and we are considering the infinite-time problem. We observe that (8.87) reduces to (8.45) for the single-input case ($m = 1$), where $R^{-1} = 1/\gamma$. The multivariable optimal control law $\mathbf{u}^*(\mathbf{x})$, corresponding to the single-input optimal control law $u^*(\mathbf{x})$ in (8.44), is

$$\mathbf{u}^*(\mathbf{x}) = -R^{-1}B^T P \mathbf{x} \qquad (8.88)$$

Therefore, we obtain an optimal feedback control law as the solution to the linear quadratic multivariable control problem.

EXAMPLE 8.7

Let a second-order multivariable plant to be controlled be described by

$$\dot{x}_1 = x_2 + u_1$$

$$\dot{x}_2 = -4x_1 - 4x_2 + u_2 \qquad (8.89)$$

as shown in Figure 8.12a. The performance measure to be minimized is

$$J(\mathbf{u}) = \int_0^\infty \left(2x_1^2 - 2x_1x_2 + 36x_2^2 + u_1^2 + u_2^2\right) dt \qquad (8.90)$$

From (8.89), we identify the matrices A and B as

$$A = \begin{pmatrix} 0 & 1 \\ -4 & -4 \end{pmatrix}$$

$$B = \begin{pmatrix} 1 & 0 \\ 0 & 1 \end{pmatrix} = I \qquad (8.91)$$

Moreover, we identify Q and R from (8.90) as

$$Q = \begin{pmatrix} 2 & -1 \\ -1 & 36 \end{pmatrix}$$

$$R = \begin{pmatrix} 1 & 0 \\ 0 & 1 \end{pmatrix} = I \qquad (8.92)$$

We solve for the 2 by 2 matrix P in (8.87) by substituting (8.91) and (8.92) into

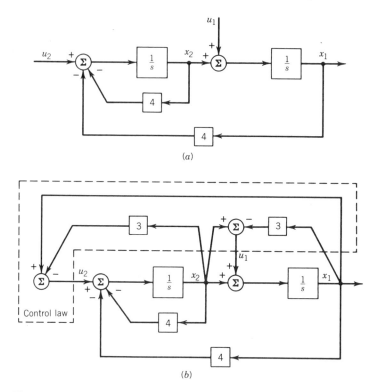

Figure 8.12 Multivariable optimal control for Example 8.7.

(8.87) to obtain

$$\begin{pmatrix} 2 & -1 \\ -1 & 36 \end{pmatrix} + \begin{pmatrix} 0 & 1 \\ -4 & -4 \end{pmatrix}^T \begin{pmatrix} p_{11} & p_{12} \\ p_{12} & p_{22} \end{pmatrix} + \begin{pmatrix} p_{11} & p_{12} \\ p_{12} & p_{22} \end{pmatrix} \begin{pmatrix} 0 & 1 \\ -4 & -4 \end{pmatrix}$$

$$- \begin{pmatrix} p_{11} & p_{12} \\ p_{12} & p_{22} \end{pmatrix} \begin{pmatrix} 1 & 0 \\ 0 & 1 \end{pmatrix} \begin{pmatrix} 1 & 0 \\ 0 & 1 \end{pmatrix}^{-1} \begin{pmatrix} 1 & 0 \\ 0 & 1 \end{pmatrix}^T \begin{pmatrix} p_{11} & p_{12} \\ p_{12} & p_{22} \end{pmatrix} = \begin{pmatrix} 0 & 0 \\ 0 & 0 \end{pmatrix} \quad (8.93)$$

Solving (8.93) gives

$$P = \begin{pmatrix} 3 & -1 \\ -1 & 3 \end{pmatrix} \quad (8.94)$$

Using (8.91), (8.92), and (8.94) in (8.88) yields the multivariable optimal control as

$$\mathbf{u}^*(\mathbf{x}) = -R^{-1}B^T P\mathbf{x}$$

$$= -\begin{pmatrix} 1 & 0 \\ 0 & 1 \end{pmatrix}^{-1} \begin{pmatrix} 1 & 0 \\ 0 & 1 \end{pmatrix}^T \begin{pmatrix} 3 & -1 \\ -1 & 3 \end{pmatrix} \begin{pmatrix} x_1 \\ x_2 \end{pmatrix}$$

$$= \begin{pmatrix} -3x_1 + x_2 \\ x_1 - 3x_2 \end{pmatrix} \quad (8.95)$$

Figure 8.12*b* shows this optimal feedback controller.

8.5 REDUCED-ORDER OBSERVERS

The distinguishing feature of this chapter has been the assumption that all state variables are available for use in a feedback controller. In this section we show how to reconstruct any inaccessible state variables by using reduced-order observers [6]. The reduced-order observer accepts as its input both the plant output **y** and the plant input **u**. A further problem is the need to guarantee that misalignments in initial conditions between the observer variables and the corresponding plant variables are quickly corrected. We show that eigenvalues of the observer can be selected arbitrarily to resolve this problem.

Let a multivariable plant be described by

$$\dot{\mathbf{x}} = A\mathbf{x} + B\mathbf{u}$$

$$\mathbf{y} = C\mathbf{x} \tag{8.96}$$

where the r-vector **y** has components available for feedback. We can select a set of state variables (or perform a linear transformation) that results in the elements of **y** as individual state variables. Let the remaining $n - r$ state variables form the elements of a vector **z**, where

$$\mathbf{z} = E\mathbf{x} \tag{8.97}$$

Using (8.96) and (8.97), we can form

$$\dot{\mathbf{y}} = F_1\mathbf{y} + F_2\mathbf{z} + G_1\mathbf{u}$$

$$\dot{\mathbf{z}} = F_3\mathbf{y} + F_4\mathbf{z} + G_2\mathbf{u} \tag{8.98}$$

Let T be an arbitrary $n - r$ by r constant matrix. Defining **w** as the difference between **z** and $T\mathbf{y}$, we may express $\dot{\mathbf{w}}$ from (8.98) as

$$\dot{\mathbf{w}} = \dot{\mathbf{z}} - T\dot{\mathbf{y}}$$

$$= (F_4 - TF_2)\mathbf{z} + (F_3 - TF_1)\mathbf{y} + (G_2 - TG_1)\mathbf{u}$$

$$= (F_4 - TF_2)\mathbf{w} + (F_3 - TF_1 + F_4T - TF_2T)\mathbf{y} + (G_2 - TG_1)\mathbf{u} \tag{8.99}$$

Since **z** contains state variables that are not accessible, we do not know $\mathbf{z}(t_0)$ and, hence, $\mathbf{w}(t_0)$. However, we can select T such that the eigenvalues of $F_4 - TF_2$ in (8.134) yield a sufficiently fast settling time for the observer state **w**. Once (8.99) is solved for the approximation $\hat{\mathbf{w}}(t)$, obtained by using estimated initial conditions, we can then form an approximation $\hat{\mathbf{z}}$ for the inaccessible state variables in **z** as

$$\hat{\mathbf{z}} = \hat{\mathbf{w}} + T\mathbf{y} \tag{8.100}$$

Finally, we note that the combined plant-observer is of order $n + (n - r) =$

$2n - r$, where the eigenvalues of the $(n - r)$th-order observer are selected arbitrarily.

EXAMPLE 8.8

Consider the linear plant in Example 8.6 described by

$$\dot{x}_1 = x_2$$

$$\dot{x}_2 = -2x_2 + Ku$$

$$y = x_1 \qquad (8.101)$$

Recall that in Example 8.6 we compared the series compensation design shown in Figure 8.10 with a feedback controller obtained by optimization theory. A basic restriction in the optimal controller design was that the state variable x_2 was inaccessible. Therefore, let us define

$$z = x_2 \qquad (8.102)$$

and then determine a first-order observer to yield an approximation $\hat{x}_2$ for x_2. Corresponding to (8.98), we use (8.101) and (8.102) to form

$$\dot{y} = z$$

$$\dot{z} = -2z + Ku \qquad (8.103)$$

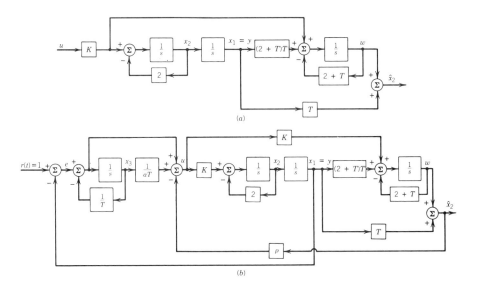

Figure 8.13 Reduced-order observer for Example 8.8.

Setting $w = z - Ty$, we write

$$\dot{w} = \dot{z} - T\dot{y}$$

$$= (-2z + Ku) - Tz$$

$$= (-2 - T)z - (-2 - T)Ty + (-2 - T)Ty + Ku$$

$$= (-2 - T)w - (2 + T)Ty + Ku \qquad (8.104)$$

Figure 8.13a shows the combined plant-observer with the approximation for x_2 formed as in (8.100). The student is asked in Problem 8.18 to verify that the transfer function from u to $\hat{x}_2$ is indeed $K/(s + 2)$. Finally, Figure 8.13b shows the closed-loop feedback system resulting when the optimal feedback control law in (8.82) is applied by using x_1 from the plant and $\hat{x}_2$ from the observer.

EXAMPLE 8.9

Suppose only x_1 is accessible in the multivariable plant of Example 8.7 described by

$$\dot{x}_1 = x_2 + u_1$$

$$\dot{x}_2 = -4x_1 - 4x_2 + u_2$$

$$y = x_1 \qquad (8.105)$$

We set $z = x_2$ and form a first-order observer to yield the approximation $\hat{x}_2$. Corresponding to (8.98), we form

$$\dot{y} = z + u_1$$

$$\dot{z} = -4y - 4z + u_2 \qquad (8.106)$$

Again, with $w = z - Ty$, we have

$$\dot{w} = \dot{z} - T\dot{y}$$

$$= (-4y - 4z + u_2) - T(z + u_1)$$

$$= (-4 - T)z - 4y - Tu_1 + u_2$$

$$= (-4 - T)z - (-4 - T)Ty + (-4 - T)Ty - 4y - Tu_1 + u_2$$

$$= (-4 - T)w - (4 + 4T + T^2)y - Tu_1 + u_2 \qquad (8.107)$$

Figure 8.14a shows the combined multivariable plant and first-order observer.

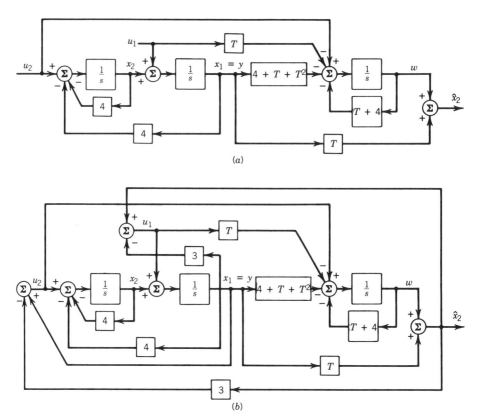

Figure 8.14 Reduced-order observer for Example 8.9.

Using this observer output as an approximation for x_2, we show the optimal control law of Example 8.7 in Figure 8.14b.

8.6 ADVANCED TOPICS IN CONTROL

The introductory nature of this book prohibits us from including details of a number of advanced control systems concepts. However, we can mention these advanced topics briefly and encourage their study in research programs or in further courses, perhaps at the graduate level. First, the need for work in microprocessor control has increased dramatically in recent years because of the rapid expansion in the development and use of microprocessors and minicomputers [7,8]. Robotics has become another extremely popular area for control systems applications that feature design including electromechanical sensors, plants, controllers, and actuators [9]. Stochastic optimal control is an area that focuses on the combined design of state variable estimators and associated feedback controllers for operations in environments having significant noise disturbances [10]. When only random time curves of system inputs and outputs are available, it becomes necessary to perform system identification by using techniques based on either continuous-time functions or discrete-time series analysis [11–13]. Finally, decentralized control algorithms are being studied for the hierarchical control of large-scale systems [14,15] and adaptive control algorithms for systems having models that are only partially known [16,17]. Interested students are encouraged to pursue these advanced topics in the current technical literature [7–17].

SUMMARY

In this final chapter, we looked at design procedures based on the modern state variable approach for linear plants with complete state feedback. Initially, we reformulated the design problem in terms of the minimization of a performance measure and then derived the Euler-Lagrange equations as necessary conditions for optimality. We showed that the solution of these equations yields the optimal open-loop control function $u^*(t)$.

We considered the linear quadratic control problem in which the plant is linear and the performance measure is the integral of a quadratic function of the plant state $\mathbf{x}$ and input u. We derived the matrix Riccati equation and showed that the optimal closed-loop controller $u^*(\mathbf{x})$ is a linear function of the plant state variables. Moreover, for the infinite-time problem, we showed that the feedback gains are constant when the plant is a time-invariant system. We presented the extension of these optimal control results to the multivariable case.

In making design comparisons, we showed the equivalence between SVFB control (pole placement) and optimization theory for linear plants by constructing the quadratic performance measure that is minimized by a given SVFB control law. However, since series compensation techniques require only the plant output for feedback, we are unable to determine the corresponding quadratic performance measures for optimal control directly. Using a reduced-order observer to approximate inaccessible state variables permits us to generate estimates of the complete state for feedback and, hence, to complete the desired comparison. We concluded with a discussion of selected advanced topics in control theory.

REFERENCES

1. D. E. Kirk. *Optimal Control Theory*. Englewood Cliffs, N.J.: Prentice-Hall, 1970.
2. B. D. O. Anderson and J. B. Moore. *Linear Optimal Control*. Englewood Cliffs, N.J.: Prentice-Hall, 1971.

3. M. Athans and P. L. Falb. *Optimal Control: An Introduction to the Theory and Its Applications.* New York: McGraw-Hill, 1966.

4. A. P. Sage and C. C. White, III. *Optimum Systems Control,* 2nd ed. Englewood Cliffs, N.J.: Prentice-Hall, 1977.

5. R. E. Kalman. "When Is a Linear Control System Optimal?" *Transactions ASME Ser. D: J. Basic Eng.,* Vol. 86 (March 1964), pp. 1–10.

6. D. G. Luenberger. *Introduction to Dynamic Systems: Theory, Models, and Applications.* New York: Wiley, 1979.

7. G. F. Franklin and J. D. Powell. *Digital Control of Dynamic Systems.* Reading, Mass.: Addison-Wesley, 1980.

8. B. C. Kuo. *Digital Control Systems.* New York: Holt, Rinehart and Winston, 1980.

9. Special Issue of the *Proceedings of the IEEE.* "Robotics and Factories of the Future" (July 1983).

10. P. S. Maybeck. *Stochastic Models, Estimation, and Control,* Vol. 1. New York: McGraw-Hill, 1979.

11. D. Graupe. *Identification of Systems,* 2nd ed. Huntington, N.Y.: Krieger, 1976.

12. T. Kailath (guest editor). "Special Issue on System Identification and Time-Series Analysis." *IEEE Transactions on Automatic Control,* Vol. AC-19, No. 6 (December 1974).

13. G. E. P. Box and G. M. Jenkins. *Time Series Analysis, Forecasting and Control.* San Francisco: Holden-Day, 1970.

14. M. Athans (guest editor). Special Issue on "Large-Scale Systems and Decentralized Control." *IEEE Transactions on Automatic Control,* Vol. AC-23, No. 2 (April 1978).

15. A. P. Sage. *Methodology of Large-Scale Systems.* New York: McGraw-Hill, 1977.

16. I. D. Landau. *Adaptive Control—The Model Reference Approach.* New York: Dekker, 1979.

17. G. C. Goodwin and K. S. Sin. *Adaptive Filtering Prediction and Control.* Englewood Cliffs, N.J.: Prentice-Hall, 1984.

PROBLEMS

(§8.1) **8.1** Let the closed-loop system in Figure 8.15 have $x(0) = 0$. Determine the optimal feedback gain k that minimizes each of the following

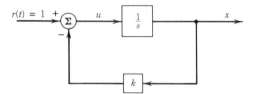

Figure 8.15 The closed-loop system for optimization in Problem 8.1.

performance measures. Evaluate $J(u^*)$ and explain the practical implication of minimizing $J(u)$ in each case.

a. $J(u) = \int_0^\infty u^2\, dt$

b. $J(u) = \int_0^1 tu^2\, dt$

c. $J(u) = \int_0^\infty [(2 - x)^2 + u^2]\, dt$

(§ 8.2) **8.2** Let a linear plant be described by

$$\dot{x} = -x + u$$

with $x(0) = 1$. If the performance measure is given by

$$J(u) = \int_0^\infty (x^2 + u^2)\, dt$$

express the Euler-Lagrange equations and identify the necessary boundary conditions. Solve these equations for the optimal control $u^*(t)$.

(§ 8.2) **8.3** For the plant and performance measure in Problem 8.2, let $u = kx$. Sketch the curve of $J(k)$ versus k. Compare the value of k obtained by setting $dJ/dk = 0$ with the one obtained as the ratio of $u^*(t)$ to $x^*(t)$ in Problem 8.2.

(§ 8.2) **8.4** Repeat the instructions of Problem 8.2 for the plant described by

$$\dot{x} = -2x + u$$

with $x(0) = 10$ and the performance measure given by

$$J(u) = \int_0^\infty (2x^2 + u^2)\, dt$$

(§ 8.2) **8.5** For the linear plant in Example 8.2 described by

$$\dot{x} = -3x + 4u$$

with $x(0) = x_0$, what infinite-time quadratic performance measure is minimized with negative unity feedback ($k = -1$)? In other words, find γ such that

$$J(u) = \int_0^\infty (x^2 + \gamma u^2)\, dt$$

is minimized for $k = -1$.

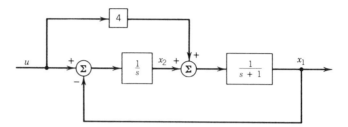

Figure 8.16 The linear plant of Problem 8.6.

(§ 8.2)   **8.6**   Express (but do not solve) the Euler-Lagrange equations for the linear plant shown in Figure 8.16 with a performance measure given by

$$J(u) = \int_0^\infty \left(x_1^2 + 2x_2^2 + 4u^2 \right) dt$$

(§ 8.2)   **8.7**   Let a second-order linear plant be described by

$$\dot{x}_1 = x_2$$

$$\dot{x}_2 = -3x_1 - 4x_2 + u$$

with $x_1(0) = 1$ and $x_2(0) = 0$. If the performance measure is given as

$$J(u) = \int_0^\infty \left(x_1^2 + x_2^2 + u^2 \right) dt$$

solve the Euler-lagrange equations to obtain $\mathbf{x}^*(t)$, $\mathbf{\lambda}^*(t)$, and $u^*(t)$.

(§ 8.2)   **8.8**   Determine optimal feedback gains k_1 and k_2 for the plant and performance measure given in Problem 8.7.

(§ 8.2)   **8.9**   Show that the time-varying feedback control law given by

$$u^*(x, t) = -\left[\tanh (1 - t) \right] x$$

is obtained from the solution to (8.43) when the linear plant is described by

$$\dot{x} = u$$

and the performance measure to be minimized is given by

$$J(u) = \int_0^1 (x^2 + u^2) dt$$

(§ 8.3) **8.10** Let a linear plant be described by

$$\dot{x}_1 = x_2$$

$$\dot{x}_2 = -12x_1 - 7x_2 + u$$

Find the quadratic performance measure that is minimized by the optimal feedback control given by

$$u^*(\mathbf{x}) = -2x_1 - x_2$$

for arbitrary initial conditions on the state $\mathbf{x}(t)$.

(§ 8.3) **8.11** Repeat the instructions of Problem 8.10 if the plant is described by

$$\dot{x}_1 = x_2$$

$$\dot{x}_2 = x_3$$

$$\dot{x}_3 = -6x_1 - 11x_2 - 6x_3 + u$$

and the optimal feedback control is

$$u^*(\mathbf{x}) = -x_1 - x_2 - x_3$$

(§ 8.3) **8.12** Find the quadratic performance measure that is minimized if the linear plant in Problem 3.29 has an optimal feedback control specified by the SVFB control determined there.

(§ 8.3) **8.13** What first-order linear plant is being considered if the feedback control

$$u^*(x) = -5x$$

minimizes the performance measure given by

$$J(u) = \int_0^\infty (x^2 + u^2)\, dt?$$

Assume the initial state $x(0)$ is arbitrary.

(§ 8.3) **8.14** What second-order linear plant, expressed in direct phase variable form, yields the feedback control

$$u^*(\mathbf{x}) = -x_1 - 2x_2$$

to minimize

$$J(u) = \int_0^\infty (x_1^2 + x_2^2 + 2u^2)\, dt?$$

Assume an arbitrary initial state $\mathbf{x}(0)$.

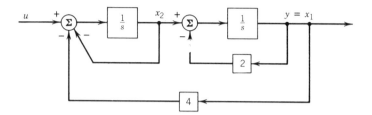

Figure 8.17 The linear plant of Problem 8.17 for which a reduced-order observer is required.

(§8.4) **8.15** Determine the optimal feedback control that minimizes

$$J(u) = \int_0^\infty \left(x_1^2 + x_2^2 + u_1^2 + u_2^2 \right) dt$$

if the linear multivariable plant is given by

$$\dot{x}_1 = -2x_1 + x_2 + u_1$$

$$\dot{x}_2 = -x_1 - x_2 + u_2$$

Assume an arbitrary initial state x(0).

(§8.4) **8.16** What quadratic performance measure is minimized by the feedback control

$$\mathbf{u}^*(\mathbf{x}) = \begin{pmatrix} -x_1 - 2x_2 \\ -2x_1 - x_2 \end{pmatrix}$$

for the linear multivariable plant given in Problem 8.15?

(§8.5) **8.17** Determine a first-order observer to approximate x_2 for the linear plant in Figure 8.17. Let this observer have an eigenvalue of -4. Show that any misalignment in initial conditions between $x_2(0)$ and $\hat{x}_2(0)$ decays as e^{-4t}.

(§8.5) **8.18** Verify that the transfer function from u to $\hat{x}_2$ in Figure 8.13a is given by

$$\frac{\hat{X}_2(s)}{U(s)} = \frac{K}{s + 2}$$

Moreover, show that $\hat{x}_2$ is not controllable from the input u.

(§8.5) **8.19** Let $y = x_1$ for the linear plant given in Problem 8.11. Determine a second-order observer to approximate x_2 and x_3. Let both eigenvalues of this reduced-order observer be located at -2.

(§ 8.5) **8.20** Consider the linear plant described by

$$\dot{x}_1 = -3x_1 + u$$

$$\dot{x}_2 = -2x_1 - x_2 + u$$

$$y = x_1$$

Show that it is impossible to construct an observer to approximate the inaccessible state variable x_2. Why?

ANSWERS TO SELECTED PROBLEMS

Chapter 1

1.1   Natural system with inherent feedback (**b, f**); natural system with human-devised controller (**e, g**); human-devised system with human-devised controller (**a, c, d, h**).

1.2   **d.**   Referring to Figure 1.2*b*, the plant block contains the ship dynamics, sensor(s) are direction sensors, the controller is a combined comparator and rudder actuator, the system input is the desired course, and typical disturbances are unexpected currents.

1.3   **a.**   Referring to Figure 1.2*a*, the open-loop controllers can be identified as timers and the plant as the lighting devices with light intensity (brightness) as the plant output.

      b.   Referring to Figure 1.2*b*, the system input is a setting for the desired light intensity, the controller sensors are optical sensors and comparative devices, the plant is composed of the mercury vapor lamps with light intensity (brightness) as the output.

1.5   Referring to Figure 1.2*b*, the system input is the desired drilling mud density, a controller/sensor senses mud density and adds barite, the plant is the mud circulation system, and the output is the actual drilling mud density.

1.13   **a.**   Nonlinear, continuous time, time invariant, deterministic
      b.   Linear, continuous time, time invariant, stochastic
      c.   Linear, discrete time, time invariant, deterministic
      d.   Nonlinear, continuous time, time varying, deterministic

1.14 $ML^2 d^2\theta/dt^2 + B\,d\theta/dt + MgL\theta = f(t)L$

1.15 a. True

b. False—linear system stability does not depend on the system's input.

c. True (for perfect sensors)

Chapter 2

2.1 New data provide an impartial test for model validation purposes.

2.2 a. $K/[(1 + R_1C_1s)(1 + R_2C_2s)]$

b. $1/(1 + R_1C_1s + R_2C_2s + R_1C_2s + R_1R_2C_1C_2s^2)$

2.3 $KR_2/[R_1 + R_2 + R_1R_2C(1 - K)s]$

2.4 $-[M_2s^2 + (B_1 + B_2)s + (K_1 + K_2)]/\Delta(s), \quad \Delta(s) = [M_2s^2 + (B_1 + B_2)s + (K_1 + K_2)][M_1s^2 + B_1s + K_1] - (B_1s + K_1)^2$

2.7 a. $1/(Js^2 + Bs + K)$, where $J = J_1 + (N_1/N_2)^2 J_2$, $B = B_1 + (N_1/N_2)^2 B_2$, and $K = (N_1/N_2)^2 K_2$

b. $1/(Js^2 + Bs)$, where $J = J_1 + (N_1/N_2)^2 J_2 + (N_1/N_3)^2 J_3$ and $B = (N_1/N_2)^2 B_2$

2.10 $0.1/[s(5 + 0.1s)(0.25s + 0.12)]$

2.11 $0.2/[s(5 + 0.2s)(0.2s + 1) + 0.04]$

2.13 $5.3/\{s[(5 + 0.1s)(0.25s + 0.12) + 1] + 5.3\}$

2.15 $[G_1(1 + G_3) + G_3 + G_1G_2G_3]/[1 + G_1 + G_3 + G_1G_3]$

2.16 $(AG + A)/(1 - GH - AC - AB - ABH + ACGH)$

2.17 $(G_1G_2G_3G_4 + G_1G_2G_5 + G_1G_3G_4G_6 + G_1G_5G_6)/\Delta(s)$, where $\Delta(s) = 1 + G_1G_2H_1 + G_1G_2G_3G_4H_4 + G_4H_3 + G_1G_3G_4G_6H_4 + G_1G_5G_6H_4 + G_2G_3H_2 + G_1G_2G_5H_4 + G_1G_2G_3G_6H_1H_2$

2.19 a. Define six nodes (1 to 6) for the associated signal flow graph. Node 1 is the input node and has a unity-gain branch outgoing to Node 2. Node 6, the output node, has a unity-gain branch incoming from Node 5. Other unidirectional branches have gains: F(2 to 3), P(3 to 4), C(4 to 5), A(3 to 2), G(4 to 3), B(5 to 4), and D(4 to 2).

b. Define seven nodes (1 to 7) for the associated signal flow graph. Node 1 is the input node and has a unity-gain branch outgoing to Node 2. Node 7, the output node, has a unity-gain branch incoming from Node 6. Other unidirectional branches have gains: A(3 to 4), C(4 to 5), E(5 to 6), B(4 to 3), G(4 to 3), D(5 to 4), F(6 to 5), and two unity-gain branches (2 to 4 and 3 to 5).

2.20 a. $(v_i - v_{C_2})/R_1 = (v_{C_2} - v_0)/R_2 + C_2\,dv_{C_2}/dt$

$(v_{C_2} - v_0)/R_2 + C_1\,dv_{C_1}/dt = 0$

$v_i - v_0 = v_{C_1}$

b. $A = \begin{pmatrix} -1/(R_2C_1) & -1/(R_2C_1) \\ -1/(R_2C_2) & -[1/(R_1C_2) + 1/(R_2C_2)] \end{pmatrix}$

$b = \begin{pmatrix} 1/(R_2C_1) \\ 1/(R_1C_2) + 1/(R_2C_2) \end{pmatrix}$ $c = \begin{pmatrix} -1 \\ 0 \end{pmatrix}$

c. $\dfrac{s^2 + [1/(R_1C_2) + 1/(R_2C_2)]s + 1/(R_1R_2C_1C_2)}{s^2 + [1/(R_2C_1) + 1/(R_1C_2) + 1/(R_2C_2)]s + 1/(R_1R_2C_1C_2)}$

2.21 a. $\dot{x} = \begin{pmatrix} -(R_1 + R_2)/(R_1R_2C_1) & -1/(R_2C_1) \\ -1/(R_2C_2) & -1/(R_2C_2) \end{pmatrix} x$

$+ \begin{pmatrix} (R_1 + R_2)/(R_1R_2C_1) \\ 1/(R_2C_2) \end{pmatrix} u$

$y = \begin{pmatrix} -1 \\ -1 \end{pmatrix}^T x + (1)v_i$

b. Define $\alpha_0 = 1/(R_1R_2C_1C_2)$ and $\alpha_1 = \alpha_0(R_1C_1 + R_1C_2 + R_2C_2)$. $V_0(s)/V_i(s) = s^2/(s^2 + \alpha_1 s + \alpha_0)$

$\dot{x} = \begin{pmatrix} 0 & 1 \\ -\alpha_0 & -\alpha_1 \end{pmatrix} x + \begin{pmatrix} 0 \\ 1 \end{pmatrix} u$

$y = \begin{pmatrix} -\alpha_0 \\ -\alpha_1 \end{pmatrix}^T x + (1)v_i$

2.23 a. $A = \begin{pmatrix} 0 & 1 & 0 \\ 0 & 0 & 1 \\ -6 & -11 & -6 \end{pmatrix}$ $b = \begin{pmatrix} 0 \\ 0 \\ 1 \end{pmatrix}$ $c = \begin{pmatrix} 100 \\ 45 \\ 5 \end{pmatrix}$

b. $A = \begin{pmatrix} -6 & 1 & 0 \\ -11 & 0 & 1 \\ -6 & 0 & 0 \end{pmatrix}$ $b = \begin{pmatrix} 5 \\ 45 \\ 100 \end{pmatrix}$ $c = \begin{pmatrix} 1 \\ 0 \\ 0 \end{pmatrix}$

c. $A = \begin{pmatrix} -1 & 0 & 0 \\ 0 & -2 & 0 \\ 0 & 0 & -3 \end{pmatrix}$ $b = \begin{pmatrix} 1 \\ 1 \\ 1 \end{pmatrix}$ $c = \begin{pmatrix} 30 \\ -30 \\ 5 \end{pmatrix}$

d. $A = \begin{pmatrix} -3 & 2 & 5 \\ 0 & -2 & 5 \\ 0 & 0 & -1 \end{pmatrix}$ $b = \begin{pmatrix} 0 \\ 0 \\ 1 \end{pmatrix}$ $c = \begin{pmatrix} 2 \\ 2 \\ 5 \end{pmatrix}$

2.26 $M = (b, Ab, A^2b) = \begin{pmatrix} 4 & 2 & 10 \\ 1 & 5 & 7 \\ 0 & 1 & 2 \end{pmatrix}$, det $M = 18$, controllable

$Q = (c, A^Tc, (A^T)^2c) = \begin{pmatrix} 0 & 1 & -4 \\ 1 & -4 & 13 \\ -5 & 15 & -45 \end{pmatrix}$, det $Q = 0$, not observable

2.28 $Y(s)/U(s) = (s^3 + 6s^2 + 13s + 3)/(s^3 + 5s^2 + 6s + 1)$

2.30 a. $\dot{x} = \begin{pmatrix} 0 & 1 \\ -6 & -5 \end{pmatrix} x + \begin{pmatrix} 0 \\ 1 \end{pmatrix} u, \quad y = \begin{pmatrix} 4 \\ 1 \end{pmatrix}^T x$

b. $\dot{x}^* = \begin{pmatrix} -2 & 0 \\ 0 & -3 \end{pmatrix} x^* + \begin{pmatrix} 1 \\ 1 \end{pmatrix} u, \quad y = \begin{pmatrix} 2 \\ -1 \end{pmatrix}^T x^*$

c. $P = \begin{pmatrix} 1 & -1 \\ -2 & 3 \end{pmatrix}$

Chapter 3

3.1   **a.**   Steady-state errors
        b.   Steady-state errors, time response shaping
        c.   Steady-state errors, system sensitivity
        d.   Steady-state errors, bandwidth considerations, relative stability

3.8   **a.** $0, 10, 0, 0$
        b. $2, \infty, \infty, 20$
        c. $1, \infty, 80, 0$
        d. $2, \infty, \infty, 1$

3.9   **a.** $\frac{1}{11}, 0, 0, 0$
        b. $\infty, 0.1, \infty, 2$
        c. $\infty, 0, \beta/80, 0$

3.10   **a.**   No, poles on $j\omega$ axis
         b. -10
         c.   No, poles in right half of the s-plane
         d. 0

3.11   1

3.12   **a.**   0
         b.   1

3.14 $-\mathbf{c}^T A \mathbf{b}, 0, 0$

3.17   **a.** $-2\alpha[s(s + \alpha)^2(10s + 1)]/\{[s(s + \alpha)^2(10s + 1) + K(s + 1)](s + \alpha)\}$
         b. $\alpha K s(1 + 2s)/\{[s^2(s + 10)(1 + \alpha s) + K(1 + 2s)](1 + \alpha s)\}$

3.18 $-\alpha K/[(K + 1)s + (3K + \alpha K + 1)]$

3.23 $(2s^2 + 2\zeta\omega_n s)/(s^2 + 2\zeta\omega_n s + \omega_n^2)$

3.25   **a.** $1/(15k_P)$
         b.   0, assuming asymptotic stability

3.28   **a.**   (1). $Y(s)/R(s) = Ks/[s^2 + (7 + Kk_2)s + (12 + Kk_1)]$
         c.   (1). $Y(s)/R(s) = Ks/[s^2 + (7 + Kk_1 + Kk_2)s + (12 + 4Kk_1 + 3Kk_2)]$

3.29 $K = \frac{1}{4}$, $\mathbf{k} = \begin{pmatrix} 24 \\ 28 \\ 4 \end{pmatrix}$, where $x_1 = \theta_m/4$, $x_2 = \dot{x}_1$, and $x_3 = \dot{x}_2$

3.30 $\mathbf{k}^* = P^T\mathbf{k}$

Chapter 4

4.1   Curve A: 35%, 0.4 sec, 0.4 sec, 2.5 sec
        Curve B: 0%, 2.1 sec, 0.7 sec, 2.7 sec

4.4   **a.**   Underdamped, 0.1, 1
        b.   Overdamped, 2, 10
        c.   Underdamped, 0.5, 10
        d.   Critically damped, 1, 2

4.5   **a.** 0.729, 3.16 sec, 72.9%
      c. 0.163, 0.36 sec, 16.3%

4.6   **a.** $1 - 1.005e^{-0.1t} \sin(0.995t + 84.3°)$
      b. $1 + 1.077e^{-2.68t} - 0.077e^{-37.32t}$
      c. $1 - 1.15e^{-0.5t} \sin(8.66t + 60°)$
      d. $1 - e^{-2t} - 2te^{-2t}$

4.7 $0 \le K_A \le 0.093$

4.8   **a.** 35.1%
      b. 2.86
      c. 3.74, 4.81, 7.81

4.9   89.5%, 0.068

4.10 0.0612

4.12 **a.** 72.9%
       b. 30.9%
       c. 35.6%

4.13 **a.** $10e^{-t} + 20e^{-2t} + 50e^{-3t}$
       b. $3 + 2e^{-10t}$
       c. $5e^{-2t} - 2\sqrt{5} \sin(2t - 63.4°)$

4.14 **a.** Poles: $0, -1, -2$; residues: $1, -5, 0.8$
       b. Poles: $-3, -2$; residues: $2, 5$
       c. Poles: $0, -2 + j3, -2 - j3$; residues: $1, 0.1\underline{/150°}, 0.1\underline{/-150°}$

4.15 **b.** $1 - 1.21e^{-5.59t} \sin(8.29t + 56.0°)$

4.16 **a.** $1 - 1.512e^{-7.5t} \sin(6.61t + 41.4°)$
       b. $1 - 1.068e^{-7.5t} \sin(6.61t + 110.8°)$

4.17 **a.** 2
       b. 4, 6
       c. β_1 can be any value except 1.5, $\beta_2 = 6$

4.18 **a.** 4
       b. 10, 12
       c. Any β_1 and β_2 such that $4\beta_1 + 3\beta_2 = 6$

4.19 (1) $\begin{pmatrix} 3e^{-2t} - 2e^{-3t} & e^{-2t} - e^{-3t} \\ -6e^{-2t} + 6e^{-3t} & -2e^{-2t} + 3e^{-3t} \end{pmatrix}$

4.29 $6.25 - 16.67e^{-t} + 12.5e^{-2t} - 2.08e^{-4t}$

4.30 αe^{2t}, where $\alpha = (4/3)e^2/(e^4 - 1)$

4.31 Yes (det $M = -1$), yes (det $Q = -2$)

Chapter 5

5.1   **a.** Asymptotically stable
      b. Marginally stable
      c, d. Unstable

5.2   **a.**   Integral in (5.3) equals $\frac{1}{2}$

    b. $1/[(s + 1)(s + 2)]$, poles at -1 and -2

    c. $A = \begin{pmatrix} 0 & 1 \\ -2 & -3 \end{pmatrix}$, eigenvalues of A are -1 and -2

5.3   **a.**   Unstable: 2 RHP poles, 2 LHP poles

    b.   Unstable: 2 RHP poles, 3 LHP poles

    c.   Unstable: 2 RHP poles, 3 LHP poles (includes a quad)

    d.   Unstable: 1 LHP pole, 4 $j\omega$-axis poles (doubles)

5.4   2, $K > 1$

5.5   **a.**   No, fails Hurwitz test

    b.   Yes

    c. $1 < K < 5$

5.6   **a.** $K > 3.39$

    b. $-0.5 < K < 3.31$

5.7 $K > 505$

5.8   **a.** $K > 32/3$

    b. $32/3, \sqrt{5/3}$

5.9 $K > 1 - 2T, T < \frac{1}{2}$

5.10   **a.** $K^* = 30, \omega^* = \sqrt{5}, \sigma_I = -2$

    b. $K^* = 2, \omega^* = \sqrt{2}$

    c. $K^* = 30, \omega^* = \sqrt{14}, \sigma_I = 1$

5.11   For $0 < K < 6$: 3 RHP poles, $6 < K < 11.4$: 2 RHP poles, $K > 11.4$: no RHP poles

5.13 $K^* = 90/7, \omega^* = \sqrt{20/7}, \sigma_I = 3.5$

5.14   **a.**   No $j\omega$-axis crossings

    b. $K^* = 6, \omega^* = \sqrt{3}, \sigma_I = -0.5$

5.15   **b.** $0°$

    c. $3/2$

5.16 $s_B = 3.15, \sigma_I = 3.5$

5.17 $K^* = 2.5, \omega^* = \sqrt{5}$

5.19   **a.**   Basic

    b.   Modified

    c.   Modified

5.20 $K^* = 2.4, \omega^* = \sqrt{11}$

5.21   **a.** $\sigma_I = \frac{1}{2}$

    b. $K^* = 4$

    c. $\omega^* = \sqrt{6}$

    d. $-18.4°$

5.24 $K < 5, \omega^* = \sqrt{11}$

5.26 $K^* = 6, \omega^* = \sqrt{2}$ and $K^* = 8, \omega^* = 0$; $s_B = 1 + \sqrt{3}, 1 - \sqrt{3}$

5.27 **a.** $K^* = 1 + \sqrt{2}$, $\omega^* = \sqrt{2\sqrt{2} - 1}$
b. $K^* = -2$, $\omega^* = 0$

5.32 No $j\omega$-axis crossings (except at origin for $b = 0$), $s_B = -34.1, -5.9$

Chapter 6

6.1 **a.** 14.0 rad/sec, 1.5
b. 6.3 rad/sec, 1.0

6.2 **a.** 4.0, 63°
b. 4.0, 27°

6.3 **a.** 2.97 rad/sec, 2.07, ∞, 28.0°

6.4 **b.** $\alpha \geq 7.75$

6.5 **a.** $0 < K < 67.6$
b. $0 < K < 90.6$
c. $0 < K < 88.7$

6.6 **a.** $0 < K \leq 12.5$
b. $7.0 < K < 11.9$
c. $K \geq 8.0$, which yields $8.0 \leq K < 11.9$ to satisfy all three.

6.9 **b.** Starting ($\omega = 0$) with a magnitude of unity and a phase of 180°, the magnitude increases somewhat for increasing ω and then decreases to zero as the phase decreases from 180° to $-90°$. The polar plot swings from the negative real axis of the $GH(s)$-plane through the second, first, and fourth quadrants.

6.11 For $K = 2$, $N = 0$, $P = 0$, $Z = N + P = 0$, stable. For $K = 10$, $N = +2$, $P = 0$, $Z = N + P = 2$, unstable. Nyquist plot has $\omega_{PC} = \sqrt{11}$ and $\text{Re}\{G(j\omega_{PC})\} = -K/5$, which implies $K^* = 5$. Thus, stable for $0 < K < 5$.

6.12 **a.** For $-\infty < K < 32/3$, $N = 0$, $P = 2$, $Z = 2$, unstable. For $K > 32/3$, $N = -2$, $P = 2$, $Z = 0$, stable ($\omega_{PC} = \sqrt{5/3}$).
b. For $-17.9 < K < 34.5$, $N = 0$, $P = 0$, $Z = 0$, stable. Otherwise, $N = +2$, $P = 0$, $Z = 2$, unstable. Values of ω_{PC} are 0.62 and 2.94 rad/sec.

6.13 For all $K > 0$, $N = 0$, $P = 2$, $Z = 2$, unstable

6.15 **a.** For $0 < K < 90/7$, $N = 0$, $P = 1$, $Z = 1$, unstable. For $K > 90/7$, $N = 2$, $P = 1$, $Z = 3$, unstable ($\omega_{PC} = \sqrt{20/7}$).
b. Two closed-loop poles cross the $j\omega$-axis at $\pm j\sqrt{20/7}$ for $K^* = 90/7$.

6.21 3.2

6.22 **b.** 2, 32°, 4.5
c. 1.3, 70°, 4

6.25 **a.** 1.7
b. 5.0

Chapter 7

7.1 $0 \le K \le 19.2,\ K \le 200$

7.3 $0 \le K \le 52.7$

7.5 For $C = 1\ \mu F$, $R_1 = R_2 = 100\ K\Omega$. For $C = 10\ \mu F$, $R_1 = R_2 = 1\ M\Omega$.

7.6 For $C = 1\ \mu F$, $R_1 = R_2 = 51\ K\Omega$. For $C = 10\ \mu F$, $R_1 = R_2 = 10\ K\Omega$.

7.7 One design: Steps 1 to 7 in Figure 7.11 yield $z = 15$ and $p = 53.3$ using the 20% overshoot design specification (with $\alpha = 5$), but a computer simulation in Step 8 indicates an actual overshoot of 27%. Using a 5% design specification (with $\alpha = 8$) yields $z = 10$ and $p = 72$, which gives a 15% overshoot, $t_r < 0.5$ sec, and $e_{ss} = 0.036$ rad. For the latter design, $R_1 = 100\ K\Omega$ and $R_2 = 16.1\ K\Omega$ for $C = 1\ \mu F$.

7.10 One design: $a = 2.7$, $T = 0.05$, $R_1 = 132\ K\Omega$, $R_2 = 77\ K\Omega$

7.14 One design: $a = 0.175$, $1/T = 0.06$, $R_1 = 140\ K\Omega$, $R_2 = 30\ K\Omega$

7.17 One phase-lead design: Use the root locus procedure in Figure 7.11 with a 10% overshoot design requirement ($\alpha = 4$) to yield $z = 5$ and $p = 25$. A computer simulation indicates an actual 28% overshoot and a rise time of approximately 0.2 sec. Further calculations show e_{ss} for a unit ramp input equal to 0.03 rad and a phase margin of 47.2° ($\omega_{GC} = 7.52$ rad/sec). Let $R_1 = 200\ K\Omega$, $R_2 = 50\ K\Omega$, and $C = 1\ \mu F$ in Figure 7.9a.

7.19 Use phase-lag design. So much additional phase is needed at the new gain crossover frequency that phase-lead design is not applicable.

7.20 Use phase-lead design. Phase-lag design is not applicable in this case. (Why not?)

Chapter 8

8.1 **a, b.** $k^* = \infty$, $J^* = 0$
  c. $k^* = 0.5$, $J^* = 2.5$

8.2 $\dot{\lambda}^* = -2x^* + \lambda^*$, $\dot{x}^* = -x^* + u^*$, $u^* = -\lambda^*/2$;
  $x(0) = 1$, $x(\infty) = 0$; $u^*(t) = -(\sqrt{2} - 1)e^{-\sqrt{2}t}$

8.3 $-(\sqrt{2} - 1)$, same as the ratio in Problem 8.2

8.5 0.4

8.7 $x_1^*(t) = [\sqrt{10}/(\sqrt{10} - 1)]e^{-t} - [1/(\sqrt{10} - 1)]e^{-\sqrt{10}t}$, $x_2^*(t) = \dot{x}_1^*(t)$

8.10 $Q = \begin{pmatrix} 52 & 28 - p_{11} \\ 28 - p_{11} & 11 \end{pmatrix}$, where $4 < p_{11} < 28 + \sqrt{572}$

8.13 $a/b = 2.4$; If one selects $b = 1$, then $a = 2.4$

8.14 $\dot{w} = -4w - 22y + u$, $\hat{z} = \hat{x}_2 = w + 3y$

8.20 From Theorem 2.2, $Q = (\mathbf{c}, A^T\mathbf{c}) = \begin{pmatrix} 1 & -3 \\ 0 & 0 \end{pmatrix}$, $\det Q = 0$; plant is not observable.

INDEX